T0351035

EARTHQUAKE ENGINEERING FOR CONCRETE DAMS

EARTHQUAKE ENGINEERING FOR CONCRETE DAMS

Analysis, Design, and Evaluation

Anil K. Chopra
University of California at Berkeley
California
USA

WILEY Blackwell

This edition first published 2020

Registered Offices
John Wiley & Sons, Inc., 111 River Street, Hoboken, NJ 07030, USA
9600 Garsington Road, Oxford, OX4 2DQ, UK

Editorial Office
9600 Garsington Road, Oxford, OX4 2DQ, UK

For details of our global editorial offices, customer services, and more information about Wiley products visit us at www.wiley.com.

Wiley also publishes its books in a variety of electronic formats and by print-on-demand. Some content that appears in standard print versions of this book may not be available in other formats.

Library of Congress Cataloging-in-Publication Data

Names: Chopra, Anil K., author.
Title: Earthquake engineering for concrete dams : analysis, design, and evaluation / Anil K. Chopra, University of California, Berkeley.
Description: Hoboken, NJ, USA : Wiley-Blackwell, 2020. | Includes bibliographical references and index.
Identifiers: LCCN 2019032036 (print) | LCCN 2019032037 (ebook) | ISBN 9781119056034 (hardback) | ISBN 9781119056041 (adobe pdf) | ISBN 9781119056096 (epub)
Subjects: LCSH: Concrete dams–Earthquake effects. | Dam safety. | Earthquake engineering. | Earthquake hazard analysis.
Classification: LCC TA654.6 .C4663 2019 (print) | LCC TA654.6 (ebook) | DDC 627/.80289–dc23
LC record available at https://lccn.loc.gov/2019032036
LC ebook record available at https://lccn.loc.gov/2019032037

Cover Photograph: Mauvoisin Dam, a 250-m-high arch dam, is located in the Swiss Alps. It is perhaps the best instrumented dam in the world

Cover Design: Wiley
Cover Image: © Axpo Power AG

Copy editing and Proofreading: Clare Romeo/Line by Line Services

Set in 10/12pt WarnockPro by SPi Global, Chennai, India
Printed and bound in Singapore by Markono Print Media Pte Ltd

10 9 8 7 6 5 4 3 2 1

This book is dedicated to the memory of my mentors:

Ray W. Clough

Joseph Penzien

Nathan M. Newmark

Anestis S. Veletsos

Emilio Rosenblueth

Contents

PART III: DESIGN AND EVALUATION

Preface

Concerns about the seismic safety of concrete dams has been growing during the past few decades, partly because the population at risk in locations downstream of major dams continues to expand, but also because it is increasingly evident that the design concepts used at the time most existing dams were built were inadequate. During this time span the knowledge of the complex nature and intensity of ground motions has been increasing rapidly, as thousands of recordings have now been accumulated. It is now widely recognized that ground motions intense enough to cause structural damage should be anticipated at many dam sites, and it has become apparent that the seismic designs of most dams did not fully recognize the hazard.

The structural damage sustained by Koyna Dam during an earthquake in 1967 was of profound significance to the development of earthquake engineering for concrete dams. A modern dam, designed according to analysis procedures and design criteria that represented "standard" practice worldwide at the time, had been damaged by ground shaking that was intense, but by no means extreme. It was clear that the design forces had little resemblance to how the dam responded during the earthquake. The experience at Koyna Dam was a watershed event in the sense that it dispelled the myth – at that time – among many engineers that these massive dams are immune to earthquake damage, and motivated the development of dynamic analysis procedures for concrete gravity dams, eventually revolutionizing earthquake engineering for all types of concrete dams.

As a result, earthquake analysis and design of concrete dams has progressed from static force methods involving the use of seismic coefficients, to procedures that now recognize the dynamics of dam–water–foundation systems. It is the story of this progress that is presented in this book.

This book provides a comprehensive, integrated view of this progress currently scattered in hundreds of research publications. It was conceived as a reference book for graduate students, researchers, and professional engineers. It should help graduate students study the subject before embarking on their own research on earthquake engineering for concrete dams. Researchers in this field should gain new insights and improved understanding of the subject. Professional engineers should develop a better understanding of the limitations of the various methods of dynamic analysis used in practice, and become familiar with modern methods that overcome these limitations.

The book is organized into three parts: I. Gravity Dams; II. Arch Dams; and III. Design and Safety Evaluation. The objectives of Parts I and II are to (i) develop response spectrum analysis and response history analysis procedures for concrete dams; (ii) develop an understanding of the dynamics of dams, leading to identification of system parameters that influence their dynamic response; (iii) demonstrate the effects of dam–water–foundation interaction on earthquake response; and (iv) identify factors that must be included in earthquake analysis of concrete dams. In Part I, these topics are presented in the context of two-dimensional models, which may be appropriate for gravity dams. In Part II, they are presented for three-dimensional models, applicable to all types of dams; arch, buttress, and gravity. The objectives of Part III are to (i) examine critically the definitions of design earthquakes according to various regulatory bodies and professional organizations; (ii) present modern methods for selecting ground motions; and (iii) illustrate application of dynamic analysis procedures to the design of new dams and safety evaluation of existing dams.

The book provides a comprehensive view of the subject with many references to the published literature. However, Parts I and II are based primarily on the research of several doctoral students at the University of California, Berkeley, who graduated in the year noted:

- Partha Chakrabarti, 1973
- John F. Hall, 1980
- Gregory L. Fenves, 1984
- Ka-Lun Fok, 1985
- Liping Zhang, 1990
- Han-Chen Tan, 1995
- Arnkjell Løkke, 2018

and on the work of visiting researcher,

- Jinting Wang (2008).

This book has been influenced by my own research experience in collaboration with my doctoral students, and by my experience in consulting on many projects worldwide. Over the period 1970–1995, my research on earthquake engineering for concrete dams was supported by the National Science foundation and U.S. Army Corps of Engineers.

I remain grateful to the University of California at Berkeley for the privilege of serving on its faculty.

– Anil K. Chopra

Acknowledgments

I am grateful to several individuals who helped in preparation of this book:

- Dr. Arnkjell Løkke, whose Ph.D. research is the basis for Chapter 11, prepared many of the figures in Chapters 2–7.
- Dr. N. Simon Kwong participated in the preparation of Chapter 13 in several ways; he generated numerical results, developed figures, provided advice, and reviewed drafts. He also reviewed the first part of Chapter 12 and developed its figures.
- Professor Jinting Wang provided information on seismic design of Dagangshan Dam. Based on original reports in Chinese, he prepared the first draft of Section 14.4.
- Professor Pierre Léger advised on a broad range of topics in several ways. He was generous in sharing his vast knowledge over many long conversations and in providing many publications; his command of the published literature was truly impressive. He also reviewed Chapters 11 and 12.
- Dr. Robin K. McGuire reviewed the first part of Chapter 12 and advised on seismic risk analysis concepts and methods.
- Larry K. Nuss advised on several practical aspects of design and evaluation of concrete dams, and provided publications by several government agencies.
- Claire M. Johnson, a partner from beginning to end, prepared and edited the text, assembled the manuscript, and helped in too many ways to enumerate here.
- Clare Romeo served as the copy editor and proofreader.

1 Introduction

1.1 EARTHQUAKE EXPERIENCE: CASES WITH STRONGEST SHAKING†

As far as can be determined, no large concrete dam with full reservoir has been subjected to extremely intense ground shaking. The closest to such an event was the experience at Koyna (gravity) Dam (Figure 1.1.1), with the reservoir nearly full, during the 1967 earthquake (Chopra and Chakrabarti 1973). Ground accelerations recorded at the dam site during a nearby earthquake of magnitude 6.5 had a peak value of 0.38 g in the stream direction and strong shaking lasted for 4 sec. Significant horizontal cracking occurred through a number of taller non-overflow monoliths at or near the elevation where the downstream face changes slope; however, the dam continued to retain the reservoir even though the water level was 25 m above the cracks (Figure 1.1.2). A similar experience had occurred in 1962 at Hsinfengkiang (buttress) Dam (Figure 1.1.3) during a magnitude 6.1 earthquake in close proximity. Although not recorded, ground motions were probably quite intense causing cracking at 16 m below the crest (Figure 1.1.4); the dam continued to retain the reservoir though the water level was 3 m above the cracks.

A few dams have withstood very intense ground shaking with little or no damage because of their unusual design or low water level. Perhaps the strongest shaking experienced by a concrete dam to date was that at Lower Crystal Springs Dam, a 42-m-high curved gravity structure (Figure 1.1.5) dam with nearly full reservoir, located within 350 m of the San Andreas fault that caused the magnitude 7.9 1906 San Francisco earthquake. Built with interlocking concrete blocks, the dam was undamaged, even though its reservoir was full. However, the earthquake resistance of this dam greatly exceeds that of typical gravity dams due to its curved plan and a cross section that was designed thicker than normal in anticipation of future heightening, which was never completed; a section view of this dam is shown in Figure 1.1.6.

† The first part of this section is adapted from a National Research Council report (1990).

Earthquake Engineering for Concrete Dams: Analysis, Design, and Evaluation, First Edition. Anil K. Chopra.

Figure 1.1.1 Koyna Dam, India, constructed during 1954–1963; this dam is 103 m high and 853 m long.

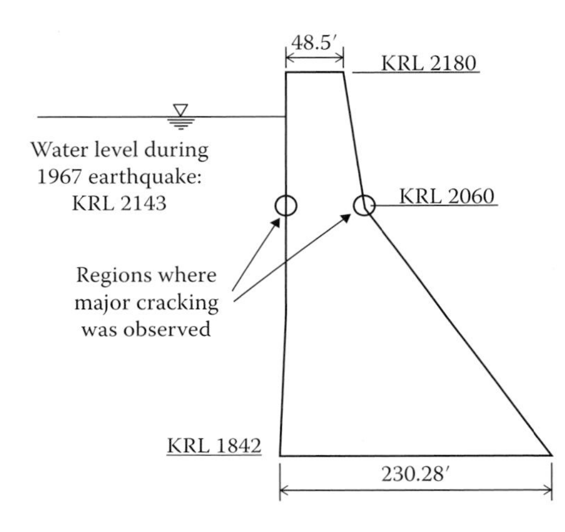

Figure 1.1.2 Cross section of Koyna Dam showing water level during 1967 earthquake and regions where principal cracking at the upstream and downstream faces was observed. Source: Adapted from National Research Council (1990).

Figure 1.1.3 Hsinfengkiang Dam, China. Completed in 1959, this dam is 105 m high and 440 m long.

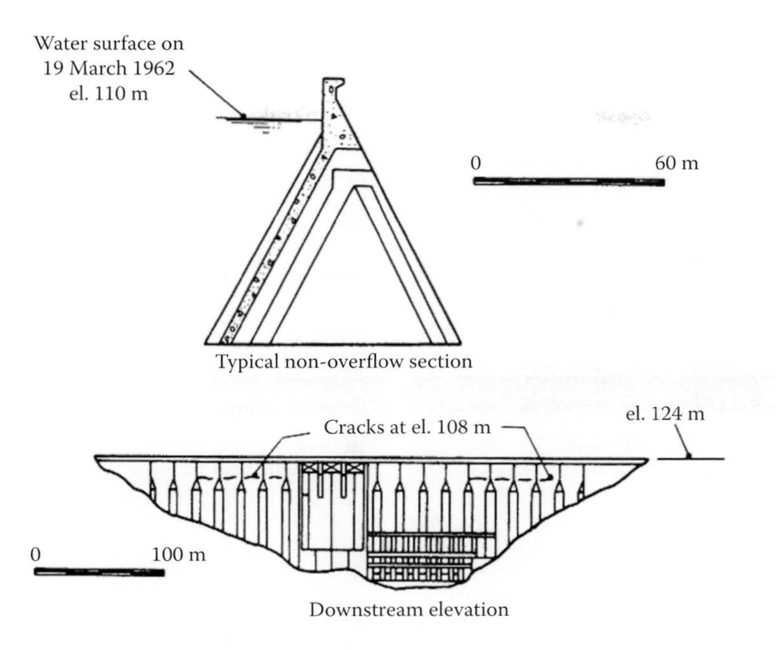

Figure 1.1.4 Cracking in Hsinfengkiang Dam, China, due to earthquake on March 19, 1962. Source: Adapted from Nuss et al. (2014).

Figure 1.1.5 Lower Crystal Springs Dam, California, USA. Built in 1888, this 45-m-high curved-gravity dam is located within 350 m of the San Andreas Fault, which is under the reservoir, oriented roughly parallel to the dam.

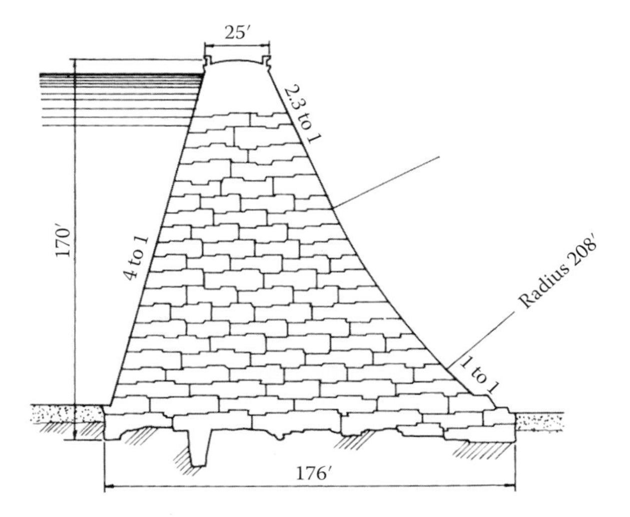

Figure 1.1.6 Section view of the Lower Crystal Springs Dam (adapted from Nuss et al. [2014] and Wieland et al. [2004]).

Another example of a concrete dam subjected to very intense shaking is the 113-m-high Pacoima (arch) Dam (Figure 1.1.7). During the 1971 magnitude 6.6 San Fernando earthquake, an accelerograph located on the left abutment ridge recorded a peak acceleration of 1.2 g in both horizontal components and 0.7 g vertical, with strong shaking lasting for 8 sec, suggesting that the excitation at the dam–foundation‡ interface – which was not recorded – must have been very intense. However, the only visible damage to the dam was a 3/8 in. opening of the contraction joint on the left thrust block and a crack in the thrust block. During the 1994 magnitude 6.7 Northridge earthquake, peak accelerations recorded ranged from 0.5 g at the base of the dam to about 2.0 g along the abutments near the crest. The damage sustained was more severe than in 1971. The contraction joint between the dam and the thrust block in the left abutment again opened, this time by 2 in. at the crest level (Figure 1.1.8), decreasing to 1/4 in. at the bottom of the joint (60 ft

Figure 1.1.7 Pacoima Dam, California, USA. Completed in 1929, this dam is 113 m high and 180 m long at the crest.

‡ The word "foundation" denotes the rock that supports the dam.

below the crest), at which point a large crack extended down diagonally through the lower part of the thrust block to meet the foundation (Figure 1.1.9). The good performance of the dam can be attributed primarily to the low water level – 45 m below the dam crest – at the time of both earthquakes. Additional information is available in Scott et al. (1995).

Figure 1.1.8 Two-inch separation between Pacoima Dam Arch (left) and the thrust block (right) on the left abutment (Scott et al. 1995).

Figure 1.1.9 Crack at the joint between the Pacoima Dam arch and the thrust block and diagonal crack in the thrust block (Scott et al. 1995).

(a)

(b)

Figure 1.1.10 Shih-Kang Dam, Taiwan, (a) before and after the Chi-Chi earthquake; (b) close-up of damaged bays. Completed in 1977, this gated spillway is 21 m high and 357 m long. (a) Two photos courtesy of C.-H. Loh, National Taiwan University, Taiwan. (b) Photo courtesy of USSD.org.

Shih-Kang Dam in Taiwan (Figure 1.1.10) – a 70-ft (21.4-m)-high, 18-bay gated spillway – located directly over a branch of the Che-Lung-Pu fault that caused the 1999 magnitude 7.6 Chi-Chi earthquake represents the first known dam failure during an earthquake. However, this failure was caused primarily by fault rupture, not ground shaking, although it was very intense, as indicated by the peak ground acceleration of 0.5 g recorded at a location 500 m from the dam. During the Chi-Chi earthquake the branch fault ruptured, with a vertical offset of 29 ft (9 m) and a horizontal offset diagonal to the dam axis of about 23 ft (7 m). As a result, bays 16–18 incurred extensive damage, but the damage to the other bays was surprisingly little; spillway piers sustained cracking, simply supported bridge girders came off their bearings, and six gates were inoperable after the earthquake.

It is clear from the preceding observations that concrete dams can be significantly damaged by ground shaking due to earthquakes. They are not as immune to damage as had commonly been presumed prior to the 1967 experience at Koyna Dam. This fact is now universally recognized, and there is much interest in the earthquake performance of concrete dams.

1.2 COMPLEXITY OF THE PROBLEM

The ability to evaluate the effects of earthquake ground motion on concrete dams is essential in order to assess the safety of existing dams, to determine the adequacy of modifications planned to improve existing dams, and to evaluate proposed designs for new dams to be constructed. However, the prediction of performance of concrete dams during earthquakes is one of the most complex and challenging problems in structural dynamics because of the following factors:

1. Dams and the impounded reservoirs† are of complicated shapes, as dictated by the topography of the site (see Figures 1.2.1 and 1.2.2).
2. The response of a dam is influenced greatly by the interaction of the motions of the dam with the impounded water and the foundation, both of which extend to large distances. Thus the mass, stiffness, material damping, radiation damping of the foundation (see Section 1.6), and the earthquake-induced hydrodynamic pressures must be considered in computing the dynamic response.
3. During intense earthquake motions, vertical contraction joints may slip or open; concrete may crack; and separation and sliding may occur at lift joints in concrete, dam–foundation interface, and fissures in foundation rock. These phenomena are highly nonlinear and extremely difficult to model realistically.
4. The response of dams is affected by variations in the intensity and frequency characteristics of the ground motion over the width and height of the canyon; however, this factor cannot be fully considered at present for lack of instrumental records to define the spatial variations of the ground motion.

Considering all these factors, analytical and computational procedures to determine the response of dam–water–foundation systems subjected to ground shaking are presented in this book. A substructure method for linear analysis of two-dimensional (2D) models, usually appropriate for gravity dams, is the subject of Chapters 2–6; and of three-dimensional (3D) models – required for arch dams, buttress dams, and gravity dams in narrow canyons – is the subject of Chapter 8. The Direct Finite-Element Method (FEM) for nonlinear analysis of 2D or 3D dam–water–foundation systems is presented in Chapter 11.

Figure 1.2.1 Olivenhain Dam, California, USA. Completed in 2003, this is a 318-ft-high roller-compacted concrete dam with a crest length of 2552 ft.

† "Reservoir" is the place of storage, not the fluid itself.

Figure 1.2.2 Morrow Point Dam, Colorado, USA, a 465-ft-high single-centered arch dam.

1.3 TRADITIONAL DESIGN PROCEDURES: GRAVITY DAMS

1.3.1 Traditional Analysis and Design

Concrete gravity dams have traditionally been designed and analyzed by very simple procedures (U.S. Army Corps of Engineers 1958; Bureau of Reclamation 1965, 1966). Earthquake effects were treated simply as static forces and were combined with the hydrostatic pressures and gravity loads. In representing the effects of horizontal ground motion – transverse to the axis of the dam – by static lateral forces, neither the dynamic response characteristics of the dam–water–foundation system nor the amplitude and frequency content of earthquake ground motion were recognized. Two types of static lateral forces were included. Forces associated with the weight of the dam were expressed as a product of a seismic coefficient – which was typically constant over the height, with a value between 0.05 to 0.10 – and the weight of the portion of the dam being considered. Water pressures, in addition to the hydrostatic pressure, were specified as the product of the seismic coefficient and a pressure coefficient that was based on assumptions of a rigid dam and incompressible water. Finally, interaction between the dam and the foundation was not considered in computing the aforementioned earthquake forces.

The traditional design criteria required that an ample safety factor be provided against overturning, sliding, and overstressing; in particular, compressive stresses should be less than one-fourth of the compressive strength. Usually tension was not permitted, and even if it was, the allowable tension was so small that the possibility of cracking of concrete was not considered.

1.3.2 Earthquake Performance of Koyna Dam

Koyna Dam in India was designed by the traditional static analysis procedure using a seismic coefficient of 0.05. Even though a "no-tension" criterion was satisfied in the design procedure, as mentioned earlier, the earthquake of December 11, 1967 caused significant horizontal cracks in the upstream and/or downstream faces of a number of the taller non-overflow monoliths near the elevation at which there is an abrupt change in slope of the downstream face (Figure 1.1.2).

To understand why the damage occurred, the dynamic response of the tallest non-overflow monolith to the recorded ground motion was computed assuming linear behavior. The results indicated large tensile stresses on both faces, with the greatest values near the elevation where the slope of the downstream face changes abruptly. These computed stresses (shown in Figure 1.3.1), which exceeded 600 psi on the upstream face and 1000 psi on the downstream face, were about two to three times the estimated tensile strength – 350 psi – of the concrete at that elevation. Hence significant cracking consistent with what was observed after the earthquake could have been anticipated. A similar analysis of the overflow monoliths indicated that cracking should not have occurred there, which is also consistent with the observed behavior.

1.3.3 Limitations of Traditional Procedures

It is apparent from the preceding discussion that the dynamic stresses that develop in gravity dams bear little resemblance to the results obtained from traditional static design procedures. In the case of Koyna Dam, no tensile stresses were expected when designing the dam for earthquake forces based on a seismic coefficient of 0.05, uniform over the height; however, the earthquake caused significant tensile cracking in the dam. This discrepancy is the result of using too small a seismic coefficient and not recognizing the amplification of acceleration over the height of the dam.

The typical design seismic coefficients, 0.05–0.10, are much smaller than the ordinates of design spectra for intense earthquake motions in the range of vibration periods for concrete

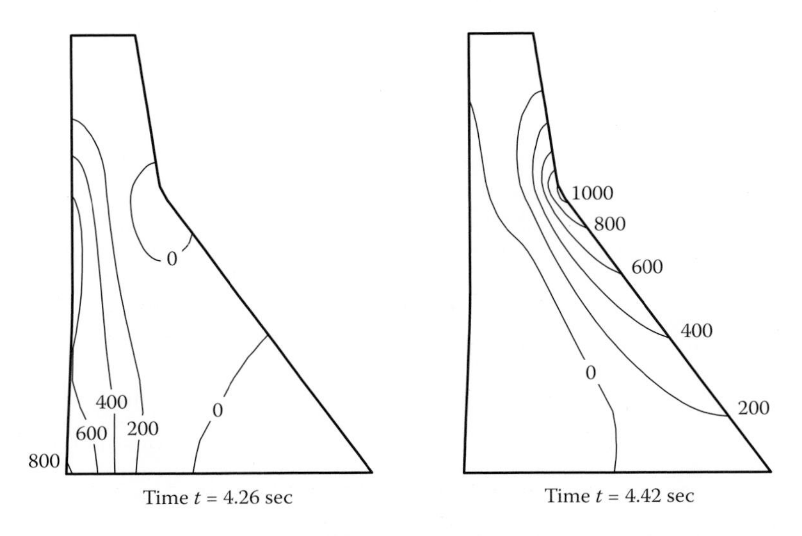

Figure 1.3.1 Maximum principal stresses in Koyna Dam at selected time instants due to transverse and vertical components of ground motion recorded during the December 11, 1967 earthquake; initial static stresses are included.

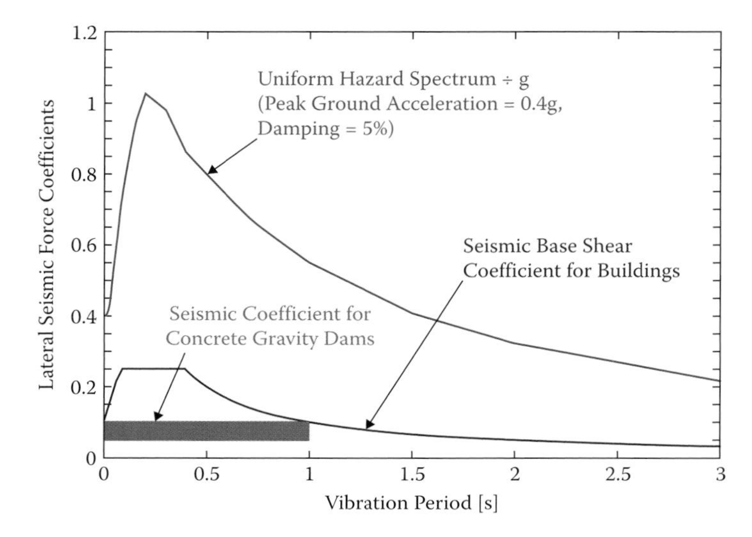

Figure 1.3.2 Comparison of uniform hazard spectrum and seismic coefficient for concrete dams and buildings. Source: Adapted from Chopra (1978).

gravity dams (Figure 1.3.2). Note that the seismic base shear coefficient values for dams are similar to those specified for multistory buildings. However, building code design provisions have been based on the premise that buildings should be able to: "(i) resist minor earthquakes without damage; (ii) resist moderate earthquakes without structural damage; and (iii) resist major earthquakes…without collapse but with some structural…damage." While these may be appropriate design objectives for buildings, major dams should be designed more conservatively, and this intended conservatism is reflected in the no-tension requirement imposed in traditional methods for designing dams. What the traditional methods fail to recognize, however, is that this requirement must be tied to the dynamic response of the dam that is influenced by its natural vibration periods and modes.

The effective modal earthquake forces may be expressed as the product of the weight of the dam per unit height and a seismic coefficient; its magnitude depends on the pseudo-acceleration spectral ordinate at the modal period and its height-wise distribution depends on the shape of the mode. The response of short-vibration-period structures, such as concrete gravity dams, is dominated by the fundamental mode of vibration, and the seismic coefficient varies over the dam height, as shown schematically in Figure 1.3.3b. In contrast, traditional analysis and design procedures ignore the dynamic amplification of response, as reflected in the response spectrum and the shape of the mode, and adopt a uniform distribution for the design coefficient (Figure 1.3.3a), resulting in an erroneous distribution of lateral forces and hence of stresses in the dam. The implications of these errors will be discussed in Chapter 7.

To eliminate these errors, it is imperative to consider the dynamics of the system subjected to realistic ground motions in estimating the earthquake response of concrete dams. In Chapters 2–6, such procedures for dynamic analysis of 2D models of gravity dams are developed. In Chapter 7, responses computed by these procedures are demonstrated to be consistent with motions of a gravity dam recorded during an earthquake and with the earthquake performance of Koyna (gravity) Dam.

The traditional design loadings for gravity dams include seismic water pressures in addition to the hydrostatic pressures, as specified by various formulas (U.S. Army Corps of Engineers 1958; Bureau of Reclamation 1966). These formulas differ somewhat in detail and in numerical

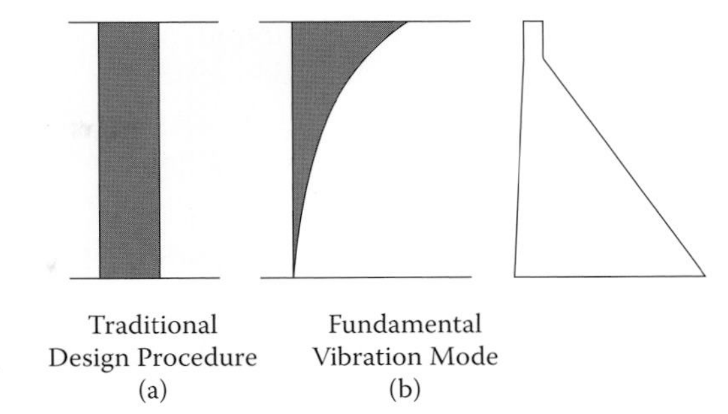

Figure 1.3.3 Distribution of seismic coefficients over dam height in traditional design and for the fundamental vibration mode. Source: Adapted from Chopra (1978).

values but not in underlying assumptions; they are all based on the classical results (Westergaard 1933; Zangar 1952) derived from analyses that assumed the dam to be rigid and water to be incompressible. One of these formulas specifies the seismic water pressure $p_e = cswH$, where c is a coefficient that varies from zero at the water surface to about 0.7 at the reservoir bottom, s is the seismic coefficient, w is the unit weight of water, and H is the total depth of water. For a seismic coefficient of 0.1, the additional water pressure at the base of the dam is about 7% of the hydrostatic pressure; and pressure values at higher elevations are even smaller. As a result, these additional water pressures have little influence on the computed stresses and hence on the geometry of the gravity section that satisfies the traditional design criteria.

On the other hand, earthquake-induced stresses in gravity dams are much larger when dam–water interaction arising from deformations of the dam and water compressibility effects are considered, as will be demonstrated in Chapters 2 and 6. It is apparent, therefore, that hydrodynamic effects are considerably underestimated because of assumptions implicit in traditional design forces.

As mentioned earlier, traditional analysis and design procedures ignore interaction between the dam and foundation. However, such interaction has very significant influence on the dynamics of the system, and, hence, on the earthquake-induced stresses. This will be demonstrated in Chapters 3 and 6.

Finally, the static overturning and sliding criteria that have been used in traditional design procedures for gravity dams have little meaning in the context of oscillatory response to earthquake motions.

1.4 TRADITIONAL DESIGN PROCEDURES: ARCH DAMS

1.4.1 Traditional Analysis and Design

Traditionally, the dynamic response of the system has not been considered in defining the earthquake forces in the design of arch dams. For example, the U.S. Bureau of Reclamation (1965) stated: “The occurrence of vibratory response of the earthquake, dam, and water is not considered, since it is believed to be a remote possibility.” Thus, the forces associated with the inertia of the dam were expressed as the product of a seismic coefficient – which was constant over the surface of the dam with a typical value of 0.10 or less – and the weight of the dam. Water pressures, in addition to the hydrostatic pressure, were specified in terms of the seismic coefficient and a pressure coefficient that was the same as for gravity dams, defined in Section 1.3.3. This pressure coefficient was based on assumptions of a rigid dam, incompressible water, and a straight dam. Generally, dynamic interaction between the dam and foundation was not considered in evaluating the aforementioned earthquake forces, but in stress analysis of arch dams the flexibility of the foundation sometimes was recognized through the use of Vogt coefficients (Bureau of Reclamation 1965).

The traditional design criteria required that the compressive stress not exceed one-fourth of the compressive strength or 1000 psi, and the tensile stress should remain below 150 psi.

1.4.2 Limitations of Traditional Procedures

As mentioned in Section 1.3.3 in the context of gravity dams, the seismic coefficient of 0.1 is much smaller than the ordinates of the pseudo-acceleration response spectra for intense ground motions (Figure 1.3.2). Thus the earthquake forces for arch dams also were greatly underestimated in traditional analysis procedures.

The effective earthquake forces on a dam due to horizontal ground motion may be expressed as the product of a seismic coefficient, which varies over the dam surface, and the weight of the dam per unit surface area. The seismic coefficient associated with earthquake forces in the first two modes of vibration of the dam (fundamental symmetric and anti-symmetric modes of a symmetric dam) varies, as shown in Figure 1.4.1. In contrast, traditional design procedures ignore the vibration properties of the dam and adopt a uniform distribution for the seismic coefficient, resulting in erroneous distribution of lateral forces and hence of stresses in the dam. A dynamic analysis procedure that eliminates such errors is developed in Chapter 8. Including dam–water–foundation interaction, this procedure is shown in Chapter 10 to produce seismic response results that are consistent with the motions of two arch dams recorded during earthquakes.

As mentioned in Section 1.3.3, the additional water pressures included in traditional design procedures for gravity dams are unrealistically small and have little influence on the computed stresses and hence on the geometry of the dam that satisfies the design criteria. This observation is equally valid for arch dams because the additional water pressures considered for arch dams are similar to those for gravity dams.

Demonstrated in Chapter 9 is the importance of two interaction mechanisms – which are ignored in traditional design – in the dynamics of arch dams. When dam–water interaction and water compressibility are properly considered, hydrodynamic effects result in significant increases in the earthquake-induced stresses in arch dams, more so than for gravity dams. Similarly, when dam–foundation interaction including foundation mass and radiation damping are properly considered, this interaction mechanism generally has a profound influence on the earthquake-induced stresses in arch dams just as in the case of gravity dams.

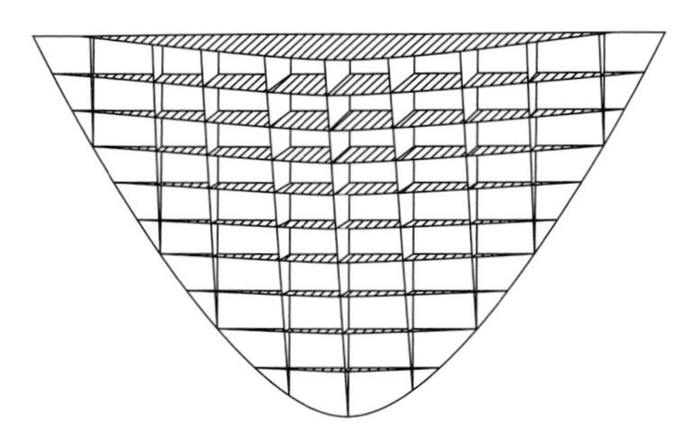

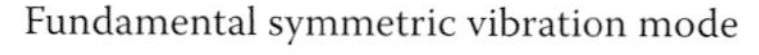

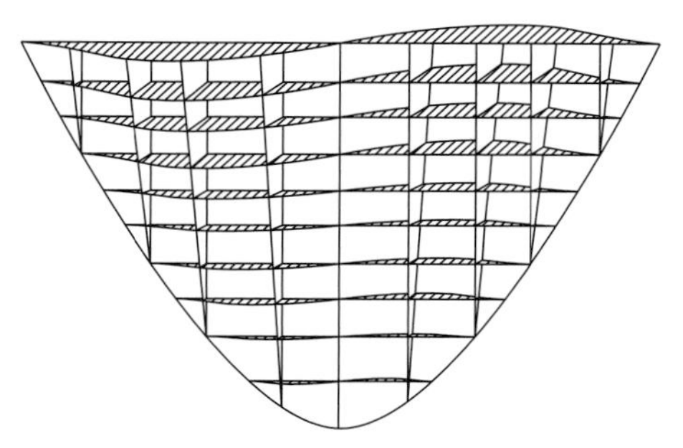

Fundamental antisymmetric vibration mode

Figure 1.4.1 Distribution of seismic coefficients over the dam surface in the first two vibration modes of an arch dam. Source: Adapted from Bureau of Reclamation (1977).

1.5 UNREALISTIC ESTIMATION OF SEISMIC DEMAND AND STRUCTURAL CAPACITY

Traditional design procedures greatly underestimate seismic demands imposed on both arch and gravity dams, as well as the capacity of these structures to resist these demands. The seismic forces associated with the mass of the dam and the hydrodynamic pressures are underestimated, as mentioned earlier. The tensile strength of concrete, which is not insignificant, is essentially ignored in the no-tension requirement in the design criteria for gravity dams and by the small value allowed for arch dams. A progressive, systematic approach to computation of seismic demands and evaluation of structural capacity is presented in Chapter 12.

Methods for designing dams must be improved in at least two major ways: (i) the tensile strength of concrete should be determined by testing cylindrical cores that are large enough – diameter equal to three or four times the size of the coarse aggregate; and (ii) seismic demands should be computed by dynamic response analysis of the dam–water–foundation system. Development of such analysis procedures is one of the main thrusts of this book.

1.6 REASONS WHY STANDARD FINITE-ELEMENT METHOD IS INADEQUATE

It is apparent from the preceding section that traditional seismic coefficient methods must be abandoned in favor of dynamic analysis procedures in order to reliably predict the earthquake-induced demands on dams. Because of the versatility of the FEM in modeling arbitrary geometries and variations of material properties, this method is suited for formulating a computational model of a concrete dam. In fact, analysis of the dam alone (no impounded water) supported on rigid foundation to ground motion specified at the base would be a standard application of the FEM. However, analysis of concrete dams is greatly complicated by the fact that the structure interacts with the water impounded in the reservoir and with the deformable foundation that supports it, and because the fluid and foundation domains extend to large distances (Figures 1.2.1 and 1.2.2).

The interaction mechanisms may be modeled in a crude way by combining finite-element models for a limited extent of the impounded water and of the foundation with a finite-element model of the dam, thus reducing the "semi-unbounded" system to a finite-sized model with rigid boundaries, which, generally, do not exist at the site (Figure 1.6.1). Such a model does not allow for radiation of hydrodynamic pressure waves in the upstream direction or stress waves in the foundation because these waves are reflected back from the rigid boundaries, thus trapping the energy in the bounded system. Thus, a significant energy loss mechanism, referred to as radiation damping, is not represented in the bounded models of the fluid and foundation domains. Developing procedures for analysis of dam–water–foundation systems that recognize the semi-unbounded geometry of the fluid and foundation domains was a major research objective during the 1970–1995 era. Research results on this challenging problem are featured prominently in this book.

While such research was in progress, an expedient solution was proposed by Clough (1980) that included in the finite-element model a limited extent of the foundation, assumed to have no mass, and modeled hydrodynamic effects by an added mass of water moving with the dam; the design ground motion defined typically at the ground surface was applied at the bottom fixed boundary of the foundation domain; see Figure 1.6.2. This modeling approach became popular in actual projects because it was easy to implement in commercial finite-element software. However, such a model solves a problem that is very different from the real problem on two

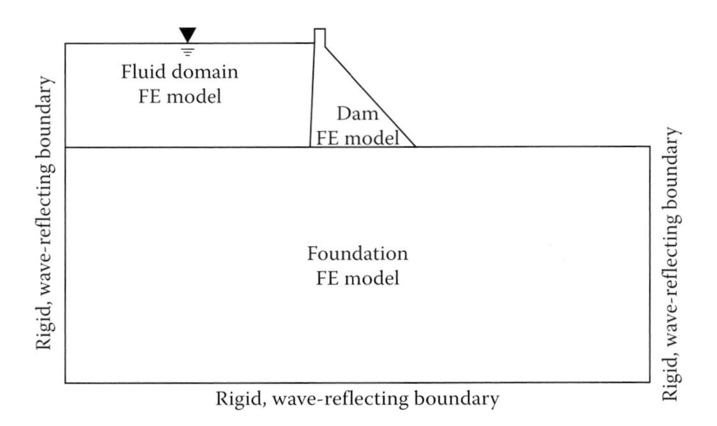

Figure 1.6.1 Standard finite-element analysis model with rigid, wave-reflecting boundaries.

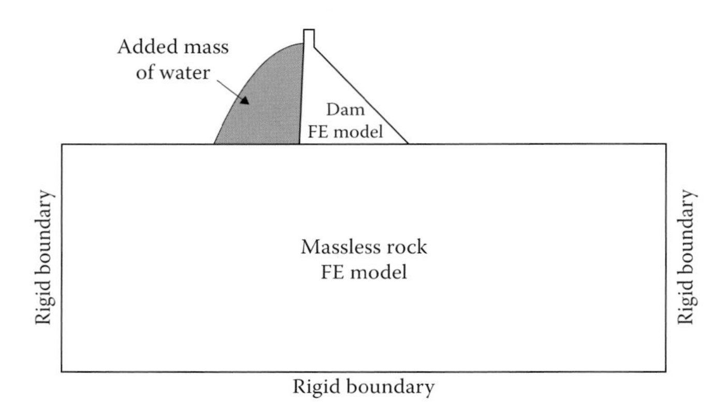

Figure 1.6.2 A popular finite-element model that assumes foundation to have no mass and models hydrodynamic effects by an added mass of water moving with the dam.

counts: (i) the assumptions of massless rock and incompressible water – implied by the added mass water model – are unrealistic, as will be demonstrated in Chapters 6 and 9; and (ii) applying ground motion specified at the ground surface to the bottom boundary of the finite-element model contradicts the recorded evidence that motions at depth generally differ significantly from surface motions.

1.7 RIGOROUS METHODS

Earthquake analysis of dams should include the following factors: (i) the semi-unbounded extent of the impounded water and foundation domains; (ii) dam–foundation interaction considering mass, flexibility, and damping of rock; and (iii) dam–water interaction considering compressibility of water and the sediments that invariably deposit at the reservoir bottom. Two approaches exist for such rigorous analyses: the *substructure method* and a *direct finite-element method*.

Presented in Chapters 5 and 8, the substructure method determines the response of idealized systems shown in Figures 1.7.1 and 1.7.2 to free-field ground motion specified at the interface between the dam and foundation; this is the motion that would have existed in the absence of the dam and impounded water. The substructure method permits different types of models for the

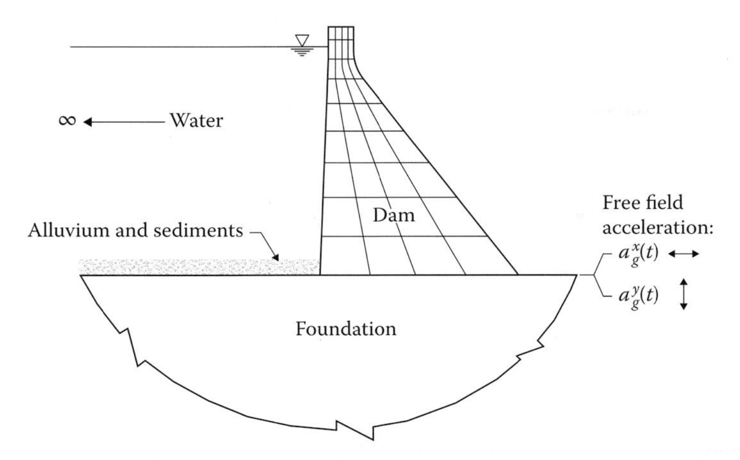

Figure 1.7.1 Gravity dam–water–foundation system.

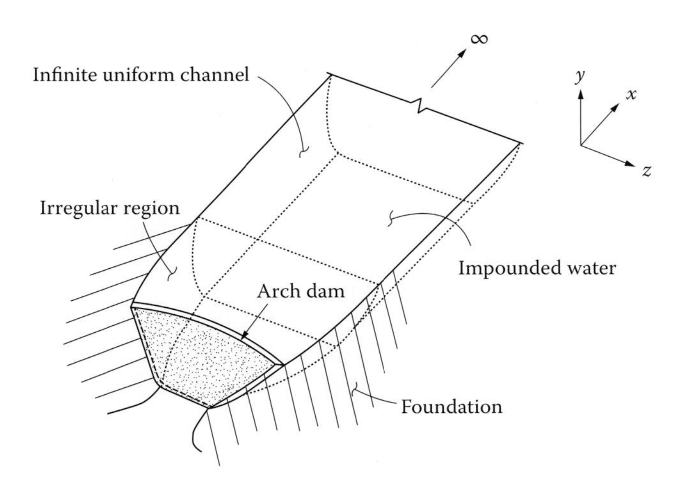

Figure 1.7.2 Arch dam–water–foundation system.

three substructures – dam, fluid domain, and foundation domain: finite-element model for the dam; and "continuum" models for the fluid and foundation domains of semi-unbounded geometry. The substructure concept permits modeling of the semi-unbounded fluid and foundation domains without truncating them to finite size and specifying the earthquake excitation directly at the dam–foundation interface.

Formulated in the frequency domain, this method is restricted to linear analysis, and requires special purpose computer programs, e.g. EAGD-84 for two-dimensional analysis of gravity dams and EACD-3D-1996 for three-dimensional analysis of arch dams. These freely available programs were developed by graduate students at the University of California, Berkeley, as a part of their research for the doctoral degree, not as commercial software programs. Thus, they lack user-friendly interfaces to facilitate input of data to define the system to be analyzed and to process response results. Despite these limitations, the aforementioned programs have been employed for seismic design of a few new dams and for seismic evaluation of several existing dams.

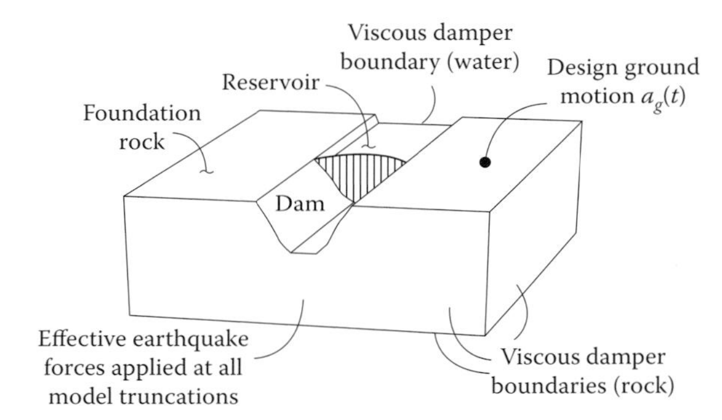

Figure 1.7.3 Finite-element model of a dam–water–foundation system with wave-absorbing boundaries.

Although linear analyses have provided great insight into the earthquake response of concrete dams, it is evident that a reliable estimate of the seismic safety of a dam can be obtained only by a nonlinear analysis if the earthquake damage is expected to be significant. The nonlinear model must recognize the possibility that the reservoir may extend to great distances upstream of the dam and the supporting rock extend to large depths and large distances in horizontal directions (Figures 1.2.1 and 1.2.2). The Direct FEM, presented in Chapter 11, overcomes the limitations of the standard FEM by introducing wave-absorbing (or non-reflecting) boundaries at two locations: (i) upstream end of the fluid domain to model its essentially semi-infinite length; and (ii) the bottom and side boundaries of the foundation domain to model its semi-unbounded geometry (Figure 1.7.3). The finite-element model of the fluid domain now includes water compressibility and reservoir bottom sediments, and the finite-element model of the foundation domain includes mass, stiffness, and material damping appropriate for the rock; water–foundation interaction is also included. Thus, the untenable assumptions of massless rock and incompressible water in the popular FEM are eliminated.

The earthquake excitation also is more realistically defined in the Direct FEM compared to the popular FEM. The excitation defined at the bottom and side boundaries of the foundation domain is determined by deconvolution of the design ground motion, typically specified on level ground at the elevation of the abutments (Figure 1.7.3). The resulting spatially varying motions cannot be input directly at wave-absorbing boundaries; instead, tractions determined from the motions are specified.

Presented in Chapter 11, the direct FEM has the great advantage over the substructure method in that is it applicable to nonlinear systems, thus permitting modeling of concrete cracking, as well as sliding and separation at contraction joints, lift joints, dam–foundation interface, and fissures in rock; however, it has the disadvantage in that it requires truncation of fluid and foundation domains, thus requiring absorbing boundaries to simulate their semi-unbounded size. This method has been developed in a form that can be implemented in any commercial finite-element code; thus, it is applicable to 3D models of all types of concrete dams: gravity, arch, and buttress. Validation of the Direct FEM applied to linear systems against the substructure method is also included in Chapter 11.

1.8 SCOPE AND ORGANIZATION

The primary goals of the book are: (i) develop dynamic analysis procedures to determine the response of concrete dams to ground shaking; (ii) identify factors that must be included in the response analyses; (iii) describe procedures for selecting ground motions for dynamic analyses;

and (iv) illustrate application of these procedures to the seismic design of new dams and safety evaluation of existing dams. Ground shaking is the only earthquake hazard that is considered; excluded are hazards such as fault rupture under the dam or its abutments, and landslides around the reservoir that may cause overtopping of the dam.

The book is organized into three parts: I. Gravity Dams: Chapters 2–7; II. Arch Dams: Chapters 8–11; and III. Design and Safety Evaluation: Chapters 12–14. Part I and Part II are developed to address the first two of the above-stated goals for gravity dams and arch dams, respectively. The various topics covered in Chapters 2–11 were mentioned in the preceding sections.

Part III includes three chapters. In Chapter 12, two levels of design earthquakes are defined and the performance requirements for the dam during each earthquake stated, among other topics. Chapter 13 is concerned with construction of the target design spectrum as well as selection of an ensemble of three-components of ground motions consistent with this spectrum. The book closes with Chapter 14 where application of modern dynamic analysis procedures to four projects are summarized.

Part I

GRAVITY DAMS

2

Fundamental Mode Response of Dams Including Dam–Water Interaction

PREVIEW

The motions of a dam during an earthquake cause dynamic pressures in the impounded water that act on the upstream face of the dam to modify the dam motions, which in turn influence the hydrodynamic pressures. It is this interaction between the dam and water that is the subject of this chapter. Considering only the fundamental mode of vibration of the dam, we first develop a procedure for analysis of dam response including dam–water interaction. Thereafter, results for dam response are presented for a wide range of parameters that characterize the dam–water system. These results provide a basis to identify the effects of dam–water interaction and their influence on the vibration properties – natural vibration period and damping ratio – and on the response of concrete gravity dams to earthquake ground motion. Also investigated are the implications of neglecting compressibility of water, an approximation that enables representation of hydrodynamic effects by inertia forces associated with an added mass of water moving with the dam. Finally, we develop an equivalent single-degree-of-freedom (SDF) system to model the response of dams including dam–water interaction that enables estimation of peak response directly from the earthquake response (or design) spectrum. Such an analysis is intended for preliminary design and safety evaluation of dams.

2.1 SYSTEM AND GROUND MOTION

The system considered consists of a monolith of a concrete gravity dam fixed (or clamped) to the horizontal surface of underlying rock, assumed to be rigid, and impounding a reservoir† of water with wave-absorptive reservoir bottom (Figure 2.1.1). We will initially study the planar

† "Reservoir" is the place of storage, not the fluid itself.

Earthquake Engineering for Concrete Dams: Analysis, Design, and Evaluation, First Edition. Anil K. Chopra.

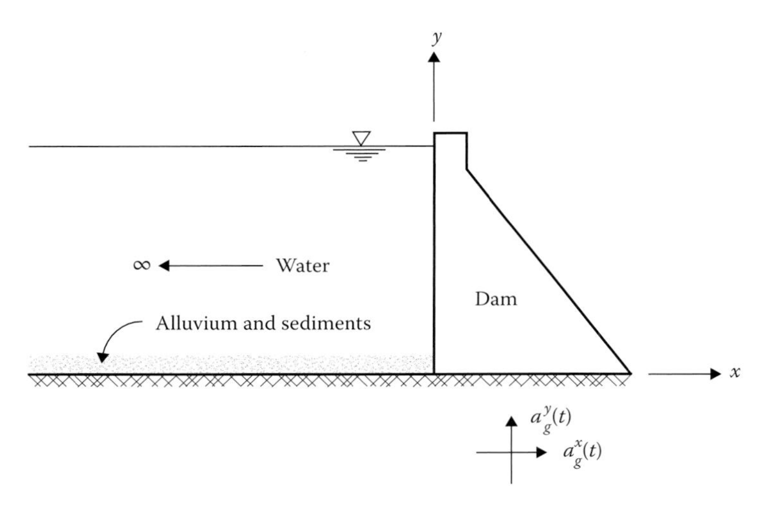

Figure 2.1.1 Dam–water system.

vibrations of an individual monolith due to earthquake excitation, a simplification that is supported by observations of monoliths vibrating somewhat independently during the earthquake response of Koyna Dam (Chopra and Chakrabarti 1972) and forced vibration tests of Pine Flat Dam (Rea et al. 1975); this simplification is discussed further in Section 5.1. The system is analyzed under the assumption of linear behavior.

The dam is idealized as a two-dimensional finite element system, thus making it possible to consider arbitrary geometry and variation of material properties. However, certain restrictions on the geometry are imposed to permit a continuum solution of the hydrodynamic wave equation in the fluid domain. For the purpose of determining hydrodynamic effects, and only for this purpose, the upstream face of the dam is assumed to be vertical. This assumption is reasonable for most concrete gravity dams, because typically the upstream face is vertical or almost vertical for most of its height, and the hydrodynamic pressure on the dam face is insensitive to small departures of the face slope from being vertical, especially if these departures are in the lower part of the dam, which is usually the case. The impounded water in the reservoir is idealized by a fluid region of constant depth and infinite length in the upstream direction.

The bottom of a reservoir upstream from a dam is generally not rigid; its flexibility could arise from flexibility of the underlying foundation[†] or deposited sediments (Figure 2.1.1). The reservoir bottom is approximately modeled by a boundary that partially absorbs incident hydrodynamic pressure waves; see Appendix 2.A for a description of this model.

The excitation for the two-dimensional dam–water system is defined by the two components of free-field ground acceleration in the plane of the monolith (or cross section) of the dam: the horizontal component $a_g^x(t)$ transverse to the longitudinal axis of the dam (i.e. in the stream direction) and the vertical component $a_g^y(t)$.

2.2 DAM RESPONSE ANALYSIS

2.2.1 Frequency Response Function

The displacements of the dam – relative to its base – vibrating in its fundamental vibration mode due to the l-component of ground motion ($l = x$ and y represents horizontal and vertical components, respectively) can be expressed as

$$r^k(x,y,t) = \phi_1^k(x,y)q_1^l(t) \qquad k = x, y; \qquad l = x, y \tag{2.2.1}$$

[†] The word "foundation" denotes the rock that supports the dam.

in which $r^x(x, y, t)$ and $r^y(x, y, t)$ are the horizontal and vertical components of displacement, respectively; $\phi_1^x(x, y)$ and $\phi_1^y(x, y)$ are the horizontal and vertical components, respectively, of the shape of the fundamental (or first) natural vibration mode of the dam fixed (or clamped) at its base to a rigid foundation with an empty reservoir; and $q_1^l(t)$ is the modal coordinate associated with this vibration mode.

Under the approximation of Eq. (2.2.1), the equation of motion for a dam supported on rigid foundation with an empty reservoir is

$$M_1 \ddot{q}_1^l + C_1 \dot{q}_1^l + K_1 q_1^l = -L_1^l a_g^l(t) \tag{2.2.2}$$

in which the generalized mass

$$M_1 = \int_A\int \left\{ m_x(x,y)\left[\phi_1^x(x,y)\right]^2 + m_y(x,y)\left[\phi_1^y(x,y)\right]^2 \right\} dx\, dy \tag{2.2.3}$$

where the integration extends over the cross-sectional area of the dam monolith; the mass density of the dam concrete $m_k(x, y) = m(x, y)$, $k = x$ and y is considered separately for the horizontal and vertical components of dam motion for convenience later in expressing the hydrodynamic effects in terms of an added mass and added damping; $C_1 = \zeta_1 \left(2M_1\omega_1\right)$; $K_1 = \omega_1^2 M_1$; ω_1, and ζ_1 are the fundamental natural frequency and the viscous damping ratio of the dam alone;

$$L_1^l = \int_A\int m_x(x,y)\phi_1^l(x,y)dx\ dy \tag{2.2.4}$$

Equation (2.2.2) can be rewritten as

$$\ddot{q}_1^l + 2\zeta_1\omega_1\dot{q}_1^l + \omega_1^2 q_1^l = -\Gamma_1^l a_g^l(t) \tag{2.2.5}$$

where

$$\Gamma_1^l = L_1^l / M_1 \tag{2.2.6}$$

For harmonic free-field ground acceleration $a_g^l(t) = e^{i\omega t}$, where ω is the exciting frequency, the modal coordinate can be expressed in terms of its complex-valued frequency response function, $q_1^l(t) = \bar{q}_1^l(\omega)e^{i\omega t}$. Upon substitution into Eq. (2.2.2) and canceling $e^{i\omega t}$ on both sides gives

$$\bar{q}_1^l(\omega) = \frac{-L_1^l}{-\omega^2 M_1 + i\omega C_1 + K_1} \tag{2.2.7}$$

We will later extend Eq. (2.2.7) to include dam–water interaction (Section 2.4) and dam–foundation interaction (Section 3.2.4).

2.2.2 Earthquake Response: Horizontal Ground Motion

In preparation for response spectrum analysis of the dam including dam–water–foundation interaction subjected only to horizontal ground motion (to be developed in Chapters 3 and 4), such analysis for the dam alone is presented first.

The response history of the modal coordinate $q_1^x(t)$ due to arbitrary ground acceleration in the x-direction can be computed from dam response to harmonic ground motion, characterized by the frequency response function (Eq. (2.2.7)), using standard Fourier synthesis techniques. Alternatively, it can be expressed in terms of $D_1(t)$, the deformation response of the first-mode single-degree-of-freedom (SDF) system, an SDF system with vibration properties – natural frequency ω_1 and damping ratio ζ_1 – of the first vibration mode of the dam. The equation of motion of this SDF system subjected to ground acceleration $a_g^x(t)$ is given by

$$\ddot{D}_1 + 2\zeta_1\omega_1\dot{D}_1 + \omega_1^2 D_1 = -a_g^x(t) \tag{2.2.8}$$

Having temporarily limited the earthquake response analysis to the x-component of ground motion, the superscript x may be dropped from q_1^x, a_g^x, and Γ_1^x. Comparing Eq. (2.2.5) to Eq. (2.2.8) gives the relation between q_1 and D_1:

$$q_1(t) = \Gamma_1 D_1(t) \tag{2.2.9}$$

where $D_1(t)$ can be determined by numerically solving Eq. (2.2.8). Substituting Eq. (2.2.9) in Eq. (2.2.1) gives the displacement history of the dam

$$r^k(x, y, t) = \Gamma_1 D_1(t)\ \phi_1^k(x, y) \qquad k = x, y \tag{2.2.10}$$

We will be especially interested in the peak value of response, or for brevity, *peak response*, defined as the maximum over time of the absolute value of the response quantity:

$$r_o \equiv \max_t |r(t)| \tag{2.2.11}$$

where the subscript "o" attached to a response quantity denotes its peak value. The peak displacements can then be expressed as

$$r_o^k(x, y) = \Gamma_1 D\left(T_1, \zeta_1\right)\ \phi_1^k(x, y) \qquad k = x, y \tag{2.2.12}$$

where $D\left(T_1,\ \zeta_1\right)$ is the ordinate of the deformation response (or design) spectrum for the x-component of ground motion evaluated at period $T_1 = 2\pi/\omega_1$, and damping ratio ζ_1; the subscript "o" that denotes peak value will subsequently be dropped to simplify notation.

The equivalent static forces associated with the peak displacements [Eq. (2.2.12)] are given by (Chopra 2017: Section 17.7)

$$f_1^k(x, y) = \Gamma_1 \frac{A\left(T_1, \zeta_1\right)}{g} w_k(x, y)\ \phi_1^k\ (x, y) \qquad k = x, y \tag{2.2.13}$$

in which $A\left(T_1,\ \zeta_1\right)$ is the ordinate of the pseudo-acceleration response (or design) spectrum, and $w_k(x, y) = gm_k(x, y)$. Because the vertical ($k = y$) component of displacements in the fundamental vibration mode, $\phi_1^k\ (x, y)$, is much smaller than their horizontal ($k = x$) component, the associated vertical forces may be dropped, leaving only the horizontal ($k = x$) component of forces in Eq. (2.2.13):

$$f_1(x, y) = \Gamma_1 \frac{A(T_1, \zeta_1)}{g} w_x(x, y)\ \phi_1^x\ (x, y) \tag{2.2.14}$$

wherein the superscript x has been dropped from $f_l^x(x, y)$ for simplicity of notation.

2.3 HYDRODYNAMIC PRESSURES

In this section we will present results for hydrodynamic pressures on the upstream face of the dam for two cases: (i) a rigid dam excited by x and y components of ground motion; and (ii) a flexible dam undergoing motion in its first mode of vibration; the three excitations are shown schematically in Figure 2.3.1. All three of these results will be utilized in deriving the fundamental mode response of the dam–water system (Section 2.4), and the hydrodynamic pressures on a rigid dam will be compared with classical solutions.

2.3.1 Governing Equation and Boundary Conditions

Assuming water to be linearly compressible and neglecting its internal viscosity, irrotational motion of the water is governed by the two-dimensional wave equation

$$\frac{\partial^2 p}{\partial x^2} + \frac{\partial^2 p}{\partial y^2} = \frac{1}{C^2}\frac{\partial^2 p}{\partial t^2} \tag{2.3.1}$$

where $p(x,\ y,\ t)$ is the hydrodynamic pressure (in excess of hydrostatic pressure), and C is the speed of pressure waves in water; $C = 4720$ fps or 1480 mps. The hydrodynamic pressure is generated by horizontal motion of the vertical upstream face of the dam and by vertical motion of

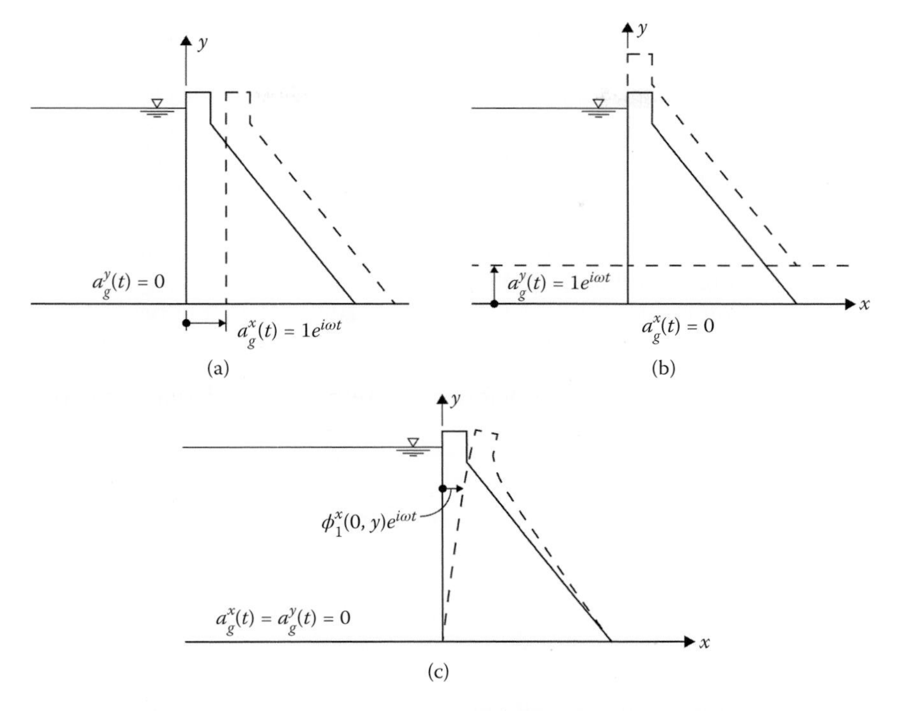

Figure 2.3.1 Acceleration excitations causing hydrodynamic pressures on the dam defined by frequency response functions: (a) $\bar{p}_0^x(0, y, \omega)$; (b) $\bar{p}_0^y(0, y, \omega)$; and (c) $\bar{p}_1(0, y, \omega)$.

the horizontal reservoir bottom. The boundary conditions for Eq. (2.3.1) governing the pressure are expressed in Eqs. (2.3.2)–(2.3.5).

The normal pressure gradient at the vertical upstream face of the dam is proportional to the horizontal acceleration of this boundary, resulting in the boundary condition for excitation cases (i) and (ii), respectively:

$$\frac{\partial p}{\partial x}(0, y, t) = -\rho a_g^x(t)\delta_{xl} \qquad l = x, y \tag{2.3.2a}$$

$$\frac{\partial p}{\partial x}(0, y, t) = -\rho \phi_1^x(0, y)\ddot{q}_1^l(t) \qquad l = x, y \tag{2.3.2b}$$

where ρ is the density of water, and δ_{kl} is the Kronecker delta function ($\delta_{xx} = \delta_{yy} = 1, \delta_{xy} = \delta_{yx} = 0$) and, contrary to the usual convention, summation is *not* implied when repeated indices appear.

Similarly, the normal pressure gradient at the horizontal bottom of the reservoir is proportional to the vertical acceleration of this boundary:

$$\frac{\partial p}{\partial y}(x, 0, t) = -\rho a_g^y(t)\delta_{yl} \qquad l = x, y \tag{2.3.3}$$

which is valid only if hydrodynamic waves are fully reflected at the boundary. This boundary condition is generalized to account for the influence of sediments at the reservoir bottom or of foundation flexibility on hydrodynamic pressures (Appendix 2)

$$\left(\frac{\partial}{\partial y} - \xi\frac{\partial}{\partial t}\right) p(x, 0, t) = -\rho a_g^y(t)\delta_{yl} \qquad l = x, y \tag{2.3.4a}$$

or

$$\frac{\partial p}{\partial y}(x, 0, t) = -\rho a_g^y(t)\delta_{yl} + \xi\frac{\partial p}{\partial t}(x, 0, t) \qquad l = x, y \tag{2.3.4b}$$

where $\xi = \rho/\rho_r C_r$, $C_r = \sqrt{(E_r/\rho_r)}$ is the compression wave velocity, E_r is the Young's modulus, and ρ_r is the density of the reservoir bottom materials. The second term on the right side in Eq. (2.3.4b) represents the modification of the vertical free-field ground acceleration due to flexibility at the reservoir bottom. Because this interactive acceleration is proportional to the time derivative of the hydrodynamic pressure, the reservoir-bottom flexibility produces a damping effect associated with partial refraction of hydrodynamic pressure waves at the reservoir bottom; ξ may be interpreted as a damping coefficient. For reservoir bottom that is rigid, $C_r = \infty$, $\xi = 0$, and the second term on the right-side of Eq. (2.3.4b) is zero, giving the boundary condition for a fully reflective reservoir bottom [Eq. (2.3.3)].

The wave reflection coefficient α, defined as the ratio of the amplitude of the reflected hydrodynamic pressure wave to the amplitude of a vertically propagating pressure wave incident on the reservoir bottom, is related to the damping coefficient, ξ (Appendix 2; Rosenblueth 1968; Hall and Chopra 1982) by

$$\alpha = \frac{1 - \xi C}{1 + \xi C} \tag{2.3.5}$$

The material properties of the sedimentary deposits at the reservoir bottom are highly variable and difficult to characterize. In contrast, the properties of the underlying rock can be better defined. Substituting them in Eq. (2.3.5) gives the corresponding value of α. For a realistic range of properties of rock, α would generally vary between 0.5 and 0.85. Researchers have attempted to measure α in the field (Ghanaat and Redpath 1995).

Neglecting the effects of waves at the free surface of water, an assumption discussed in Chopra (1967), leads to the boundary condition

$$p(x, H, t) = 0 \tag{2.3.6}$$

The hydrodynamic pressures must satisfy the boundary conditions of Eqs. (2.3.2), (2.3.4b), and (2.3.6), and the radiation condition in the upstream direction.

2.3.2 Solutions to Boundary Value Problems

The steady state hydrodynamic pressure due to unit harmonic free-field ground acceleration $a_g^l(t) = e^{i\omega t}$ can be expressed in terms of its complex frequency response function

$$p^l(x, y, t) = \bar{p}^l(x, y, \omega)e^{i\omega t} \tag{2.3.7}$$

Substituting this in Eq. (2.3.1) leads to the Helmholtz equation:

$$\frac{\partial^2 \bar{p}}{\partial x^2} + \frac{\partial^2 \bar{p}}{\partial y^2} + \frac{\omega^2}{C^2}\bar{p} = 0 \tag{2.3.8}$$

The frequency response function, $\bar{p}_0^x(x, y, \omega)$, for the hydrodynamic pressure when the excitation is the horizontal ground acceleration and the dam is rigid (Figure 2.3.1a), is the solution of Eq. (2.3.8) subject to the boundary conditions of Eqs. (2.3.2a), (2.3.4b), and (2.3.6) transformed according to Eq. (2.3.7):

$$\frac{\partial \bar{p}}{\partial x}(0, y, \omega) = -\rho \qquad \left(\frac{\partial}{\partial y} - i\omega\xi\right)\bar{p}(x, 0, \omega) = 0 \qquad \bar{p}(x, H, \omega) = 0 \tag{2.3.9}$$

The frequency response function $\bar{p}_0^y(x, 0, \omega)$ for the hydrodynamic pressure, when the excitation is the vertical ground acceleration and the dam is rigid (Figure 2.3.1b), is the solution of Eq. (2.3.8) subject to the boundary conditions of Eqs. (2.3.2a), (2.3.4b), and (2.3.6) transformed according to Eq. (2.3.7):

$$\frac{\partial \overline{p}}{\partial x}(0, y, \omega) = 0 \qquad \left(\frac{\partial}{\partial y} - i\omega\xi\right) \overline{p}(x, 0, \omega) = -\rho \qquad \overline{p}(x, H, \omega) = 0 \tag{2.3.10}$$

The frequency response function $\overline{p}_1(x, y, \omega)$ for the hydrodynamic pressure due to horizontal acceleration $\phi_1^x(0, y)e^{i\omega t}$ of the dam vibrating in its fundamental natural vibration mode (Figure 2.3.1c) is the solution of Eq. (2.3.8) subject to boundary conditions of Eqs. (2.3.2b), (2.3.4b), and (2.3.6) transformed according to Eq. (2.3.7):

$$\frac{\partial \overline{p}}{\partial x}(0, y, \omega) = -\rho\phi_1^x(0, y) \qquad \left(\frac{\partial}{\partial y} - i\omega\xi\right) \overline{p}(x, 0, \omega) = 0 \qquad \overline{p}(x, H, \omega) = 0 \tag{2.3.11}$$

The complex-valued frequency response functions $\overline{p}_0^l(x, y, \omega)$ and $\overline{p}_1(x, y, \omega)$ are obtained using standard solution methods for boundary value problems. Specialized for the upstream face of the dam, these functions are (Fenves and Chopra 1984a)

$$\overline{p}_0^x(0, y, \omega) = -2\rho H \sum_{n=1}^{\infty} \frac{\mu_n^2(\omega)}{H\left[\mu_n^2(\omega) - (\omega\xi)^2\right] + i(\omega\xi)} \frac{I_{0n}(\omega)}{\sqrt{\left[\mu_n^2(\omega) - (\omega^2/C^2)\right]}} Y_n(y, \omega) \tag{2.3.12a}$$

$$\overline{p}_0^y(0, y, \omega) = \frac{\rho}{\omega/C} \frac{1}{\cos(\omega H/C) + i\xi C \sin(\omega H/C)} \sin\frac{\omega(H - y)}{C} \tag{2.3.12b}$$

$$\overline{p}_1(0, y, \omega) = -2\rho H \sum_{n=1}^{\infty} \frac{\mu_n^2(\omega)}{H\left[\mu_n^2(\omega) - (\omega\xi)^2\right] + i(\omega\xi)} \frac{I_{1n}(\omega)}{\sqrt{\left[\mu_n^2(\omega) - (\omega^2/C^2)\right]}} Y_n(y, \omega) \tag{2.3.12c}$$

where

$$I_{0n}(\omega) = \frac{1}{H}\int_0^H Y_n(y, \omega)dy \qquad I_{1n}(\omega) = \frac{1}{H}\int_0^H \phi_1^x(0, y)Y_n(y, \omega)\, dy \tag{2.3.13}$$

The frequency response functions for hydrodynamic pressure due to horizontal motions of the upstream face of the dam, given by Eqs. (2.3.12a) and (2.3.12c) are the sum of the contributions of an infinite number of natural vibration modes of the impounded water. The complex-valued, frequency-dependent eigenvalues $\mu_n(\omega)$ satisfy Eq. (2.3.14) and the eigenfunctions $Y_n(y, \omega)$ are defined by Eq. (2.3.15):

$$e^{2i\mu_n(\omega)H} = -\frac{\mu_n(\omega) - \omega\xi}{\mu_n(\omega) + \omega\xi} \tag{2.3.14}$$

$$Y_n(y, \omega) = \frac{1}{2\mu_n(\omega)}\left\{[\mu_n(\omega) + \omega\xi]e^{i\mu_n(\omega)y} + [\mu_n(\omega) - \omega\xi]e^{-i\mu_n(\omega)y}\right\} \tag{2.3.15}$$

If the ground motion is vertical, pressure waves do not propagate upstream resulting in the much simpler frequency response function (Eq. (2.3.12b)), which is independent of the x-coordinate.

Non-absorptive Reservoir Bottom. For a rigid, non-absorptive reservoir bottom, as mentioned earlier, $\xi = 0$ and $\alpha = 1$; the eigenvalues $\mu_n(\omega)$ and eigenfunctions $Y_n(y, \omega)$ are real-valued and independent of the excitation frequency:

$$\mu_n(\omega) \equiv \mu_n = \frac{\omega_n^r}{C} \qquad \omega_n^r = \frac{2n - 1}{2}\pi\frac{C}{H} \tag{2.3.16}$$

where ω_n^r are the natural vibration frequencies of the impounded water with rigid non-absorptive reservoir bottom, and

$$Y_n(y, \omega) = \cos \mu_n y \tag{2.3.17}$$

Then Eqs. (2.3.12a) and (2.3.12b) specialize to

$$\overline{p}_0^x(0,y,\omega) = -\frac{4\rho}{\pi}\sum_{n=1}^{\infty}\frac{(-1)^{n-1}}{(2n-1)\sqrt{\mu_n^2-(\omega^2/C^2)}}\cos\mu_n y \tag{2.3.18}$$

$$\overline{p}_0^y(0,y,\omega) = \frac{\rho}{\omega/C}\frac{\sin(\omega/C)(H-y)}{\cos(\omega H/C)} \tag{2.3.19}$$

These are the same as the equations defining hydrodynamic pressures on the upstream face of a rigid dam for a non-absorptive reservoir bottom presented in Chopra (1967).

Incompressible Water. Neglecting compressibility of water is equivalent to assuming the speed C of the hydrodynamic pressure waves to be infinite. The limits of Eqs. (2.3.18) and (2.3.19) as $C \to \infty$ result in

$$\overline{p}_0^x(0,y) = -\frac{4\rho}{\pi}\sum_{n=1}^{\infty}\frac{(-1)^{n-1}}{(2n-1)\mu_n}\cos\mu_n y \tag{2.3.20}$$

$$\overline{p}_0^y(0,y) = \rho(H-y) \tag{2.3.21}$$

Observe that the hydrodynamic pressure functions $\overline{p}_0^x$ and $\overline{p}_0^y$ are now independent of the excitation frequency, and $\overline{p}_0^y$ is equal to the hydrostatic pressure.

2.3.3 Hydrodynamic Forces on Rigid Dams

The complex-valued frequency response functions $\overline{F}_0^l(\omega)$ for the hydrodynamic force on a rigid dam due to horizontal and vertical ground acceleration are computed from Eqs. (2.3.12a) and (2.3.12b) as the integral of $\overline{p}_0^l(0,y,\omega)$ over the depth of water. The real and imaginary components as well as the absolute value of $\overline{F}_0^l(\omega)$, normalized with respect to the hydrostatic force $F_{st} = \rho g H^2/2$, are plotted in Figures 2.3.2 and 2.3.3 for five different values of α as a function of the dimensionless excitation frequency ω/ω_1^r, where $\omega_1^r = \pi C/2H$ (Eq. (2.3.16)) is the first natural vibration frequency of the impounded water with non-absorptive reservoir bottom. When presented in this non-dimensional form, the plots apply to fluid domains of any depth. The real and imaginary components represent the in-phase (or 180°-out-of-phase) and 90°-out-of-phase hydrodynamic forces relative to the harmonic ground acceleration.

If the reservoir bottom is non-absorptive, i.e. $\alpha = 1$, the hydrodynamic forces due to both ground motion components are unbounded at the natural vibration frequencies ω_n^r of the impounded water. The hydrodynamic force due to vertical ground motion is in-phase or opposite-phase relative to the ground acceleration for all excitation frequencies. The hydrodynamic force due to horizontal ground motion is of opposite-phase relative to the ground acceleration for excitation frequencies less than the first natural vibration frequency ω_1^r, but a 90°-out-of-phase component exists for higher excitation frequencies.

As mentioned earlier, the hydrodynamic pressure, Eq. (2.3.12a), and hence the total force on a rigid dam due to horizontal ground motion have been expressed as an infinite series wherein each term represents the contribution of a natural vibration mode of the impounded water. If the reservoir bottom is non-absorptive, i.e. $\alpha = 1$, the contribution of the nth mode is real-valued with opposite-phase relative to the ground acceleration for excitation frequencies lower than ω_n^r, the nth natural vibration frequency; but is imaginary-valued, i.e. 90°-out-of-phase relative to the ground acceleration, for excitation frequencies higher than ω_n^r, and is unbounded when the excitation frequency is equal to ω_n^r. For excitation frequencies higher than ω_n^r the pressure wave associated with the nth mode propagates in the upstream direction of the infinitely long fluid domain resulting in radiation of energy, As the excitation frequency increases past ω_n^r, the hydrodynamic force contribution of the nth mode changes from a pressure function decaying exponentially in the upstream direction to one propagating in the upstream direction, thus reducing the real component of $\overline{F}_0^l(\omega)$ and increasing its imaginary component (Figure 2.3.2). With increasing excitation frequency, a larger number of modes are associated

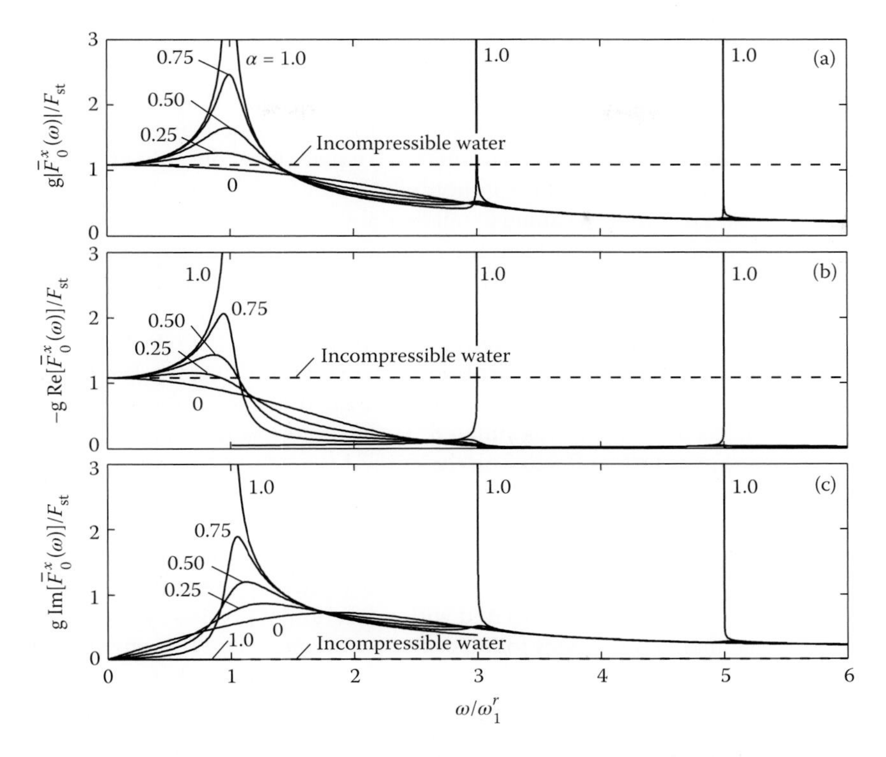

Figure 2.3.2 Hydrodynamic force on rigid dam due to horizontal ground acceleration: (a) absolute value; (b) real component; and (c) imaginary component.

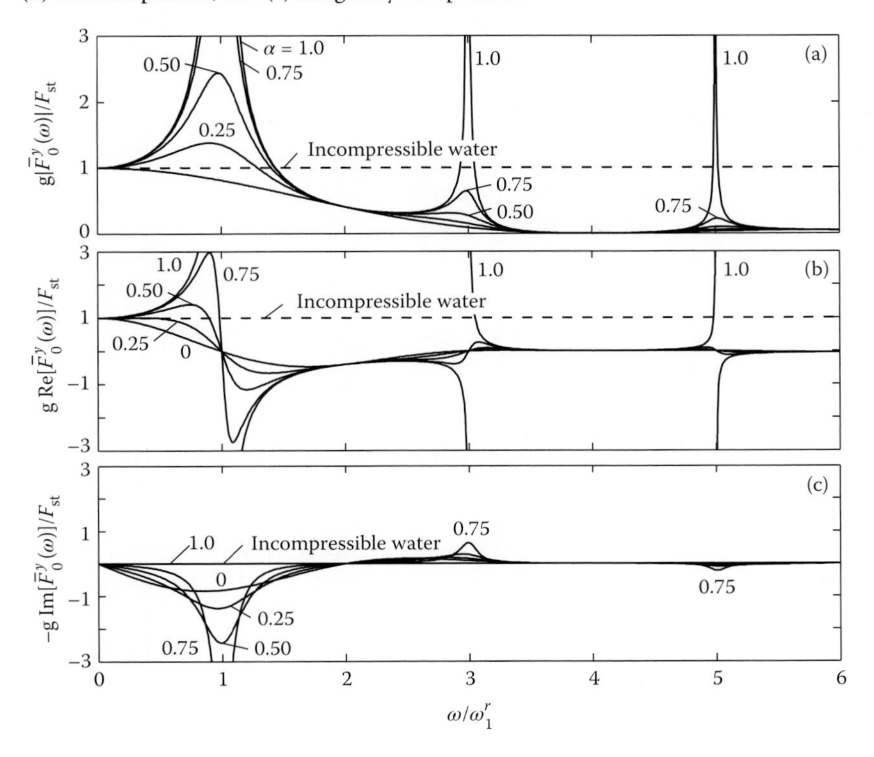

Figure 2.3.3 Hydrodynamic force on rigid dam due to vertical ground acceleration: (a) absolute value; (b) real component; and (c) imaginary component.

with the propagating pressure waves, leading to increased energy radiation and hence smaller hydrodynamic force (Figure 2.3.2a) – except for the local resonances at ω_n^r; these unbounded resonances are unrealistic artifacts of a non-absorptive boundary at the reservoir bottom.

For an absorptive reservoir bottom, the frequency-dependent eigenvalues $\mu_n(\omega)$ of the impounded water are complex-valued for all excitation frequencies. Consequently, the contribution of the nth natural vibration mode of the impounded water to the hydrodynamic force due to horizontal ground motion is complex-valued for all excitation frequencies; wherein the imaginary (or 90°-out-of-phase) component arises from the radiation of energy due to propagation of pressure waves in the upstream direction and their refraction into the reservoir bottom. This implies that if the reservoir bottom is absorptive, the hydrodynamic force contains a 90°-out-of-phase component even for excitation frequencies lower than ω_1^r (Figure 2.3.2c). Because of the additional energy loss resulting from wave absorption at the reservoir bottom, the hydrodynamic force is bounded for all excitation frequencies, the fundamental resonant peak is reduced, and the higher resonant peaks are virtually eliminated. However, the additional energy absorption into the reservoir bottom has little influence on the natural frequencies of the impounded water.

The hydrodynamic pressure due to vertical ground motion is independent of the upstream coordinate (Chopra 1967) and the pressure waves do not propagate in the upstream direction, resulting in a truly undamped system if the reservoir bottom is non-absorptive. The hydrodynamic pressure is real-valued, in-phase, or opposite-phase relative to the ground acceleration, for all excitation frequencies. Energy loss associated with refraction of pressure waves into a flexible bottom leads to an imaginary component for all excitation frequencies. This energy loss reduces the response at all frequencies and the resonant responses are now bounded.

If water compressibility is neglected, the frequency response functions for hydrodynamic pressure on a rigid dam, given by Eqs. (2.3.20) and (2.3.21), are real-valued and independent of the excitation frequency (Figures 2.3.2 and 2.3.3). The hydrodynamic force due to vertical ground motion is equal to the hydrostatic force (Figure 2.3.3), and in-phase with the ground acceleration; whereas the hydrodynamic force due to horizontal ground motion is slightly larger than the hydrostatic force (Figure 2.3.2), and has opposite-phase relative to the ground acceleration.

2.3.4 Westergaard's Results and Added Mass Analogy

In 1933 Westergaard derived an equation for the hydrodynamic pressure on the upstream face of a rigid dam due to time-harmonic horizontal ground motion, a result that for several decades profoundly influenced the treatment of hydrodynamic effects in dam analysis. The range of validity of this result will be identified in this section. His result for hydrodynamic pressure on the upstream face of the dam due to $a_g^x(t) = \cos \omega t$, expressed in the Cartesian coordinate system and notation adopted herein, is

$$\bar{p}_0^x(0, y, t) = -\frac{4\rho}{\pi} \cos \omega t \sum_{n=1}^{\infty} \frac{(-1)^{n-1}}{(2n-1)\sqrt{\mu_n^2 - (\omega^2/C^2)}} \cos \mu_n y \tag{2.3.22}$$

To evaluate this classical result, we substitute Eq. (2.3.18) in Eq. (2.3.7), and separate the real part to obtain the hydrodynamic pressure due to the excitation $a_g^x(t) = \cos \omega t$:

$$p_0^x(0, y, t) = \frac{-4\rho}{\pi} \left[\sin \omega t \left(\sum_{n=1}^{n_1-1} \frac{(-1)^{n-1}}{(2n-1)\sqrt{(\omega^2/C^2) - \mu_n^2}} \cos \mu_n y \right) \right.$$
$$\left. + \cos \omega t \left(\sum_{n=n_1}^{\infty} \frac{(-1)^{n-1}}{(2n-1)\sqrt{\mu_n^2 - (\omega^2/C^2)}} \cos \mu_n y \right) \right] \tag{2.3.23}$$

where n_1 = the minimum value of n such that $\mu_n > \omega/C$ or $\omega < \omega_n^r$. Equations (2.3.23) and (2.3.22) are identical if $n_1 = 1$, i.e. $\omega < \omega_1^r$, because then the term involving $\sin \omega t$ in Eq. (2.3.23) vanishes. Thus, Westergaard's solution is valid only if the excitation frequency ω is less than the fundamental frequency ω_1^r of the fluid domain (Chopra 1967).

Westergaard's classic paper introduced the concept that the hydrodynamic pressure acting on the upstream face of a rigid dam due to *horizontal* ground motion can be interpreted as the inertia forces associated with an added mass m_a of water moving with the dam:

$$p_0^x(0, y, t) = -m_a(y, \omega) a_g^x(t) \tag{2.3.24}$$

Comparing this with Eq. (2.3.22) and recalling that $a_g^x(t) = \cos \omega t$, the added mass is

$$m_a(y, \omega) = \frac{4\rho}{\pi} \sum_{n=1}^{\infty} \frac{(-1)^{n-1}}{(2n-1)\sqrt{\mu_n^2 - (\omega^2/C^2)}} \cos \mu_n y \tag{2.3.25}$$

Because Eq. (2.3.22) is valid only for $\omega < \omega_1^r$, the added mass analogy is also restricted to the same range of frequencies. Note that the added mass of Eq. (2.3.25) depends on the excitation frequency and is relevant only for horizontal ground motion in the stream direction.

If the compressibility of water is neglected, the added mass is given by the limit of Eq. (2.3.25) as the wave speed C approaches infinity, resulting in

$$m_a(y) = \frac{4\rho}{\pi} \sum_{n=1}^{\infty} \frac{(-1)^{n-1}}{(2n-1)\mu_n} \cos \mu_n y \tag{2.3.26}$$

Observe from Eqs. (2.3.25) and (2.3.26) that the added mass is independent of the excitation frequency only when water compressibility is neglected. This added mass may then be visualized as the mass of a body of water of width

$$b_w(y) = m_a(y) \div \rho \tag{2.3.27}$$

Moving with a rigid dam, the body of water defined by Eqs. (2.3.27) and (2.3.26) is shown in Figure 2.3.4. Also included is Westergaard's (1933) popular approximation,

$$b_w(y) = \frac{7}{8}\sqrt{H(H-y)} \tag{2.3.28}$$

Although the two results are close, neither of them is valid because they ignore compressibility of water that has an important influence on the response of dams, as will be demonstrated in Section 2.5.4. Before closing this section, we note that the above-mentioned added mass concept was restricted to horizontal ground motion in the stream direction.

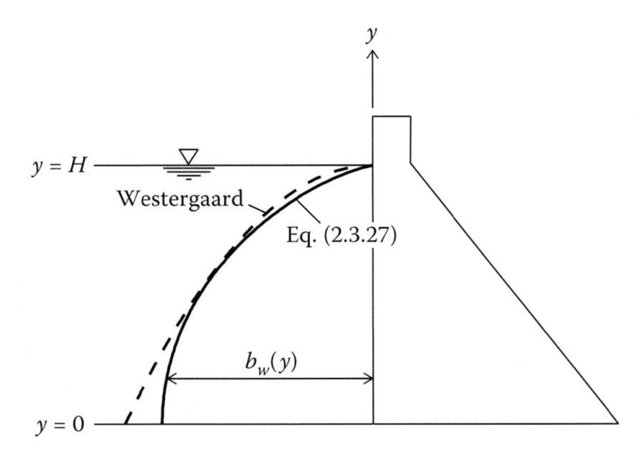

Figure 2.3.4 Body of water, assumed to be incompressible, moving with a rigid dam subjected to horizontal ground acceleration. Two results are presented: Eqs. (2.3.27) and (2.3.2).

2.4 DAM RESPONSE ANALYSIS INCLUDING DAM–WATER INTERACTION

Equation (2.2.2), which governs the fundamental modal coordinate, is extended to include the hydrodynamic pressure, $p^l(0, y, t)$, on the upstream face ($x = 0$) of the dam, resulting in

$$M_1\ddot{q}_1^l(t) + C_1\dot{q}_1^l(t) + K_1 q_1^l = -L_1^l a_g^l(t) + \int_0^H p^l(0,y,t)\phi_1^x(0,y)dy \tag{2.4.1}$$

The hydrodynamic pressure is generated by horizontal acceleration of the upstream face of the dam:

$$\ddot{r}^x(0,y,t) = a_g^x(t)\delta_{xl} + \phi_1^x(0,y)\ddot{q}_1^l(t) \qquad l = x, y \tag{2.4.2a}$$

and by vertical acceleration of the reservoir bottom:

$$\ddot{v}(x,0,t) = a_g^y(t)\delta_{yl} \qquad l = x, y \tag{2.4.2b}$$

The normal pressure gradient at the vertical upstream face of the dam is proportional to the total acceleration of this boundary, leading to the boundary condition:

$$\frac{\partial p}{\partial x}(0,y,t) = -\rho\left[a_g^x(t)\delta_{xl} + \phi_1^x(0,y)\ddot{q}_1^l(t)\right] \tag{2.4.3}$$

The boundary conditions at the reservoir bottom and free surface of water are given by Eqs. (2.3.4b) and (2.3.6), respectively. In addition to these boundary conditions, the hydrodynamic pressures must satisfy the radiation condition in the upstream direction.

The steady-state response of the dam–water system to unit harmonic free-field ground acceleration, $a_g^l(t) = e^{i\omega t}$, can be expressed in terms of complex-valued frequency response functions. Thus the modal coordinate and hydrodynamic pressure are given by

$$q_1^l(t) = \bar{q}_1^l(\omega)e^{i\omega t} \tag{2.4.4}$$

$$p^l(x,y,t) = \bar{p}^l(x,y,\omega)e^{i\omega t} \tag{2.4.5}$$

and Eq. (2.4.1) can be expressed in terms of the frequency response functions:

$$\left[-\omega^2 M_1 + i\omega C_1 + \omega_1^2 K_1\right]\bar{q}_1^l(\omega) = -L_1^l + \int_0^H \bar{p}^l(0,y,\omega)\phi_1^x(0,y)dy \tag{2.4.6}$$

Similarly, the wave Eq. (2.3.1), becomes the Helmholz equation (2.3.8), and the boundary accelerations of Eq. (2.4.2) become

$$\ddot{r}^x(0,y,t) = \left[\delta_{xl} + \phi_1^x(0,y)\,\ddot{\bar{q}}_1^l(\omega)\right]e^{i\omega t} \tag{2.4.7a}$$

$$\ddot{v}(x,0,t) = \delta_{yl}\,e^{i\omega t} \tag{2.4.7b}$$

The frequency response function $\bar{p}^l(x,y,\omega)$ is governed by Eq. (2.3.8) subject to the boundary conditions of Eqs. (2.4.3), (2.3.4b), and (2.3.6) transformed according to Eqs. (2.4.4) and (2.4.5):

$$\frac{\partial}{\partial x}\bar{p}(0,y,\omega) = -\rho\left[\delta_{xl} + \phi_1^x(0,y)\ddot{\bar{q}}_1^l(\omega)\right] \tag{2.4.8a}$$

$$\left[\frac{\partial}{\partial y} - i\omega\xi\right]\bar{p}(x,0,\omega) = -\rho\delta_{yl} \tag{2.4.8b}$$

$$\bar{p}(x,H,\omega) = 0 \tag{2.4.8c}$$

Note that the terms multiplying $-\rho$ on the right side of Eqs. (2.4.8a) and (2.4.8b) are the amplitudes of the boundary accelerations given by Eq. (2.4.7).

Using the principle of superposition, which is applicable because the governing equations and boundary conditions are linear, the frequency response function for hydrodynamic pressure can be expressed as

$$\bar{p}^l(x,y,\omega) = \bar{p}_0^l(x,y,\omega) + \bar{p}_1(x,y,\omega)\bar{\ddot{q}}_1^l(\omega) \tag{2.4.9}$$

where the frequency response functions $\bar{p}_0^l(x,y,\omega)$ and $\bar{p}_1(x,y,\omega)$ were presented in Eq. (2.3.12).

Substituting Eq. (2.4.9) with $\bar{\ddot{q}}_1^l(\omega) = -\omega^2\bar{q}_1^l(\omega)$ into Eq. (2.4.6) leads to the frequency response function for the fundamental modal coordinate when the dam is subjected to the l-component of ground motion ($l = x, y$):

$$q_1^l(\omega) = \frac{-\left[L_1^l + B_0^l(\omega)\right]}{-\omega^2\left\{M_1 + \mathrm{Re}\left[B_1(\omega)\right]\right\} + i\omega\left\{C_1 - \omega\,\mathrm{Im}\left[B_1(\omega)\right]\right\} + K_1} \tag{2.4.10}$$

in which

$$B_0^l(\omega) = -\int_0^H \bar{p}_0^l(0,y,\omega)\phi_1^x(0,y)dy \tag{2.4.11a}$$

$$B_1(\omega) = -\int_0^H \bar{p}_1(0,y,\omega)\phi_1^x(0,y)dy \tag{2.4.11b}$$

Equation (2.4.10) may be expressed in terms of the natural vibration frequency ω_1 and damping ratio ζ_1 of the dam alone:

$$q_1^l(\omega) = \frac{-\left[L_1^l + B_0^l(\omega)\right]}{-\omega^2\left\{M_1 + \mathrm{Re}\left[B_1(\omega)\right]\right\} + i\omega\left\{2\zeta_1\omega_1 M_1 - \omega\,\mathrm{Im}\left[B_1(\omega)\right]\right\} + \omega_1^2 M_1} \tag{2.4.12}$$

A comparison of Eq. (2.4.10) with Eq. (2.2.7) shows that the effects of dam–water interaction and reservoir bottom absorption are contained in the frequency-dependent hydrodynamic terms $B_0(\omega)$ and $B_1(\omega)$. The hydrodynamic effects can be interpreted as introducing an added force $B_0^l(\omega)$, and modifying the properties of the dam by an added mass represented by the real component of $B_1(\omega)$, and an added damping represented by the imaginary component $B_1(\omega)$. The added mass arises from the portion of the impounded water that reacts in phase with the motion of the dam, and the added damping arises from radiation of pressure waves in the upstream direction and from their refraction into the absorptive reservoir bottom.

2.5 DAM RESPONSE

2.5.1 System Parameters

The frequency response function $\bar{\ddot{q}}_1^l(\omega) = -\omega^2\bar{q}_1^l(\omega)$ for a dam with a fixed cross-sectional geometry and Poisson's ratio, when expressed as a function of the normalized excitation frequency ω/ω_1, depends on three system parameters: $\Omega_r = \omega_1^r/\omega_1$, the ratio of the fundamental natural vibration frequency of the impounded water to that of the dam alone; H/H_s, the ratio of water depth to the dam height; and α, the wave reflection coefficient at the reservoir bottom (Chopra 1968). We know that $\omega_1^r = \pi C/2H$ (Eq. (2.3.16)), and it can be shown that $\omega_1 = \gamma C_s/H_s$, where γ is a dimensionless factor that depends on the cross-sectional shape of the dam monolith and the Poisson's ratio of the concrete in the dam, $C_s = \sqrt{E_s/\rho_s}$, E_s is the Young's modulus, and ρ_s is the density of concrete. Therefore

$$\Omega_r = \frac{\pi}{2\gamma}\frac{C}{C_s}\frac{H_s}{H} \tag{2.5.1}$$

For fixed values of γ, C, ρ_s, and H/H_s, the frequency ratio Ω_r, is proportional to $1/\sqrt{E_s}$. Thus Ω_r decreases with increasing E_s or dam stiffness, and vice versa.

If the reservoir is empty or water is assumed to be incompressible, $\bar{\ddot{q}}_1^l(\omega)$, when expressed as a function of ω/ω_1, is independent of E_s, and α; the incompressible case implies $C = \infty$ and thus $\Omega_r = \infty$.

2.5.2 System and Cases Analyzed

The idealized monolith considered has a triangular cross section with a vertical upstream face and a downstream face with a slope of 0.8 horizontal to 1.0 vertical. The dam is assumed to be homogeneous and isotropic with linearly elastic properties for mass concrete: Poisson's ratio = 0.2, unit weight = 155 lb/cu ft, and damping ratio $\zeta_1 = 5\%$. The Young's modulus for mass concrete is varied over a wide range by specifying three values for the frequency ratio $\Omega_r = 0.80$, 1.0, and 2.0, which, for the selected system properties, correspond to $E_s = 3.94$, 2.52, and 0.63 million psi, respectively. The first two cover the range of values representative of mass concrete, and the third is unrealistically low, chosen to help identify the condition under which water compressibility can be neglected. The unit weight of water is 62.4 lb/cu ft and the velocity of pressure waves in water is $C = 4720$ ft/sec. Two values for the depth of water are considered: an empty reservoir ($H/H_s = 0$) and a full reservoir ($H/H_s = 1$). The wave reflection coefficient is varied over a wide range; the values considered are: $\alpha = 1.0$, 0.75, 0.50, 0.25, and 0.

The response of the dam to harmonic horizontal and vertical ground motion is determined by numerically evaluating Eq. (2.4.10) wherein the fundamental natural vibration frequency ω_1, and mode shape $\phi_1^k(x, y)$ were determined using a finite element idealization of the dam, and the integrals involved in M_1, L_1^l, $B_0^l(\omega)$, and $B_1(\omega)$ were computed in discretized form.

The complex-valued frequency response function, $\bar{\ddot{q}}_1^l(\omega)$, is a dimensionless response factor representing the ratio of horizontal acceleration at the dam crest to unit free-field ground acceleration in the $l(= x$ or $y)$ direction. For each case mentioned above, the absolute value of this complex-valued response factor is plotted against the normalized excitation frequency ω/ω_1. When presented in this form, the results apply to dams of all heights with the idealized triangular cross section, and chosen Ω_r, H/H_s, and α values.

2.5.3 Dam–Water Interaction Effects

Frequency response functions for dams subjected to horizontal and vertical ground motions are presented in Figures 2.5.1–2.5.4 for two selected values of $\Omega_r = 0.67$ and 1.0. Each plot contains response curves for the dam with full reservoir for five values of α and the response curve for the dam alone, i.e. with an empty reservoir. The latter is the familiar response curve for a SDF system with frequency-independent mass, stiffness, and damping parameters. However, dam–water interaction including water compressibility introduces frequency-dependent terms in Eq. (2.4.10), resulting in complicated shapes for the response curves.

The frequency response function due to horizontal ground motion displays strongly resonant behavior with large amplification over an especially narrow frequency band because of dam–water interaction and water compressibility. The single resonant peak in the response of the dam without water may become two resonant peaks for a full reservoir if the reservoir bottom is non-absorptive, a behavior that develops for systems with smaller Ω_r or stiffer dams (Figure 2.5.1). With increasing wave absorption at the reservoir bottom, i.e. decreasing α, the first resonant peak is reduced, whereas the second peak is increased, and for a small enough α the two peaks coalesce, resulting in a single resonant peak at an intermediate resonant frequency. For systems with

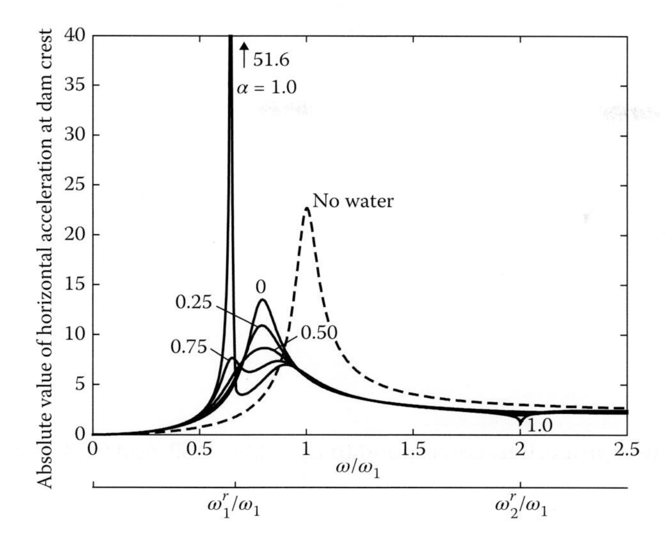

Figure 2.5.1 Dam response to harmonic horizontal ground motion; frequency ratio, $\Omega_r = 0.67$, i.e. $E_s = 5.67$ million psi; $\alpha = 1.0$, 0.75, 0.50, 0.25, and 0; response of dam alone is also shown.

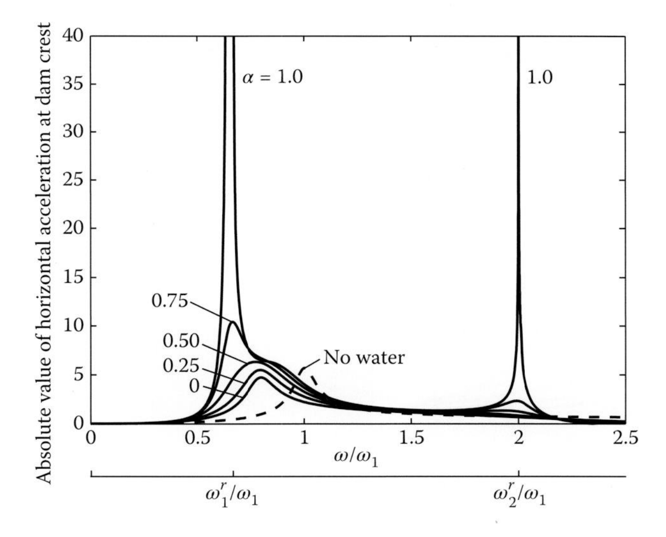

Figure 2.5.2 Dam response to harmonic vertical ground motion; frequency ratio, $\Omega_r = 0.67$, i.e. $E_s = 5.67$ million psi; $\alpha = 1.0$, 0.75, 0.50, 0.25, and 0; response of dam alone is also shown.

the larger Ω_r value, or relatively flexible dams, only a single resonant peak develops for all values of α (Figure 2.5.3). For such systems, as α decreases, increased absorption of energy through the reservoir bottom further reduces the resonant amplitude, with little change in the resonant frequency. The fundamental resonant frequency of the dam including hydrodynamic effects is lower than both the natural frequency ω_1 of the dam alone and the fundamental natural frequency ω_1^r of the impounded water.

The response function due to horizontal ground motion is especially complicated if the pressure waves are fully reflected at the reservoir bottom, i.e. $\alpha = 1$, because at excitation frequencies equal to ω_n^r, the natural vibration frequencies of the impounded water, the added mass

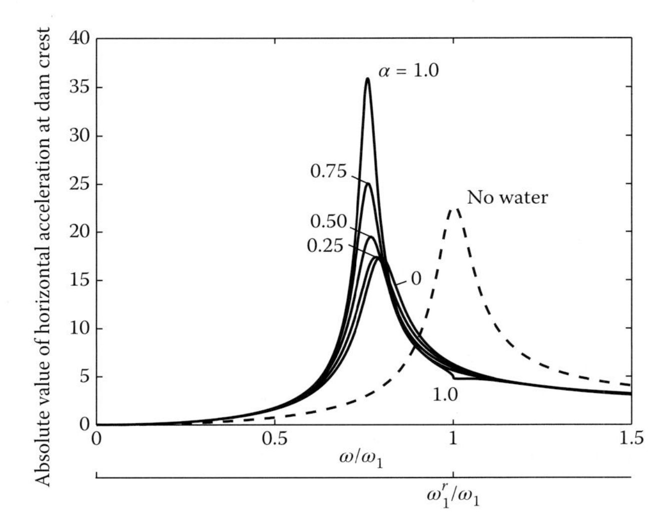

Figure 2.5.3 Dam response to harmonic horizontal ground motion; frequency ratio, $\Omega_r = 1.0$, i.e. $E_s = 2.52$ million psi; $\alpha = 1.0$, 0.75, 0.50, 0.25, and 0; response of dam alone is also shown.

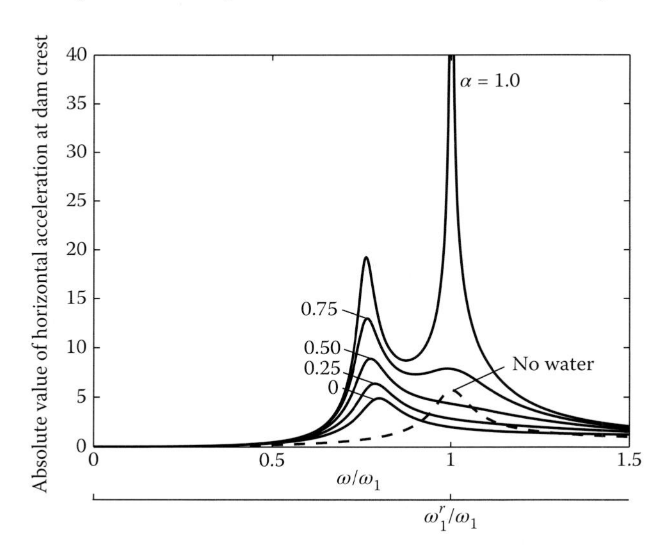

Figure 2.5.4 Dam response to harmonic vertical ground motion; frequency ratio, $\Omega_r = 1.0$, i.e. $E_s = 2.52$ million psi; $\alpha = 1.0$, 0.75, 0.50, 0.25, and 0; response of dam alone is also shown.

and force are both unbounded. When determined by a limiting process, however, the response function due to horizontal ground motion has bounded values at ω_n^r (Chopra 1968), which appear as local dips in the response curve (Figures 2.5.1 and 2.5.3).

The response function due to vertical ground motion also displays the first resonance at a frequency lower than the natural frequency ω_1 of the dam alone with complicated behavior in the frequency range between ω_1 and ω_1^r that is dominated by the unbounded response values at excitation frequencies equal to ω_n^r (Figures 2.5.2 and 2.5.4). These unbounded peaks are not the result of resonance in the usual sense, which is associated with the denominator in Eq. (2.4.10) attaining a minimum, but are caused by the unbounded added force. Reservoir bottom absorption reduces the added force associated with both ground motion components and the added mass to bounded values at ω_n^r (Section 2.3.3). Consequently, the dips at ω_n^r in the response function due to horizontal ground motion are eliminated; and the unbounded values at ω_n^r in the response function due to vertical ground motion are reduced to bounded peaks, which disappear for the smaller values of α.

2.5.4 Implications of Ignoring Water Compressibility

Earthquake analysis of dams is greatly simplified if compressibility of water is ignored, because then hydrodynamic effects can be modeled by a frequency-independent added mass (Section 2.3.4). Here, we answer the important question: can water compressibility be ignored in the earthquake analysis of concrete gravity dams?

Frequency response curves of the dam without water, when presented using normalized scales as in Figures 2.5.5 and 2.5.6, are independent of the modulus of elasticity, E_s, of concrete. Similarly, the response curves including hydrodynamic effects do not vary with E_s if water compressibility is ignored (Chopra 1968). However, the Ω_r parameter, or correspondingly the E_s value, affects the response functions when water compressibility is included.

The response of the dam is greatly influenced by the frequency ratio, Ω_r. i.e. the relative frequencies of the two interacting systems, as demonstrated by Figures 2.5.1 and 2.5.3. The percentage decrease in the fundamental resonant frequency of the dam due to dam–water interaction is larger for the smaller values of Ω_r, i.e. larger values of E_s. The response value at resonance as well as the shape of the response curve in the neighborhood of the natural frequencies of the dam and of the impounded water also depend significantly on the frequency ratio Ω_r.

With increasing Ω_r, or decreasing E_s, the effects of water compressibility on response become smaller and the response curve approaches the result for incompressible water. For systems with $\Omega_r = 2.0$, the effects of water compressibility are insignificant in the response to horizontal ground motion (Figures 2.5.5 and 2.5.7) but are still noticeable in the response to vertical ground motion (Figures 2.5.6 and 2.5.8). These results confirm the earlier conclusion that

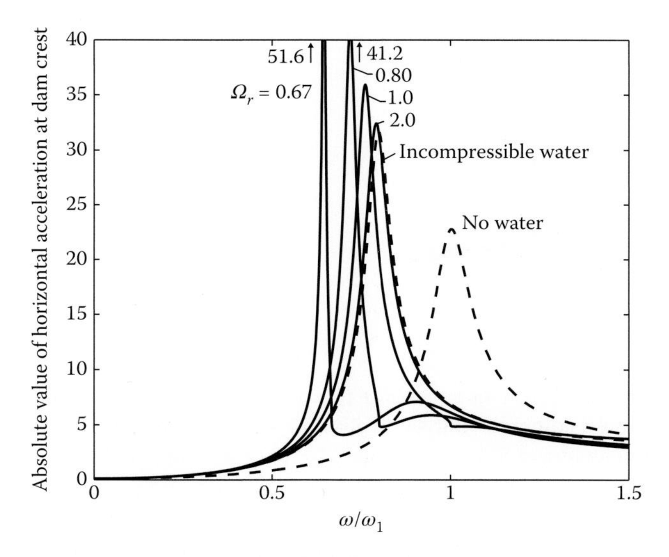

Figure 2.5.5 Influence of frequency ratio, Ω_r, on dam response to harmonic horizontal ground motion. Results are presented (1) including water compressibility with $\alpha = 1$ for $\Omega_r = 0.67$, 0.80, 1.0, and 2.0 ($E_s = 5.67$, 3.94, 2.52, and 0.63 million psi); and (2) assuming water to be incompressible.

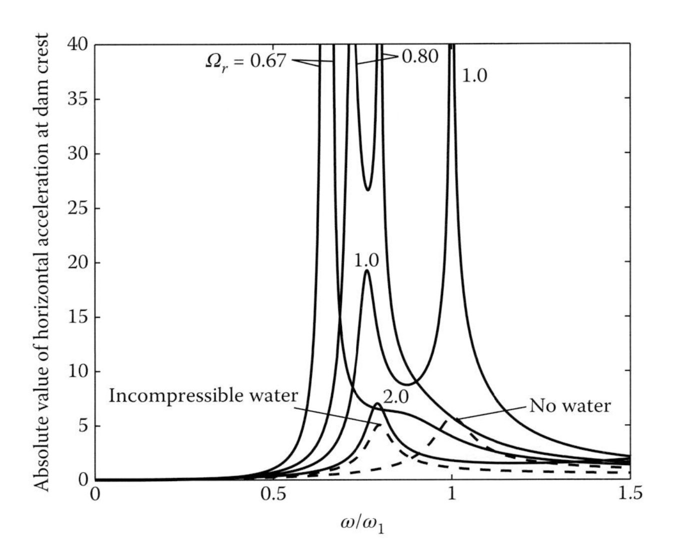

Figure 2.5.6 Influence of frequency ratio, Ω_r, on dam response to harmonic vertical ground motion. Results are presented (1) including water compressibility with $\alpha = 1$ for $\Omega_r = 0.67$, 0.80, 1.0, and 2.0 ($E_s = 5.67$, 3.94, 2.52, and 0.63 million psi); and (2) assuming water to be incompressible.

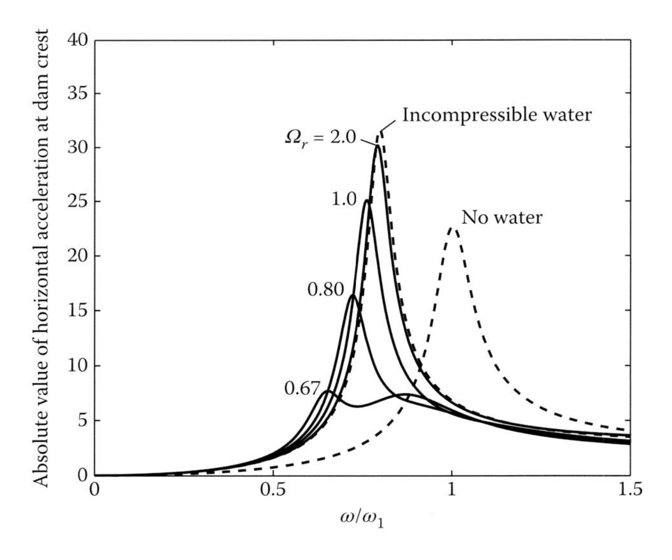

Figure 2.5.7 Influence of frequency ratio, Ω_r, on dam response to harmonic horizontal ground motion. Results are presented (1) including water compressibility with $\alpha = 0.75$ for $\Omega_r = 0.67$, 0.80, 1.0, and 2.0 ($E_s = 5.67$, 3.94, 2.52, and 0.63 million psi); and (2) assuming water to be incompressible.

the effects of water compressibility become insignificant for systems with $\Omega_r > 2$ or $\omega_1 < 2\omega_1^r$ (Chopra 1968), which for the system considered implies $E_s < 0.63$ million psi. However, E_s for mass concrete is generally in the range of 2–5 million psi. Consequently, errors in response results will be significant if water compressibility is ignored. This conclusion will be reinforced further in the context of response to earthquake excitation (Chapter 6).

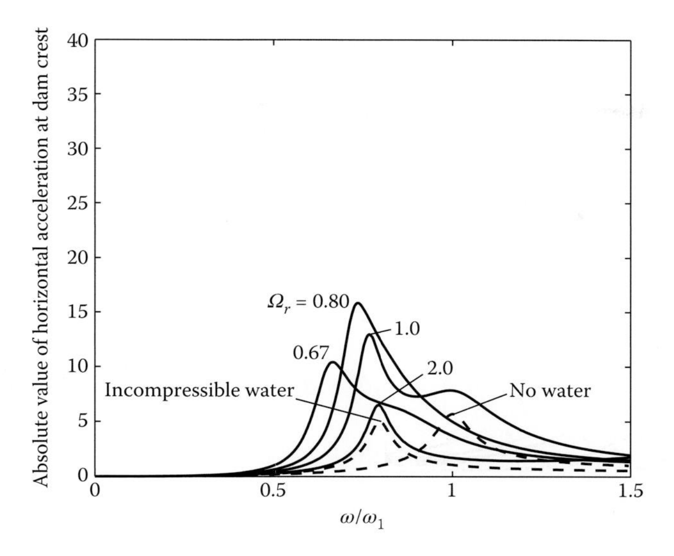

Figure 2.5.8 Influence of frequency ratio, Ω_r, on dam response to harmonic vertical ground motion. Results are presented (1) including water compressibility with $\alpha = 0.75$ for $\Omega_r = 0.67$, 0.80, 1.0, and 2.0 ($E_s = 5.67$, 3.94, 2.52, and 0.63 million psi); and (2) assuming water to be incompressible.

2.5.5 Comparison of Responses to Horizontal and Vertical Ground Motions

Comparing the response of the dam to horizontal and vertical ground motions (Figures 2.5.5 and 2.5.6) it is apparent – consistent with common view – that without impounded water the response to vertical ground motion is relatively small because the excitation term L_1^l in Eq. (2.4.10) is much smaller for vertical ground motion ($l = y$) than for horizontal ground motion ($l = x$). With impounded water in the reservoir, the response is affected by the added hydrodynamic mass, damping, and force terms in Eq. (2.4.10). Because the added hydrodynamic mass and damping are independent of the excitation direction, the response functions due to horizontal and vertical ground motions display the same resonant frequency and effective damping. However, the added force is associated with hydrodynamic pressures acting in the horizontal direction on the vertical upstream face of a rigid dam, whether the ground motion is horizontal or vertical. In the latter case, $B_0^y(\omega)$ is much larger than the small L_1^y. As a result, hydrodynamic effects (including water compressibility) cause a larger increase in response to vertical ground motion than in response to horizontal ground motion. This is apparent by comparing the response curves for the dam without water and with water associated with the two excitations (Figures 2.5.1 versus 2.5.2 and Figures 2.5.3 versus 2.5.4).

Putting this issue in historical context, the response of concrete gravity dams to the vertical component of ground motion was overestimated in the earliest studies based on the assumption of non-absorptive reservoir bottom (Chopra and Chakarabarti 1973, 1974). An absorptive reservoir bottom gives a more plausible estimate of the response to vertical ground motion and its smaller contribution to the total earthquake response of the dam (Fenves and Chopra 1983). Although not negligible, this contribution is of secondary importance. Thus, the response to only the horizontal component of ground motion is considered in developing the procedure for simplified analysis of two-dimensional dam–water–foundation systems (Sections 2.6, 3.4, and 3.5); this procedure is intended for preliminary analysis and design of dams.

2.6 EQUIVALENT SDF SYSTEM: HORIZONTAL GROUND MOTION

It was shown in Section 2.4 that including the interaction between the dam and compressible water results in the following complex-valued frequency response function for the modal coordinate due to horizontal ground motion (Eq. (2.4.10) with the superscript $l = x$ dropped):

$$q_1(\omega) = \frac{-\left[L_1 + B_0(\omega)\right]}{-\omega^2\left\{M_1 + Re\left[B_1(\omega)\right]\right\} + i\omega\left\{C_1 - \omega\, Im\left[B_1(\omega)\right]\right\} + K_1} \tag{2.6.1}$$

in which the hydrodynamic terms $B_0(\omega)$ and $B_1(\omega)$ are equivalent to those in Eq. (2.4.10) but are now defined slightly differently:

$$B_0(\omega) = \int_0^H \bar{p}_0(y,\omega)\phi_1^x(0,y)dy \tag{2.6.2a}$$

$$B_1(\omega) = \int_0^H \bar{p}_1(y,\omega)\phi_1^x(0,y)dy \tag{2.6.2b}$$

where $\bar{p}_0(y,\omega)$ and $\bar{p}_1(y,\omega)$ are frequency response functions for the hydrodynamic pressure on the upstream face due to horizontal ground acceleration of a rigid dam, and acceleration of a dam in its fundamental mode of vibration, respectively:

$$\bar{p}_j(y,\omega) = 2\rho H \sum_{n=1}^{\infty} \frac{\mu_n^2(\omega)}{H\left[\mu_n^2(\omega) - (\omega\xi)^2\right] + i(\omega\xi)} \frac{I_{jn}(\omega)}{\sqrt{\mu_n^2(\omega) - (\omega^2/C^2)}} Y_n(y,\omega), \quad j = 0, 1 \tag{2.6.3}$$

where

$$I_{jn}(\omega) = \frac{1}{H}\int_0^H f_j(y)Y_n(y,\omega)dy \quad j = 0, 1 \tag{2.6.4}$$

in which $f_0(y) = 1$ and $f_1(y) = \phi_l^x(0,y)$. For convenience later in defining the added hydrodynamic mass, the preceding pressure functions are defined for the positive x-direction upstream, thus giving algebraic signs opposite those of the corresponding equations in Section 2.3, where the positive x-direction was downstream.

Including dam–water interaction, the frequency response function $\bar{q}_1(\omega)$ for the modal coordinate associated with the fundamental vibration mode of the dam, Eq. (2.6.1), is a complicated function of excitation frequency ω that contains frequency-dependent hydrodynamic terms. To develop a simplified analytical procedure, the dam–water system will be modeled by an equivalent SDF system with frequency-independent values for the hydrodynamic terms. Such a procedure, developed by Chopra (1978) for dam–water systems with non-absorptive reservoir bottom and later extended to include reservoir bottom absorption (Fenves and Chopra 1985c), is presented next.

2.6.1 Modified Natural Frequency and Damping Ratio

The properties of the equivalent SDF system are defined as those of the dam with an empty reservoir modified by an added mass and an added damping that represent the hydrodynamic effects of the impounded water. The mass density $\tilde{m}_k(x,y)$, $k = x, y$, of the equivalent SDF system is defined as

$$\tilde{m}_x(x,y) = m_x(x,y) + m_a(y)\delta(x) \tag{2.6.5a}$$

$$\tilde{m}_y(x,y) = m_y(x,y) \tag{2.6.5b}$$

where $\delta(x)$ is the Dirac delta function. Because the hydrodynamic pressure on a vertical upstream face acts in the horizontal direction, the "added mass" $m_a(y)$ applies only to the horizontal displacement of the dam and is concentrated at the upstream face of the dam. The frequency-independent "added mass" is defined as

$$m_a(y) = \frac{\overline{p}_1(y, \tilde{\omega}_r)}{\phi_1^x(0, y)} \tag{2.6.6}$$

where the natural vibration frequency $\tilde{\omega}_r$ of the equivalent SDF system approximates the fundamental resonant frequency of the dam–water system. If the reservoir bottom is absorptive, $m_a(y)$ is complex valued. Thus, it is not a mass quantity in the usual sense; only its real-valued component contributes to an added mass, whereas the imaginary-valued component implies an added damping. Furthermore, this added mass representing hydrodynamic effects on a flexible dam differs from the one in Eq. (2.3.25), which was determined assuming the dam to be rigid. It is inappropriate to use the latter in dynamic analysis of dams because they are flexible structures.

The complex-valued frequency response function for the modal coordinate of the equivalent SDF system will be of the same form as Eq. (2.2.7), but L_1, M_1, and C_1 will be different; thus

$$\overline{\tilde{q}}_1(\omega) = \frac{-\tilde{L}_1}{-\omega^2 \tilde{M}_1 + i\omega \tilde{C}_1 + K_1} \tag{2.6.7}$$

wherein the superscript $l = x$ has been dropped for notational simplicity, and the over-tilde is included to indicate that the equivalent SDF system models the dam response including dam–water interaction; $\tilde{M}_1$, $\tilde{C}_1$, and $\tilde{L}_1$ are obtained by substituting Eq. (2.6.5) in the standard definitions of generalized mass, damping, and force (Chopra 2017, Chapter 17):

$$\tilde{M}_1 = M_1 + \mathrm{Re}\left\{ \int_0^H m_a(y) \left[\phi_1^x(0, y)\right]^2 dy \right\} \tag{2.6.8a}$$

$$\tilde{C}_1 = C_1 - \omega \mathrm{Im}\left\{ \int_0^H m_a(y) \left[\phi_1^x(0, y)\right]^2 dy \right\} \tag{2.6.8b}$$

$$\tilde{L}_1 = L_1 + \int_0^H m_a(y) \phi_1^x(0, y) dy \tag{2.6.8c}$$

and M_1, C_1, and L_1 for the dam alone were defined in Section 2.2.1. Substituting Eq. (2.6.6) into Eq. (2.6.8) leads to

$$\tilde{M}_1 = M_1 + \mathrm{Re}\left\{ \int_0^H \overline{p}_1\left(y, \tilde{\omega}_r\right) \phi_1^x(0, y) dy \right\} \tag{2.6.9a}$$

$$\tilde{C}_1 = C_1 - \omega\, \mathrm{Im}\left\{ \int_0^H \overline{p}_1\left(y, \tilde{\omega}_r\right) \phi_1^x(0, y) dy \right\} \tag{2.6.9b}$$

$$\tilde{L}_1 = L_1 + \int_0^H \overline{p}_1(y, \tilde{\omega}_r) dy \tag{2.6.9c}$$

We will demonstrate that if the excitation frequency ω that appears in $\tilde{C}_1$ is replaced by the resonant frequency $\tilde{\omega}_r$, Eq. (2.6.7) will give the same resonant response as the rigorous Eq. (2.6.1). Using the solutions for $p_0(y, \omega)$ and $p_1(y, \omega)$ in Eq. (2.6.3), replacing ω by $\tilde{\omega}_r$ in C_1, and recalling Eq. (2.6.2) for the definition of $B_0(\omega)$ and $B_1(\omega)$, Eq. (2.6.9) becomes

$$\tilde{M}_1 = M_1 + \mathrm{Re}\left[B_1(\tilde{\omega}_r)\right] \tag{2.6.10a}$$

$$\tilde{C}_1 = C_1 - \tilde{\omega}_r \mathrm{Im}\left[B_1(\tilde{\omega}_r)\right] \tag{2.6.10b}$$

$$\tilde{L}_1 = L_1 + B_0\left(\tilde{\omega}_r\right) \tag{2.6.10c}$$

with the generalized mass, damping and force of the equivalent SDF system given by Eq. (2.6.10), a comparison of Eqs. (2.6.7) and (2.6.1) shows that $\overline{\tilde{q}}_1(\tilde{\omega}_r) = \overline{q}_1(\tilde{\omega}_r)$, i.e. the equivalent SDF system gives the exact value for the fundamental resonant response of the dam–water system.

The natural vibration frequency, $\tilde{\omega}_r$, of the equivalent SDF system is $\tilde{\omega}_r = \sqrt{K_1/\tilde{M}_1}$; substituting $K_1 = \omega_1^2 M_1$ and Eq. (2.6.10a) for $\tilde{M}_1$, gives

$$\tilde{\omega}_r = \frac{\omega_1}{\sqrt{1 + \frac{\text{Re}[B_1(\tilde{\omega}_r)]}{M_1}}} \tag{2.6.11}$$

Because $\tilde{\omega}_r$ appears on both sides, this equation must be solved iteratively for $\tilde{\omega}_r$. Hydrodynamic effects always reduce the natural vibration frequency because $\text{Re}[B_1(\omega)] > 0$ for all excitation frequencies. Does Eq. (2.6.11), determined from the properties of the equivalent SDF system, give the correct reduction in frequency of the dam due to dam–water interaction? The fundamental resonant frequency of the dam–water system is approximately given by the excitation frequency that makes the real-valued component of the denominator in the exact fundamental mode response, Eq. (2.6.1), equal to zero. This argument also leads to Eq. (2.6.11), which demonstrates that the mass of the equivalent SDF system defined in Eqs. (2.6.5) and (2.6.6) reduces the fundamental resonant frequency of the dam due to hydrodynamic effects by the exact amount.

The damping ratio of the equivalent SDF system $\tilde{\xi}_r = \tilde{C}_1/2\tilde{M}_1\tilde{\omega}_r$ is determined by substituting Eqs. (2.6.10b) and (2.6.10a) and utilizing Eq. (2.6.11):

$$\tilde{\xi}_r = \frac{\tilde{\omega}_r}{\omega_1}\zeta_1 + \zeta_r \tag{2.6.12}$$

in which the added damping due to upstream propagation of hydrodynamic waves and their absorption at the reservoir bottom is represented by the frequency-independent damping ratio, ζ_r, defined as

$$\zeta_r = -\frac{1}{2}\frac{1}{M_1}\left(\frac{\tilde{\omega}_r}{\omega_1}\right)^2 \text{Im}[B_1(\tilde{\omega}_r)] \tag{2.6.13}$$

The damping ratio ζ_r is non-negative because $\text{Im}[B_1(\omega)] \leq 0$ for all excitation frequencies.

For dam–water systems with non-absorptive reservoir bottom ($\alpha = 1$), $\bar{p}_1(y, \omega_r)$ is real-valued (Section 2.3). Thus, $m_a(y)$ is real-valued and Eq. (2.6.11) reduces to the earlier result (Chopra 1978) for the natural vibration frequency of the equivalent SDF system; and the added damping ratio, ζ_r, is zero, so Eq. (2.6.13) reduces to the earlier expression (Chopra 1978) for the damping ratio of the equivalent SDF system. An absorptive reservoir bottom ($\alpha < 1$) results in complex-valued $m_a(y)$, which modifies the resonant frequency and increases the damping because now hydrodynamic pressure waves propagate upstream and refract into the absorptive reservoir bottom at the excitation frequency, $\tilde{\omega}_r$.

2.6.2 Evaluation of Equivalent SDF System

The effectiveness of the equivalent SDF system in representing the fundamental mode response of dams with impounded water is demonstrated in Figure 2.6.1. The exact and equivalent-SDF-system responses of an idealized concrete gravity dam monolith with the triangular cross section (described in Section 2.5.2) to harmonic horizontal ground motion were computed by numerically evaluating Eqs. (2.6.1) and (2.6.7), respectively. The absolute value of the complex-valued frequency response function for horizontal acceleration at the dam crest is plotted against the normalized excitation frequency parameter, ω/ω_1, so the results are valid for dams of any height, H_s. Figure 2.6.1 demonstrates that the equivalent SDF system provides a good approximation of the fundamental mode response of the dam with impounded water for a wide

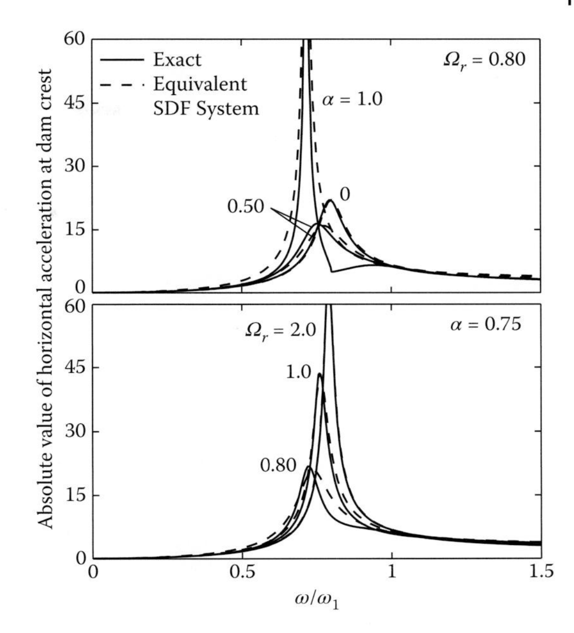

Figure 2.6.1 Comparison of exact and equivalent SDF system response of dams on rigid foundation with impounded water due to harmonic horizontal ground motion; $\zeta_1 = 2\%$.

range of values of the frequency ratio Ω_r – hence of the concrete modulus, E_s – and of the wave reflection coefficient, α, at the reservoir bottom. The approximation of the frequency bandwidth of the resonant peak is more accurate if the reservoir bottom is absorptive because additional energy is lost at this boundary, which eliminates the sharp peaks in the response curves.

The exact value of the fundamental resonant period, obtained from the resonant peak of $\overline{q}_1(\omega)$, computed from Eq. (2.6.1), is compared in Figure 2.6.2 with the natural vibration period, $\tilde{T}_r = 2\pi/\tilde{\omega}_r$, of the equivalent SDF system in which $\tilde{\omega}_r$ is computed from Eq. (2.6.11). It is apparent that the natural vibration period of the equivalent SDF system provides a very accurate approximation of the fundamental resonant period of the dam with impounded water if the reservoir bottom is non-absorptive, but it is slightly less accurate if the reservoir bottom is absorptive.

2.6.3 Hydrodynamic Effects on Natural Frequency and Damping Ratio

Figure 2.6.2 demonstrates that dam–water interaction lengthens the vibration period, with this effect being especially small for H/H_s, less than 0.5, but increasing rapidly with water depth (Chakrabarti and Chopra 1974). Furthermore, the vibration period ratio, $\tilde{T}_r/T_1$, increases as the frequency ratio, Ω_r, decreases (i.e. the modulus of elasticity, E_s, of the concrete increases) because of interaction between the closely-spaced fundamental vibration frequencies of the dam and water (Section 2.5.3); these observations first appeared in Chopra (1968). As the reservoir bottom becomes more absorptive, i.e. as the wave reflection coefficient α decreases, the fundamental resonant period is reduced from its value for a non-absorptive reservoir bottom. This occurs because reservoir bottom absorption reduces the hydrodynamic terms (Section 2.3.3), thus reducing the value of the added mass. The wave reflection coefficient, α, has little influence on the fundamental resonant period for larger values of Ω_r, i.e. smaller values of E_s. However, the $\tilde{T}_r/T_1$ ratio is relatively insensitive to E_s if the reservoir bottom is absorptive with $\alpha \leq 0.5$.

The effects of reservoir bottom absorption on the added damping ratio ζ_r (Figure 2.6.3), and thus, on the damping ratio, $\tilde{\zeta}_r$, of the equivalent SDF system (Figure 2.6.4), are more complicated than its effects on the vibration period. As the wave reflection coefficient, α, decreases from

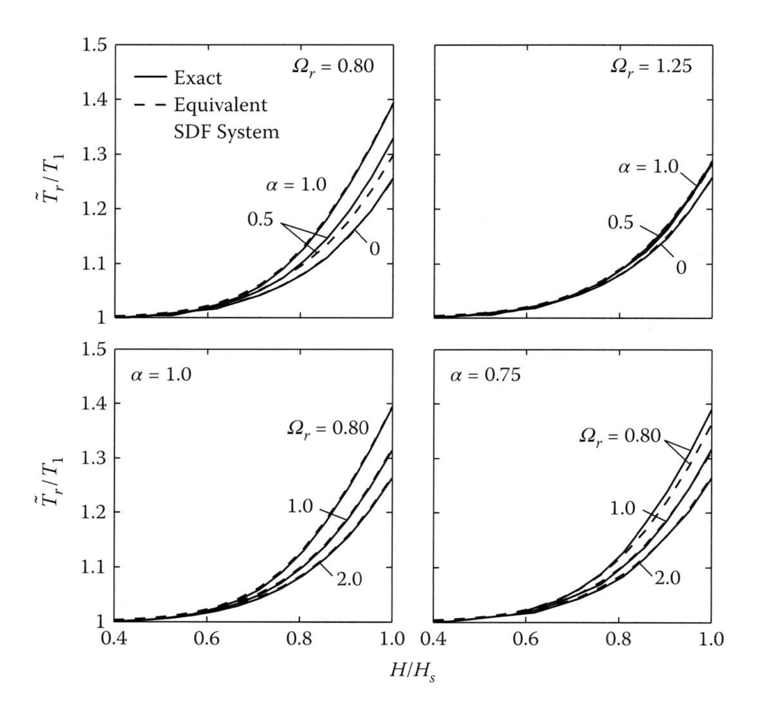

Figure 2.6.2 Comparison of exact and approximate (equivalent SDF system) values of the ratio of fundamental vibration periods of the dam on rigid foundation with and without impounded water. Results presented are for various values of the frequency ratio Ω_r and the wave reflection coefficient α.

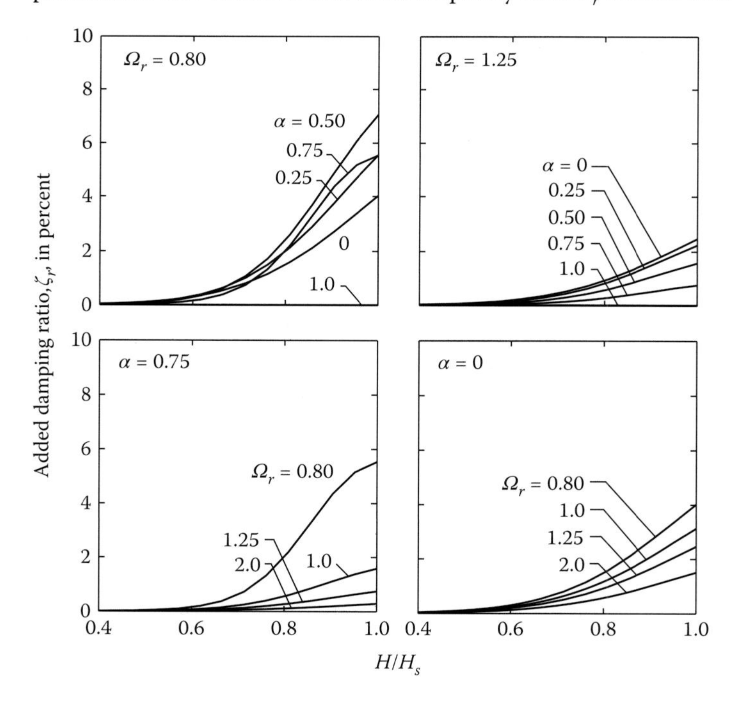

Figure 2.6.3 Added damping ratio ζ_r due to dam–water interaction and reservoir bottom absorption. Results presented are for various values of the frequency ratio Ω_r and the wave reflection coefficient α.

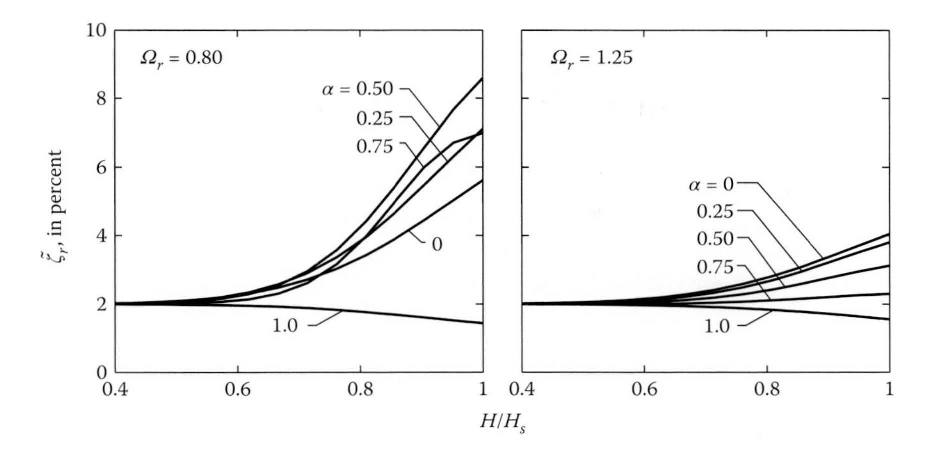

Figure 2.6.4 Damping ratio $\tilde{\zeta}_r$ of the equivalent SDF system representing dams on rigid foundation with impounded water; $\zeta_1 = 2\%$.

unity, ζ_r increases monotonically from zero for larger values of Ω_r, i.e. smaller values of E_s, but the trends are more complicated for smaller values of Ω_r, i.e. larger values of E_s. This latter, unexpected behavior in ζ_r results from the previously observed effects of reservoir bottom absorption on the natural vibration frequency, $\tilde{\omega}_r$, of the equivalent SDF system (Eq. 2.6.11), which is the frequency at which the added damping, ζ_r, is evaluated (Eq. 2.6.13). The added damping ratio depends on the relative values of $\tilde{\omega}_r$ and ω_1^r; recall that the latter is the fundamental natural vibration frequency of the impounded water. As Ω_r decreases (i.e. E_s increases, implying that the dam becomes stiffer), $\tilde{\omega}_r$ approaches ω_1^r, and the imaginary-valued component of the hydrodynamic term, $B_1(\tilde{\omega}_r)$, increases as α decreases from unity to zero, thus increasing ζ_r. Figure 2.6.3 also shows that the wave reflection coefficient, α, has a larger effect on the added damping for smaller values of Ω_r than for larger Ω_r. If the reservoir bottom is absorptive ($\alpha < 1$), the added damping ratio ζ_r increases as Ω_r decreases, with the rate of increase becoming smaller as α decreases.

Considering that $\tilde{\omega}_r$ is less than ω_1, Eq. (2.6.12) indicates that dam–water interaction reduces the effectiveness of the structural damping. Unless this reduction is compensated by the added damping ζ_r due to reservoir bottom absorption, the overall damping ratio, ζ_r, will be less than ζ_1 (Figure 2.6.4).

2.6.4 Peak Response

Because the equivalent SDF system accurately predicts the response of dam–water system to horizontal ground motion over a complete range of excitation frequencies and for a wide range of parameters characterizing the properties of the dam, water, and reservoir bottom materials, it is applicable to the analysis of dam response to arbitrary ground motion. In particular, the peak displacements of the dam and equivalent static lateral forces can be expressed by appropriately modifying Eqs. (2.2.12) and (2.2.14), specialized for horizontal ground motion:

$$r^k(x,y) = \tilde{\Gamma}_1 D\left(\tilde{T}_r, \tilde{\zeta}_r\right)\, \phi_1^k(x,y) \qquad k = x, y \tag{2.6.14}$$

$$f_1(x,y) = \tilde{\Gamma}_1 \frac{A\left(\tilde{T}_r, \tilde{\zeta}_r\right)}{g} \tilde{w}_x(x,y)\, \phi_1^x(x,y) \tag{2.6.15}$$

In these equations, $\tilde{\Gamma}_1 = \tilde{L}_1/\tilde{M}_1$; $D\left(\tilde{T}_r, \tilde{\zeta}_r\right)$ and $A\left(\tilde{T}_r, \tilde{\zeta}_r\right)$ are the ordinates of the deformation and pseudo-acceleration response spectra at the natural vibration period $\tilde{T}_r$ and damping ratio $\tilde{\zeta}_r$ of the equivalent SDF system, and $\tilde{w}(x, y) = g\tilde{m}_x(x, y)$ with $\tilde{m}_x(x, y)$ defined by Eqs. (2.6.5a) and (2.6.6). Substituting these equations in Eq. (2.6.15) gives the final expression for the equivalent static lateral forces (Chopra 1978):

$$f_1(x, y) = \tilde{\Gamma}_1 \frac{A\left(\tilde{T}_r, \tilde{\zeta}_r\right)}{g} \left[w_x(x, y)\phi_1^x(x, y) + g\overline{p}_1\left(y, \tilde{\omega}_r\right) \delta(x)\right] \tag{2.6.16}$$

Comparing Eq. (2.6.16) with Eq. (2.2.12), we note that dam–water interaction introduces hydrodynamic pressures $\overline{p}_1\left(y, \tilde{\omega}_r\right)$ at the upstream face, increases Γ_1, and modifies the period and damping ratio where the spectral ordinate is determined.

APPENDIX 2: WAVE-ABSORPTIVE RESERVOIR BOTTOM

A2.1 Reservoir Bottom Sediments

The normal pressure gradient at the horizontal bottom of the reservoir is proportional to the vertical acceleration of the boundary. This is the specified free-field excitation $a_g^y(t)$ if the boundary is rigid, resulting in Eq. (2.3.3), repeated here for convenience:

$$\frac{\partial p}{\partial y}(x, 0, t) = -\rho a_g^y(t)\delta_{yl} \qquad l = x, y \tag{A2.1}$$

This boundary condition is modified in this appendix to include the flexibility of sediments deposited at the reservoir bottom.

Flexibility of the reservoir bottom modifies the free-field acceleration $a_g^y(t)$ by an unknown interactive acceleration $\ddot{v}(x, 0, t)$, and the boundary condition of Eq. (A2.1) becomes

$$\frac{\partial p}{\partial y}(x, 0, t) = -\rho \left[a_g^y(t)\delta_{yl} + \ddot{v}(x, 0, t)\right] \tag{A2.2}$$

For steady-state harmonic excitation due to $a_g^y(t) = 1e^{i\omega t}$, Eq. (A2.2) becomes

$$\frac{\partial \overline{p}}{\partial y}(x, 0, \omega) = -\rho \left[\delta_{yl} - \omega^2\overline{v}(x, 0, \omega)\right] \tag{A2.3}$$

where $\overline{p}(x, 0, \omega)$ and $\overline{v}(x, 0, \omega)$ are complex frequency response functions for $p(x, 0, t)$ and $v(x, 0, t)$, respectively.

The reservoir bottom is represented approximately by a one-dimensional model – independent of the x-coordinate – that does not explicitly consider the thickness of the sediment layer. The frequency response function $\overline{v}(x, 0, \omega)$ for the vertical displacement of the reservoir bottom (i.e. the surface of the sediment layer) due to interaction between the impounded water and the reservoir bottom materials can be expressed in terms of the hydrodynamic pressure at the reservoir bottom:

$$\overline{v}(x, 0, \omega) = -\,\mathcal{C}(\omega)\overline{p}(x, 0, \omega) \tag{A2.4}$$

The compliance function $\mathcal{C}(\omega)$ for the reservoir bottom is defined as the harmonic displacement at the reservoir bottom due to unit harmonic pressure $p(x, 0, t) = 1e^{i\omega t}$ at the reservoir bottom.

The compliance function $\mathcal{C}(\omega)$ can be derived by solving the one-dimensional Helmholtz equation:

$$\frac{\partial^2\overline{v}}{\partial y'^2} + \frac{\omega^2}{C_r^2}\overline{v} = 0 \tag{A2.5}$$

that governs the steady-state vibration of the model of the reservoir bottom materials, where $\overline{v}\left(y', \omega\right)$ is the frequency response function for vertical displacement in the layer of reservoir bottom materials, $C_r = \sqrt{(E_r/\rho_r)}$ is the compression wave speed, E_r is the modulus of elasticity, and ρ_r is the density of the reservoir bottom materials. The equilibrium condition at the surface of the layer of reservoir bottom materials ($y' = 0$) is that the pressure in the fluid equals the normal stress; thus

$$1e^{i\omega t} = E_r \frac{\partial}{\partial y'} \overline{v}(0, \omega) e^{i\omega t} \tag{A2.6}$$

The solution of Eq. (A2.5) subject to the equilibrium condition of Eq. (A2.6) and the radiation condition in the negative y'-direction gives

$$\overline{v}\,(y', \omega) = -i \left(\frac{1}{\rho_r C_r} \frac{1}{\omega} \right) \exp \left[i \frac{\omega}{C_r} y' \right] \tag{A2.7}$$

By definition, $\mathcal{C}(\omega) = \overline{v}(0, \omega)$; therefore, the compliance function for the reservoir bottom is given by

$$\mathcal{C}(\omega) = -i \left(\frac{1}{\rho_r C_r} \frac{1}{\omega} \right) \tag{A2.8}$$

The compliance function $\mathcal{C}(\omega)$ is imaginary-valued for all excitation frequencies, so the reservoir bottom materials, as modeled, introduces an additional mechanism for energy loss. Because the thickness of the sediment layer is not recognized explicitly, this compliance function is applied at the surface of the underlying foundation ($y = 0$).

The substitution of Eqs. (A2.4) and (A2.8) into Eq. (A2.3) gives the boundary condition at the absorptive reservoir bottom:

$$\left[\frac{\partial}{\partial y} - i\omega\xi \right] \overline{p}(x, 0, \omega) = -\rho \delta_{yl} \qquad l = x, y \tag{A2.9}$$

where the damping coefficient $\xi = \rho / \rho_r C_r$. This boundary condition for time-harmonic motion takes the following form for transient motion:

$$\left(\frac{\partial}{\partial y} - \xi \frac{\partial}{\partial t} \right) p(x, 0, t) = -\rho a_g^y(t) \delta_{yl}, \qquad l = x, y \tag{A2.10}$$

which is identical to Eq. (2.3.4a).

We next relate the wave reflection coefficient, α, which is the ratio of the amplitude of the reflected hydrodynamic pressure wave to the amplitude of a vertically propagating pressure wave incident on the reservoir bottom, to the damping coefficient ξ. Consider a downward traveling wave in the fluid domain that strikes the fluid-sediment boundary. Hydrodynamic pressures are governed by the one-dimensional version of Eq. (2.3.8):

$$\frac{\partial^2 \overline{p}}{\partial y^2} + \frac{\omega^2}{C^2} \overline{p} = 0 \tag{A2.11}$$

The general solution of this equation is

$$\overline{p}(y, \omega) = A(\omega) \exp \left(i \frac{\omega}{C} y \right) + B(\omega) \exp \left(-i \frac{\omega}{C} y \right) \tag{A2.12}$$

where $A(\omega)$ is the amplitude of the hydrodynamic pressure wave incident to the reservoir bottom, and $B(\omega)$ is the amplitude of the reflected wave.

An equation for the ratio $B(\omega)/A(\omega)$, termed the reflection coefficient, α, can be obtained by substituting Eq. (A2.12) into the boundary condition of Eq. (A2.9) with $a_g^y = 0$, i.e. zero free-field acceleration:

$$\alpha = \frac{1 - \xi C}{1 + \xi C} \tag{A2.13}$$

which is independent of the excitation frequency.

The wave reflection coefficient α is a more physically meaningful description than is the damping coefficient ξ of the behavior of hydrodynamic pressure waves at the reservoir bottom. Although the wave reflection coefficient depends on a pressure wave's angle of incidence at the reservoir bottom, the value α for vertically incident waves, as given in Eq. (A2.13), is used here as a metric to characterize the absorptiveness of the reservoir bottom materials. The wave reflection coefficient, α, may range within the limiting values of 1 and -1. For rigid reservoir bottom materials, $C_r = \infty$ and $\xi = 0$, resulting in $\alpha = 1$, i.e. full reflection or no absorption of hydrodynamic pressures waves. For very soft reservoir bottom materials, C_r approaches zero and ξ tends to ∞, resulting in $\alpha = -1$. It is believed that α values from 1 to 0 would cover the wide range of materials encountered at the bottom of actual reservoirs (Ghanaat and Redpath 1995); $\alpha = 0$ indicates no reflection of hydrodynamic pressure waves.

A2.2 Application to Flexible Foundation

The preceding theory leading up to Eqs. (A2.10) and (A2.13) is also applicable to modeling of wave refraction in the underlying foundation. Such a derivation was presented even earlier by Hall and Chopra (1982).

The wave-reflection coefficient, α, at the water–foundation boundary can be computed from Eq. (A2.13), wherein $\xi = \rho/\rho_r C_r$ and $C_r = \sqrt{(E_r/\rho_r)}$; both C_r and ξ are determined from known values of Young's modulus, E_r, and mass density, ρ_r, for rock.

A2.3 Comments on the Absorbing Boundary

Wave absorption – or, alternatively, wave refraction – at the reservoir bottom is represented only approximately by the boundary condition of Eq. (A2.10). A hydrodynamic pressure wave impinging on the reservoir bottom results in a reflected hydrodynamic pressure wave in the water and two refracted waves, dilatational and rotational, in the sediments or underlying rock. The angle of reflection is equal to the angle of incidence and the angles of refraction of the two refracted waves are given by Snell's law. Although the boundary condition given by Eq. (A2.10) allows for proper reflection of hydrodynamic pressure waves for any angle of incidence, the only refracted waves allowed in the foundation or in the sediments are downward, vertically propagating waves.

3
Fundamental Mode Response of Dams Including Dam–Water–Foundation Interaction

PREVIEW

It was assumed in Chapter 2 that the foundation[†] is rigid and the excitation at the base of the dam was specified by the free-field horizontal motion of the ground surface. This is the motion that would be recorded at the site if the dam were not present. However, the forces associated with a vibrating dam that are transmitted to deformable rock modify the free-field horizontal motion of the foundation and may cause a rocking component in addition to the horizontal component. Viewing this phenomenon from a different point of view, a flexibly supported dam differs from the rigidly supported structure in two aspects: first, its natural vibration period is longer, and, second, part of its vibrational energy is dissipated into the supporting medium by radiation of waves and by hysteretic action in the medium itself. It is this dynamic interaction between the dam and foundation – henceforth referred to as dam–foundation interaction – that is the subject of the first part of this chapter.

Considering only the fundamental mode of vibration of the dam, as in Chapter 2, we first develop a procedure for analysis of dam response including dam–foundation interaction. Response results determined by this analysis procedure enable us to identify the effects of dam–foundation interaction and how they influence the vibration properties – natural vibration period and damping ratio – and response of concrete gravity dams to earthquake ground motion. Also investigated are the implications of ignoring the mass of foundation rock, an assumption that is obviously unrealistic but has been popular in professional practice. Thereafter, we develop an equivalent SDF system to model the response of dams including dam–foundation interaction that enables estimation of the peak response directly from the earthquake response (or design) spectrum. Finally, the equivalent SDF system and the response spectrum analysis procedure is extended to include combined dam–water–foundation interaction.

[†] The word "foundation" denotes the rock that supports the dam.

Earthquake Engineering for Concrete Dams: Analysis, Design, and Evaluation, First Edition. Anil K. Chopra.

3.1 SYSTEM AND GROUND MOTION

The system considered consists of a concrete gravity dam monolith supported on the horizontal surface of underlying flexible foundation, and impounding a reservoir of water with alluvium and sediments deposited at the reservoir bottom (Figure 3.1.1). The system is analyzed under the assumption of linear behavior for the concrete dam, the impounded water, and the foundation.

As mentioned in Section 2.1, minimal restrictions on the geometry of the dam cross section are imposed to permit a continuum solution for hydrodynamic pressures. In addition, the surfaces of the foundation and reservoir bottom are assumed to be horizontal. Idealization of the water impounded in the reservoir and a reservoir bottom that partially absorbs incident hydrodynamic waves is as described in Section 2.1. The foundation underlying the dam is idealized as a homogeneous, isotropic, viscoelastic half-plane. This idealization is motivated by the fact that most concrete dams are founded on competent rock and that similar rocks usually extend to large depths, i.e. there is no obvious "rigid" boundary such as a soil–rock interface at shallow depths. It is for this reason that the foundation is idealized as a semi-unbounded domain. Only rigid-body motions of the base of the dam are permitted in the simplified analysis procedure developed in this chapter, an assumption that will be justified in the next paragraph, and later relaxed when a rigorous procedure for earthquake analysis of dam–water–foundation systems is developed in Chapter 5. As in the later part of Chapter 2, the earthquake excitation considered in the simplified analysis of the system is the horizontal free-field ground acceleration, $a_g(t)$, in the plane of the dam monolith, i.e. transverse to the dam axis; for notational convenience, the superscript x has been temporarily dropped from $a_g^x(t)$.

A rigorous procedure for earthquake analysis of concrete gravity dams, which will be presented in Chapter 5, allows for a deformable base of the dam in contact with the foundation, as permitted by a finite element idealization of the dam. Computed by this procedure, the horizontal acceleration at the crest of the idealized triangular dam monolith (described in Section 2.5.2) relative to the horizontal free-field ground acceleration, is plotted in Figure 3.1.2 for two values of the moduli ratio E_f/E_s, in which E_f and E_s are the moduli of elasticity for the foundation rock and dam concrete, respectively. Also presented in Figure 3.1.2 is the dam response computed under the assumption that the base of the dam is rigid, i.e. motions of the base due to foundation-rock flexibility are restricted to the rigid-body modes (horizontal translation and rocking). It is apparent that base deformation has little effect on the response of the dam at lower excitation frequencies – zero to well beyond the fundamental resonant frequency of the system. Thus, the response of the dam in its fundamental mode of vibration may be computed assuming that the base of the dam is rigid.

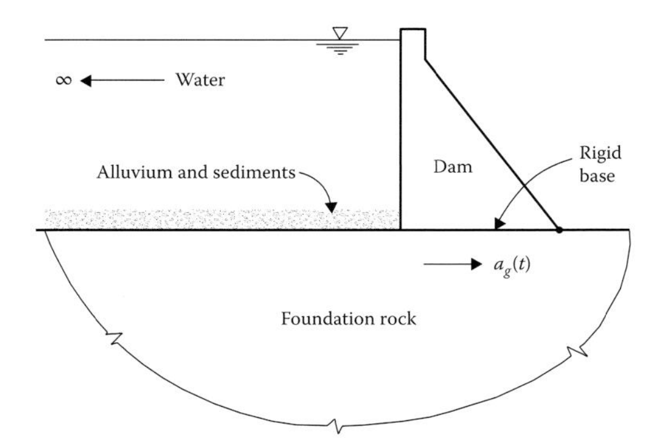

Figure 3.1.1 Dam–water–foundation system.

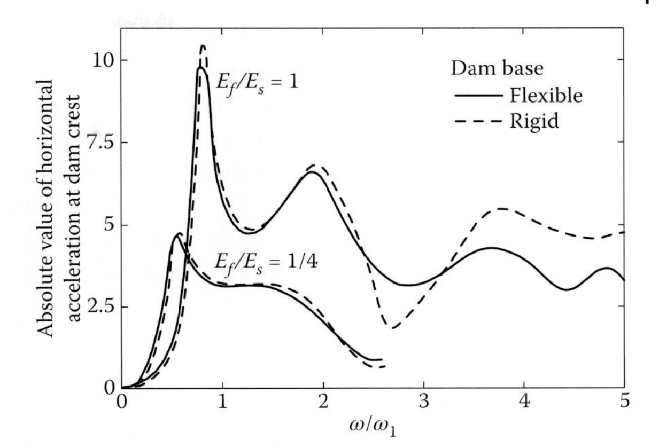

Figure 3.1.2 Effect of rigid base assumption on response of dams including dam–foundation interaction to harmonic horizontal ground motion; reservoir is empty, fundamental mode damping ratio, $\zeta_1 = 5\%$, constant hysteretic damping factor for the foundation, $\eta_f = 0.10$.

3.2 DAM RESPONSE ANALYSIS INCLUDING DAM–FOUNDATION INTERACTION

3.2.1 Governing Equations: Dam Substructure

Shown in Figure 3.2.1, the displacements of the dam – relative to the rigid base – vibrating in its fundamental mode of vibration are given by Eq. (2.2.1), repeated for convenience.

$$r^k(x, y, t) = \phi_1^k(x, y)q_1(t), \qquad k = x, y \tag{3.2.1}$$

in which $r^x(x, y, t)$ and $r^y(x, y, t)$ are the horizontal and vertical components of displacement, respectively; $\phi_1^x(x, y)$ and $\phi_1^y(x, y)$, are the horizontal and vertical components, respectively, of the shape of the fundamental (or first) natural vibration mode of the dam clamped (or fixed) at its base to a rigid foundation, with an empty reservoir; and $q_1(t)$ is the modal coordinate associated with this vibration mode.

Under the influence of horizontal ground motion, the base of the structure displaces horizontally by an amount that is generally different from the free-field displacement – this difference is denoted by u_0 – and it rotates by an amount θ. The displaced configuration of the dam–foundation system is defined by the displacements u_0 and θ of the base and the modal coordinate q_1 that characterizes the deformations of the dam, Figure 3.2.1.

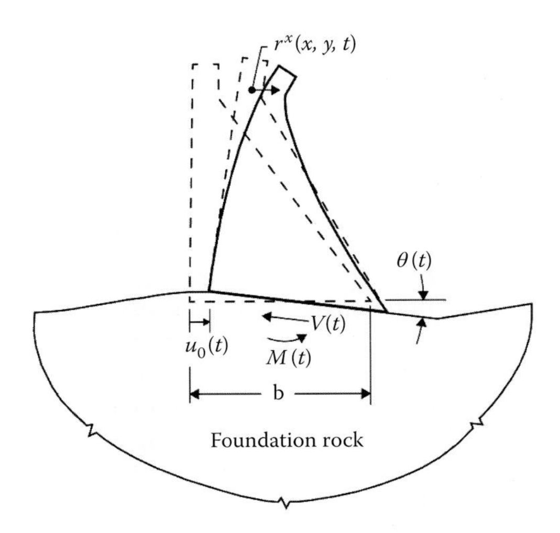

Figure 3.2.1 Displaced configuration of dam with rigid base supported on flexible foundation.

The equations of motion governing these three degrees of freedom (DOFs) are presented without derivation, which is available in Fenves and Chopra (1984a, Appendix D). Equation (2.2.2) governing the motion of the dam vibrating in its fundamental mode of vibration, defined by modal coordinate $q_1(t)$, is modified to include the displacements $u_0(t)$ and $\theta(t)$ of the dam base and the associated base shear, $V(t)$, and base moment, $M(t)$, due to interaction between the dam and foundation (Figure 3.2.1), resulting in the first equation below. Equilibrium of all horizontal forces provides the second equation, and equilibrium of moments of all forces about the centroid of the base of the dam gives the third equation. The three equations of motion are

$$M_1\ddot{q}_1 + C_1\dot{q}_1 + K_1 q_1 + L_1\ddot{u}_0 + L_1^{\theta}\ddot{\theta} = -L_1 a_g(t) \tag{3.2.2a}$$

$$L_1\ddot{q}_1 + m_t\ddot{u}_0 + L_{\theta}^{x}\ddot{\theta} + V = -m_t a_g(t) \tag{3.2.2b}$$

$$L_1^{\theta}\ddot{q}_1 + L_{\theta}^{x}\ddot{u}_0 + I_t\ddot{\theta} + M = -L_{\theta}^{x} a_g(t) \tag{3.2.2c}$$

in which M_1, C_1, K_1, and L_1 were defined in Section 2.2.1; and m_t denotes the total mass of the dam monolith:

$$m_t = \iint m(x, y)\,dx\,dy \tag{3.2.3a}$$

where $m(x, y)$ is the mass density of the dam concrete, and I_t denotes the mass moment of inertia of the dam monolith about the centroid of its base:

$$I_t = \iint \left\{ m_x(x, y)y^2 + m_y(x, y)\left(\frac{1}{2}b - x\right)^2 \right\} dx\,dy \tag{3.2.3b}$$

in which b is the breadth of the dam base; and $m_k(x, y) = m(x, y)$, $k = x, y$;

$$L_1^{\theta} = \iint \left\{ m_x(x, y)y\phi_1^x(x, y) + m_y(x, y)\left(\frac{1}{2}b - x\right)\phi_1^y(x, y) \right\} dx\,dy; \tag{3.2.3c}$$

and

$$L_{\theta}^{x} = \iint m_x(x, y)\,y\,dx\,dy \tag{3.2.3d}$$

3.2.2 Governing Equations: Foundation Substructure

For harmonic free-field ground acceleration, $a_g(t) = e^{i\omega t}$, the interaction forces and displacements at the dam base can be expressed in terms of their complex frequency response functions; thus $V(t) = \bar{V}(\omega)^{i\omega t}$, $M(t) = \bar{M}(\omega)^{i\omega t}$, $u_0(t) = \bar{u}_0(\omega)^{i\omega t}$, and $\theta(t) = \bar{\theta}(\omega)^{i\omega t}$. The frequency response functions for the interaction displacements and interaction forces at the dam base are related by the complex-valued, dynamic (frequency-dependent) stiffness matrix:

$$\begin{bmatrix} K_{VV}(\omega) & K_{VM}(\omega) \\ K_{MV}(\omega) & K_{MM}(\omega) \end{bmatrix} \begin{Bmatrix} \bar{u}_0(\omega) \\ \bar{\theta}(\omega)b \end{Bmatrix} = \begin{Bmatrix} \bar{V}(\omega) \\ \bar{M}(\omega)/b \end{Bmatrix} \tag{3.2.4}$$

Presented in Dasgupta and Chopra (1979) is a procedure for evaluating the dynamic flexibility functions at uniformly spaced nodal points on the surface of a homogeneous, isotropic, viscoelastic half-plane from the solution of two boundary value problems with prescribed surface stresses between adjacent nodal points. The flexibility functions are assembled and inverted to obtain the dynamic stiffness matrix of the foundation – defined with reference to the DOFs at the uniformly spaced nodal points at the dam base in a finite element idealization of the dam (Section 5.3) – which is transformed by the constraint matrix for a rigid base to obtain the dynamic stiffness matrix of Eq. (3.2.4).

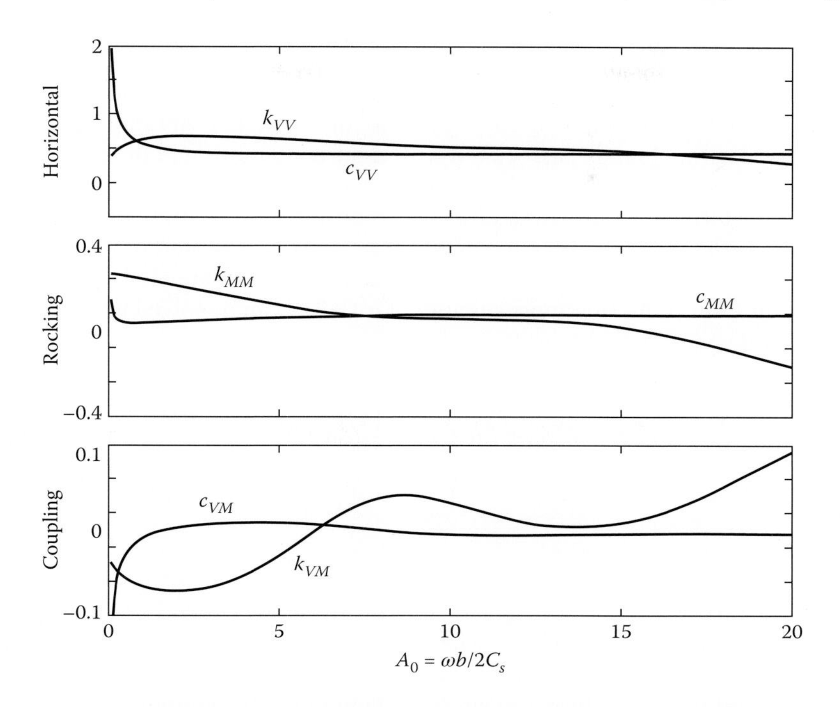

Figure 3.2.2 Horizontal, rocking, and coupling dynamic stiffness functions $K_{ij} = E_f(k_{ij} + iA_0c_{ij})$; k_{ij} and c_{ij} are plotted against dimensionless frequency $A_0 = \omega b/2C_s$; $\eta_f = 0.04$.

The dynamic stiffnesses, K_{VV}, K_{VM}, and K_{MM}, are smooth, slowly varying functions of frequency (Figure 3.2.2); note that coupling terms are equal: $K_{VM}(\omega) = K_{MV}(\omega)$. Unlike the impounded water, the half-plane does not exhibit resonant behavior (compare Figures 3.2.2 and 2.3.2a).

3.2.3 Governing Equations: Dam–Foundation System

For harmonic free-field ground acceleration, $a_g(t) = e^{i\omega t}$, the displacements and forces can be expressed in terms of their complex-valued frequency response functions: $q_1(t) = \overline{q}_1(\omega)e^{i\omega t}$, $u_0(t) = \overline{u}_0(\omega)^{i\omega t}$, $\theta(t) = \overline{\theta}(\omega)^{i\omega t}$, $V(t) = \overline{V}(\omega)^{i\omega t}$, and $M(t) = \overline{M}(\omega)^{i\omega t}$. Substituting these functions in Eq. (3.2.2) and canceling the $e^{i\omega t}$ terms gives

$$\left(-\omega^2 M_1 + i\omega C_1 + K_1\right)\overline{q}_1(\omega) - \omega^2 L_1 \overline{u}_0(\omega) - \omega^2 L_1^\theta\,\overline{\theta}(\omega) = -L_1$$

$$-\omega^2 L_1 \overline{q}_1(\omega) - \omega^2 m_t \overline{u}_0(\omega) - \omega^2 L_\theta^x\,\overline{\theta}(\omega) + \overline{V}(\omega) = -m_t$$

$$-\omega^2 L_1^\theta\,\overline{q}_1(\omega) - \omega^2 L_\theta^x\,\overline{u}_0(\omega) - \omega^2 I_t\,\overline{\theta}(\omega) + \overline{M}(\omega) = -L_\theta^x \tag{3.2.5}$$

and then using Eq. (3.2.4) to express interaction forces in terms of interaction displacements leads to

$$\begin{bmatrix} -\omega^2 M_1 + i\omega C_1 + K_1 & -\omega^2 L_1 & -\omega^2 L_1^\theta \\ -\omega^2 L_1 & -\omega^2 m_t + K_{VV}(\omega) & -\omega^2 L_\theta^x + K_{MV}(\omega)b \\ -\omega^2 L_1^\theta & -\omega^2 L_\theta^x + K_{MV}(\omega)b & -\omega^2 I_t + K_{MM}(\omega)b^2 \end{bmatrix} \begin{Bmatrix} \overline{q}_1(\omega) \\ \overline{u}_0(\omega) \\ \overline{\theta}(\omega) \end{Bmatrix} = -\begin{Bmatrix} L_1 \\ m_t \\ L_\theta^x \end{Bmatrix} \tag{3.2.6}$$

3.2.4 Dam Response Analysis

The solution of Eq. (3.2.6) for the frequency response function, $\overline{q}_1(\omega)$, for the fundamental modal coordinate is complicated by the implicit "contributions" of the higher vibration modes of the dam to the inertial terms m_t, L_θ^x, and I_t associated with rigid-body motion of the dam permitted by foundation flexibility. Numerical results demonstrate that the dam response is accurately represented if these inertial terms are approximated by the contribution of the fundamental vibration mode: $m_t \approx m_1^*$, $L_\theta^x \approx m_1^* h_1^*$, and $I_t \approx m_1^* \left(h_1^*\right)^2$, in which $m_1^* = \left(L_1\right)^2/M_1$ and $h_1^* = L_1^\theta/L_1$ are the effective mass and effective height, respectively, of the dam in its fundamental vibration mode (Chopra 2017, Section 17.6). With this approximation, Eq. (3.2.6) can be solved for $\overline{q}_1(\omega)$ (Fenves and Chopra 1984a, Appendix E) to obtain

$$\overline{q}_1(\omega) = \frac{-L_1}{-\omega^2 M_1 + i\omega C_1 + K_1 - \omega^2 M_1 \left(1 + i2\zeta_1 \omega/\omega_1\right) \mathcal{F}(\omega)} \tag{3.2.7}$$

in which

$$\mathcal{F}(\omega) = \frac{\dfrac{K_{VV}(\omega)}{m_1^*\omega_1^2} + \dfrac{K_{MM}(\omega)}{m_1^*\omega_1^2}\left(\dfrac{b}{h_1^*}\right)^2 - 2\dfrac{K_{VM}(\omega)}{m_1^*\omega_1^2}\left(\dfrac{b}{h_1^*}\right)}{\dfrac{K_{VV}(\omega)}{m_1^*\omega_1^2}\dfrac{K_{MM}(\omega)}{m_1^*\omega_1^2}\left(\dfrac{b}{h_1^*}\right)^2 - \left[\dfrac{K_{VM}(\omega)}{m_1^*\omega_1^2}\left(\dfrac{b}{h_1^*}\right)\right]^2} \tag{3.2.8}$$

The term $\mathcal{F}(\omega)$ in the denominator of Eq. (3.2.7) contains the effects of dam–foundation interaction; this is evident by comparing Eq. (3.2.7) with Eq. (2.2.7), specialized for horizontal ground motion ($l = x$). This complex-valued term reduces the effective stiffness of the dam because of foundation flexibility and hence modifies the effective damping of the system because of material damping in the foundation and radiation of waves away from the vibrating dam into the half-space. As will be demonstrated later, the effective damping of the dam may increase or decrease because of dam–foundation interaction.

3.3 DAM–FOUNDATION INTERACTION

3.3.1 Interaction Effects

Computed from Eq. (3.2.7), the absolute value of the relative (to the moving base of the dam) acceleration at the crest of the idealized triangular dam monolith is plotted as a function of the frequency of the harmonic free-field ground acceleration of unit amplitude (Figure 3.3.1). When presented in normalized form, these frequency response functions do not depend separately on E_s and E_f, but only on the ratio E_f/E_s. Results are presented for four values of $E_f/E_s = \infty$, 2, 1, and 1/4, and a hysteretic damping factor $\eta_f = 0.04$. The first represents rigid foundation, and the last case is a very flexible foundation: the elastic modulus for the foundation rock is one-fourth of the modulus for dam concrete.

Dam–foundation interaction affects the response of the dam in a simpler manner than does dam–water interaction because, as mentioned earlier, unlike impounded water in the reservoir, the half-plane foundation model does not exhibit any resonant frequencies. As the E_f/E_s ratio decreases, which for a fixed value of the concrete modulus implies an increasingly flexible foundation, the fundamental resonant frequency of the dam decreases, the amplitude of the resonant response of the dam decreases, and the frequency bandwidth at resonance increases, implying an increase in the apparent damping of the structure (Figure 3.3.1). These effects due to foundation flexibility and damping, both material and radiation, have been examined extensively for concrete gravity dams (Fenves and Chopra 1985a,b) and for buildings (Veletsos 1977; Veletsos and Meek 1974).

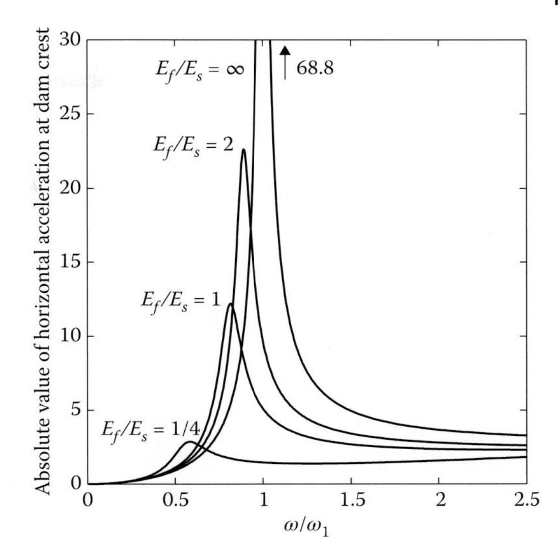

Figure 3.3.1 Influence of dam–foundation interaction on dam response to horizontal ground motion; $\zeta_1 = 2\%$ and $\eta_f = 0.4$; reservoir is empty.

Can coupling terms in the dynamic stiffness matrix be ignored? The coupling term, $K_{VM}(\omega)$ and $K_{MV}(\omega)$, which are usually neglected in the analysis of multistory buildings (Veletsos 1977; Veletsos and Meek 1974), have a significant influence on the fundamental mode response of dams, as shown in Figure 3.3.2 for the idealized triangular dam monolith. The additional radiation damping associated with the imaginary part of the coupling term is apparently significant for a squat, heavy structure such as a concrete gravity dam because, unlike slender multistory buildings, its base translational motion is comparable to the rotational motion.

3.3.2 Implications of Ignoring Foundation Mass

Earthquake analysis of dams is greatly simplified if the mass of foundation rock is ignored because then only flexibility of the foundation needs to be considered, which can be computed by analyzing a bounded-size model of the foundation as part of a standard finite element analysis. Can this seemingly unrealistic assumption be justified?

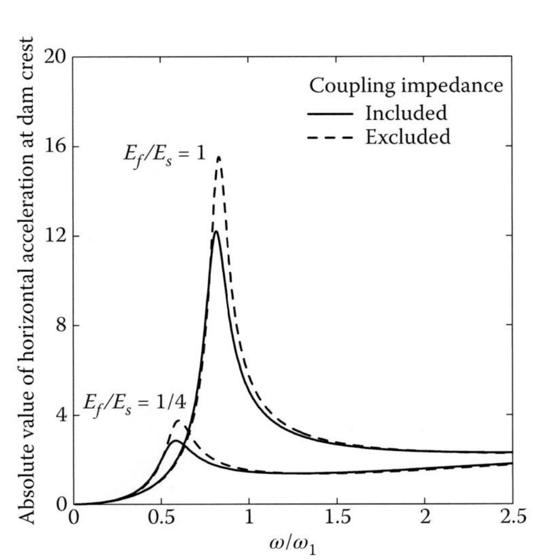

Figure 3.3.2 Effect of coupling term in foundation stiffness matrix on response of dams to harmonic horizontal ground motion; $\zeta_1 = 2\%$, $\eta_f = 0.04$.

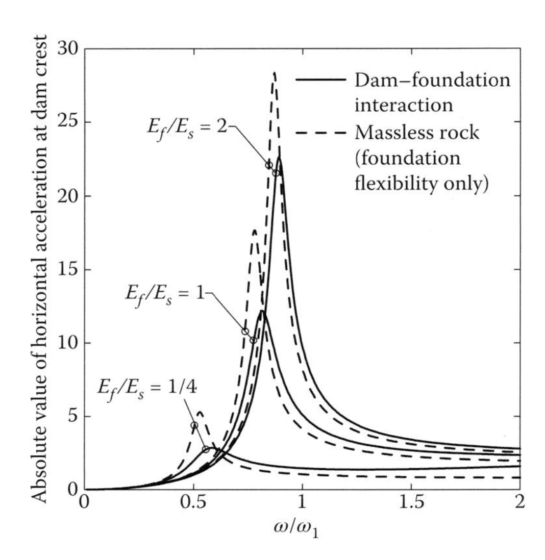

Figure 3.3.3 Influence of foundation model on dam response to horizontal ground motion. Results are presented for two cases: (i) including all effects of dam–foundation interaction; and (ii) ignoring mass of foundation rock, i.e. considering foundation flexibility only; $\zeta_1 = 2\%$, $\eta_f = 0.04$.

Presented in Figure 3.3.3 are frequency response curves for the idealized triangular dam monolith computed for two cases: (i) all effects of dam–foundation interaction are included; and (ii) foundation rock is assumed to be massless, implying that flexibility of the foundation is included but foundation mass, material damping, and radiation damping are all ignored. It is seen that foundation flexibility roughly accounts for the frequency shift due to dam–foundation interaction. However, the response amplitude at the resonant frequency is overestimated if the foundation rock is assumed massless, because then foundation material and radiation damping is also ignored. This overestimation of response is large, especially for the smaller values of E_f/E_s.

These results suggest that the earthquake response of the dam will be overestimated if the mass of foundation rock is ignored. We will demonstrate in Chapter 6 that this overestimation can be unacceptably large.

3.4 EQUIVALENT SDF SYSTEM: DAM–FOUNDATION SYSTEM

The fundamental mode response of the dam including dam–foundation interaction can be modeled by a rigidly supported equivalent single-degree-of-freedom (SDF) system following the procedure developed for building–foundation systems (Veletsos 1977; Veletsos and Meek 1974). The properties of this system are defined to recognize the reduction in stiffness and change in damping of the dam due to dam–foundation interaction. These modifications are contained in the foundation dynamic flexibility term, $\mathcal{F}(\omega)$ [Eq. (3.2.8)] that appears in the frequency response function $\overline{q}_1(\omega)$ for the modal coordinate associated with the fundamental vibration mode of the dam [Eq. (3.2.7)].

3.4.1 Modified Natural Frequency and Damping Ratio

The natural vibration frequency $\tilde{\omega}_f$ and damping ratio $\tilde{\zeta}_f$ of the equivalent SDF system are determined in such a way that the resonant frequency and resonant response amplitude of this system are equal to the corresponding quantities for the dam–foundation system (Figure 3.2.1) when subjected to the same free-field harmonic ground motion.

The natural vibration frequency, $\tilde{\omega}_f$, of the equivalent SDF system that models the fundamental mode response of dam–foundation systems is approximately given by the excitation frequency that makes the real-valued component of the denominator in Eq. (3.2.7) equal to zero; see Appendix 3:

$$\tilde{\omega}_f = \frac{\omega_1}{\sqrt{\left\{1 + \mathrm{Re}\left[\mathcal{F}\left(\tilde{\omega}_f\right)\right]\right\}}} \tag{3.4.1}$$

Because $\tilde{\omega}_f$ appears on both sides of Eq. (3.4.1), it must be solved iteratively. The natural frequency, $\tilde{\omega}_f$, of the equivalent SDF system is less than the natural frequency, ω_1, of the dam on rigid foundation because Re[$\mathcal{F}(\omega)$] > 0 for all excitation frequencies. In passing we make two observations: First, Eq. (3.4.1) is similar to the expression derived earlier for building–foundation systems (Bielak 1976; Veletsos 1977). Second, dropping the coupling stiffness terms in Eq. (3.2.8) and substituting it into Eq. (3.4.1) gives the same expression as the one in Veletsos (1977) for building–foundation systems.

The frequency response function for the equivalent SDF system with natural vibration frequency $\tilde{\omega}_f$ and damping ratio $\tilde{\zeta}_f$ can be shown to be (see Appendix 3)

$$\tilde{\bar{q}}_1(\omega) = \left(\frac{\tilde{\omega}_f}{\omega_1}\right)^2 \frac{-L_1}{-\omega^2 M_1 + i\omega\left(2\tilde{\zeta}_f M_1 \tilde{\omega}_f\right) + \tilde{\omega}_f^2 M_1} \tag{3.4.2}$$

The damping ratio $\tilde{\zeta}_f$ is determined by equating the resonant response of the equivalent SDF system, determined from Eq. (3.4.2), to the response of the dam–foundation system to harmonic (free-field) excitation at frequency, $\tilde{\omega}_f$, determined from Eq. (3.2.7): $\tilde{\bar{q}}_1\left(\tilde{\omega}_f\right) = \bar{q}_1\left(\tilde{\omega}_f\right)$. It can be shown that the damping ratio of the equivalent SDF system is (see Appendix 3)

$$\tilde{\zeta}_f = \left(\frac{\tilde{\omega}_f}{\omega_1}\right)^3 \zeta_1 + \zeta_f \tag{3.4.3}$$

in which the added damping due to dam–foundation interaction is represented by ζ_f, defined as

$$\zeta_f = -\frac{1}{2}\left(\frac{\tilde{\omega}_f}{\omega_1}\right)^2 \mathrm{Im}\left[\mathcal{F}\left(\tilde{\omega}_f\right)\right] \tag{3.4.4}$$

The damping ratio, ζ_f, is positive because Im[$\mathcal{F}(\omega)$] < 0 for all excitation frequencies. It includes the energy dissipation due to material damping in the rock and energy loss due to radiation of elastic waves away from the vibrating dam into the half-space. In passing we note that Eq. (3.4.3) for the damping of dam–foundation systems has the same form as for building–foundation systems (Veletsos 1977; Veletsos and Meek 1974).

A comparison of Eq. (3.4.2) with Eq. (2.2.7) demonstrates that the response of the dam–foundation system to harmonic free-field ground motion is $\left(\tilde{\omega}_f/\omega_1\right)^2$ times the response of the rigidly supported SDF system – with natural vibration frequency $\tilde{\omega}_f$ and damping ratio $\tilde{\zeta}_f$ – subjected to the same excitation. Thus, as will be shown in Section 3.4.3, the earthquake response of the dam–foundation system can be readily determined once $\tilde{\omega}_f$ and $\tilde{\zeta}_f$ are known.

3.4.2 Evaluation of Equivalent SDF System

The frequency response function $\tilde{\bar{q}}_1(\omega)$ for the equivalent SDF system, computed from Eq. (3.4.2) – with the natural vibration frequency, $\tilde{\omega}_f$, and damping ratio, $\tilde{\zeta}_f$, given by Eqs. (3.4.1) and (3.4.3), respectively – is presented in Figure 3.4.1, wherein it is compared with the exact function $\bar{q}_1(\omega)$ of Eq. (3.2.7) for the idealized triangular monolith (first presented in Figure 3.3.1). These results demonstrate that, over a wide range of excitation frequencies, the

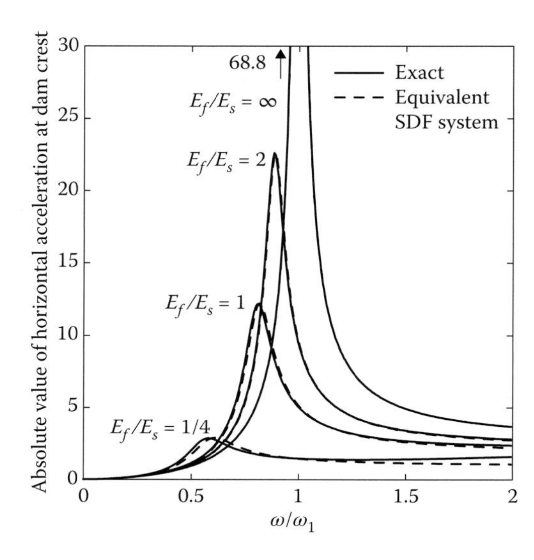

Figure 3.4.1 Comparison of exact and equivalent SDF system responses of dams to horizontal ground motion. Results are presented for various values of E_f/E_s, the ratio of elastic moduli of foundation rock and dam concrete; reservoir is empty; $\zeta_1 = 2\%$, $\eta_f = 0.04$.

equivalent SDF system accurately represents the fundamental mode response of dams including dam–foundation interaction effects.

The exact value of the fundamental resonant period of the idealized triangular monolith, given by the resonant peak of $\overline{q}_1(\omega)$ is compared in Figure 3.4.2 with the natural vibration period, $\tilde{T}_f = 2\pi/\tilde{\omega}_f$, of the equivalent SDF system. A measure of the lengthening of the fundamental resonant period of the dam due to dam–foundation interaction, the period ratio $\tilde{T}_f/T_1$, is shown for a range of E_f/E_s values. It is apparent that the natural vibration period of the equivalent SDF system is close to the fundamental resonant period of the dam–foundation system, but its accuracy decreases as E_f/E_s decreases, i.e. as the foundation becomes more flexible; for $E_f/E_s = 1$, the error is 1.4%, and for $E_f/E_s = 1/4$, the error is 7.3%. However, the increasing error in the vibration period $\tilde{T}_f$ has little effect on the accuracy of the SDF system response, as shown in Figure 3.4.1, because the damping due to dam–foundation interaction increases as E_f/E_s decreases, resulting in response functions that do not resonate sharply.

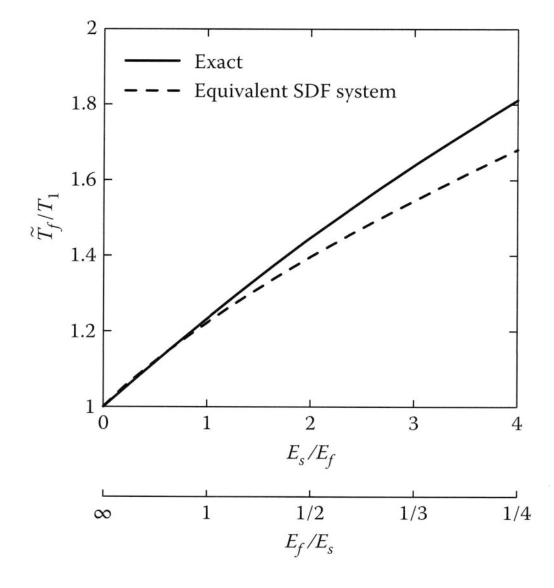

Figure 3.4.2 Comparison of exact and approximate (equivalent SDF system) values of the fundamental vibration period $\tilde{T}_f$ of dams for a range of E_f/E_s values; $\eta_f = 0.04$.

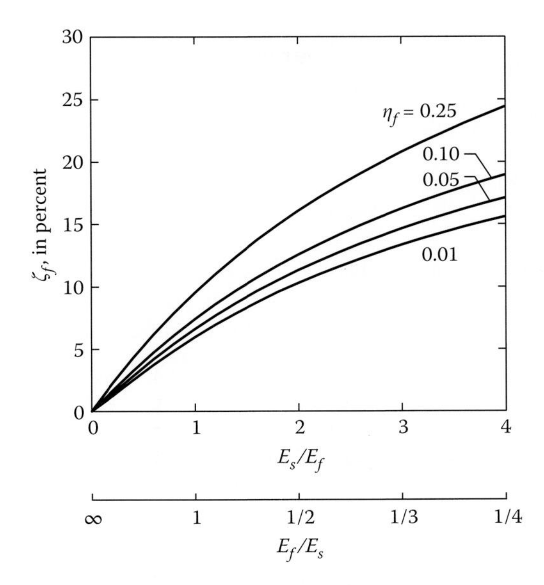

Figure 3.4.3 Added damping ratio ζ_f due to dam–foundation interaction for a range of E_f/E_s values and various values of η_f, the constant hysteretic damping factor for foundation rock.

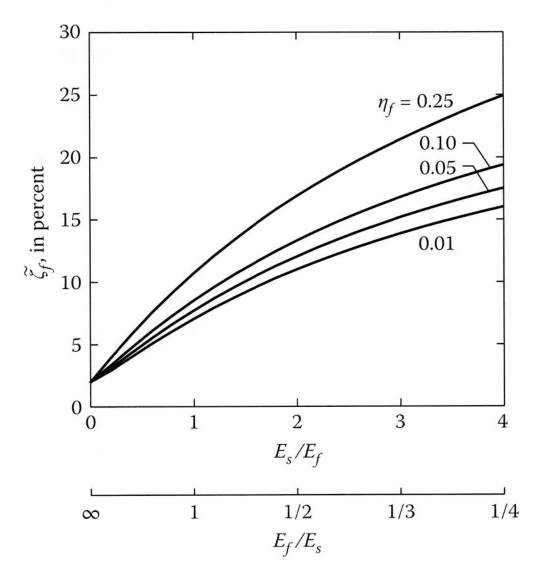

Figure 3.4.4 Damping ratio $\tilde{\zeta}_f$ of the equivalent SDF system representing dam–foundation systems for a range of E_f/E_s values and various values of η_f, the constant hysteretic damping factor for the foundation rock; $\zeta_1 = 2\%$.

The added damping ratio, ζ_f, due to dam–foundation interaction is presented in Figure 3.4.3, and the damping ratio, $\tilde{\zeta}_f$, of the equivalent SDF system is shown in Figure 3.4.4 for a range of values of the hysteretic damping factor, η_f, for the foundation. The damping ratios, ζ_f and $\tilde{\zeta}_f$, increase as the foundation becomes more flexible and with increasing damping factor η_f. Considering that $\tilde{\omega}_f$ is less than ω_1, Eq. (3.4.3) indicates that dam–foundation interaction reduces the effectiveness of the structural damping. However, unlike for slender multistory buildings (Veletsos 1977), for a wide range of E_f/E_s and η_f values, this reduction is more than compensated by the added damping due to dam–foundation interaction, which leads to an increase in the overall damping of the dam, as shown in Figure 3.4.4. The effects of dam–foundation interaction on dam response may be neglected if the moduli ratio, E_f/E_s, is relatively large. In particular, if E_f/E_s is greater than five, the vibration period is lengthened by only 5%, the added damping ratio, ζ_f, is only 1.4%, and consequently, the foundation may be treated as rigid.

3.4.3 Peak Response

Because the equivalent SDF system accurately predicts the response of dams including dam–foundation interaction over a complete range of excitation frequencies and for a wide range of ratios of elastic moduli for the dam concrete and foundation rock, it is valid for the analysis of dam response to arbitrary ground motion. In particular, the peak deformations of the dam can be expressed by comparing Eqs. (3.4.2) and (2.2.7) and modifying Eq. (2.2.12) appropriately:

$$r^k(x,y) = \left(\frac{\tilde{\omega}_f}{\omega_1}\right)^2 \Gamma_1 D\left(\tilde{T}_f, \tilde{\zeta}_f\right) \phi_1^k(x,y) \qquad k = x, y \tag{3.4.5}$$

in which the subscript "o" has been dropped for notational convenience; $D\left(\tilde{T}_f, \tilde{\xi}_f\right)$ is the ordinate of the deformation response (or design) spectrum for the earthquake ground motion evaluated at period $\tilde{T}_f = 2\pi/\tilde{\omega}_f$ and damping ratio $\tilde{\xi}_f$. The equivalent static forces – acting horizontally on the dam – associated with the peak deformations of Eq. (3.4.5) are given by Chopra 2017, Section 17.7:

$$f_1(x,y) = \Gamma_1 \frac{A\left(\tilde{T}_f, \tilde{\xi}_f\right)}{g} w_x(x,y)\phi_1^x(x,y) \tag{3.4.6}$$

in which $A\left(\tilde{T}_f, \tilde{\xi}_f\right)$ is the ordinate of the pseudo-acceleration response (or design) spectrum.

This equation is similar to the corresponding expression for a rigidly supported dam, Eq. (2.2.14), except that the spectral ordinate is determined at the modified period, $\tilde{T}_f$, and modified damping, $\tilde{\xi}_f$.

3.5 EQUIVALENT SDF SYSTEM: DAM–WATER–FOUNDATION SYSTEM

The equivalent SDF systems that represent the fundamental mode response of dams considering the separate effects of dam–water interaction and dam–foundation interaction were presented in Sections 2.6 and 3.4, respectively. In a similar manner, the fundamental mode response of dams including both types of interaction, simultaneously, has been modeled by an equivalent SDF system (Fenves and Chopra 1985d). Without a detailed derivation of this complicated analysis, only the final equations for the natural vibration frequency and damping ratio of the equivalent SDF system are presented next.

3.5.1 Modified Natural Frequency and Damping Ratio

The natural vibration frequency $\tilde{\omega}_1$ of the equivalent SDF system including both interaction effects is given by

$$\tilde{\omega}_1 = \left(\frac{\tilde{\omega}_r}{\omega_1}\right)\left(\frac{\tilde{\omega}_f}{\omega_1}\right)\omega_1 \tag{3.5.1}$$

where frequencies $\tilde{\omega}_r$ and $\tilde{\omega}_f$ approximate the fundamental resonant frequencies of two special cases: (i) the dam with impounded water on rigid foundation; and (ii) the dam–foundation system with an empty reservoir, respectively. According to Eq. (3.5.1), reductions in ω_1 due to dam–water interaction (including reservoir-bottom absorption) and due to dam–foundation interaction are determined independently and combined to obtain $\tilde{\omega}_1$ for the dam–water–foundation system.

The frequency response function for this equivalent SDF system is (see Fenves and Chopra 1985d)

$$\bar{\tilde{q}}_1(\omega) = \left(\frac{\tilde{\omega}_f}{\omega_1}\right)^2 \frac{-\tilde{L}_1}{-\omega^2\tilde{M}_1 + i\omega\left(2\tilde{\xi}_1\tilde{M}_1\tilde{\omega}_1\right) + \tilde{\omega}_1^2\tilde{M}_1} \tag{3.5.2}$$

in which $\tilde{M}_1 = M_1 + \text{Re}\left[B_1\left(\tilde{\omega}_r\right)\right]$ and $\tilde{L}_1 = L_1 + B_0\left(\tilde{\omega}_r\right)$, first defined in Section 2.6.1. A comparison of Eq. (3.5.2) with Eq. (3.4.2) shows that including hydrodynamic effects increases the generalized mass, which correspondingly reduces the natural vibration frequency of the equivalent SDF system from $\tilde{\omega}_f$ to $\tilde{\omega}_1$, and increases the generalized earthquake force. Also, the damping ratio now includes a contribution from propagation of hydrodynamic pressure waves upstream and their refraction into the absorptive reservoir-bottom, in addition to contributions from structural damping in the dam, and from radiation and material damping in the foundation domain.

The damping ratio, $\tilde{\zeta}_1$, of the equivalent SDF system is obtained by equating the resonant response of the equivalent SDF system from Eq. (3.5.2) to the exact fundamental mode response (not presented here) at the natural vibration frequency, $\tilde{\omega}_1$ (Fenves and Chopra 1985d). Simplifying the resulting equation by a series of clever approximations, leads to the final result for the effective damping ratio, $\tilde{\zeta}_1$, of the equivalent SDF system:

$$\tilde{\zeta}_1 = \left(\frac{\tilde{\omega}_r}{\tilde{\omega}_1}\right)\left(\frac{\tilde{\omega}_f}{\tilde{\omega}_1}\right)^3 \zeta_1 + \zeta_r + \zeta_f \tag{3.5.3}$$

in which ζ_r is given by Eq. (2.6.123) and ζ_f by Eq. (3.4.4). According to Eq. (3.5.3), the contributions of dam–water interaction and dam–foundation interaction to the damping of the system are obtained independently and added to structural damping, with its effectiveness properly reduced because of the interaction effects, to give the damping ratio $\tilde{\zeta}_1$ of the dam–water–foundation system.

It is important to note that the effects of dam–water interaction and dam–foundation interaction on the parameters of the equivalent SDF system – natural vibration frequency, damping, generalized mass, and force – are computed independently of each other and applied sequentially to give values for the parameters that include both interaction effects. The ability to separate the interaction effects in the computation of the natural vibration frequency and generalized mass is a consequence of the fact that dam–foundation interaction has a small influence on the added hydrodynamic mass, $m_a(y)$, and dam–water interaction does not substantially alter the effects of dam–foundation interaction. The approximations introduced to achieve this desirable result are documented in a detailed derivation (Fenves and Chopra 1985d). The main attraction in separately considering the effects of dam–water interaction and of dam–foundation interaction, at the expense of some loss in accuracy, is of course to simplify the evaluation of the properties of the equivalent SDF system that represents the fundamental mode response of dams. Such an approximate procedure is intended only for the preliminary phase of design of new dams and of safety evaluation of existing dams.

3.5.2 Evaluation of Equivalent SDF System

The effectiveness of the equivalent SDF system in representing the fundamental mode response of dams including dam–water–foundation interaction is shown in Figure 3.5.1; results are presented for the triangular dam monolith for several values of the moduli ratio, E_f/E_s, and wave reflection coefficient, α, for the reservoir bottom. The "exact" fundamental mode response was computed from Eqs. 5 and 6 in Fenves and Chopra (1985d), and the response of the equivalent SDF system was computed using Eq. (3.5.2), with the natural vibration frequency, $\tilde{\omega}_1$, and damping ratio $\tilde{\zeta}_1$ determined from Eqs. (3.5.1) and (3.5.3), respectively.

These results demonstrate that the equivalent SDF system provides a good approximation to the fundamental mode response of concrete gravity dams for a wide range of values of the moduli ratio, E_f/E_s, and wave reflection coefficient, α. The quality of the approximation is judged to be satisfactory for the preliminary design or evaluation of dams. This conclusion is especially impressive if we recall that the complicated effects of dam–water interaction, reservoir bottom absorption, and dam–foundation interaction have all been included; a number of clever but rational approximations were introduced to derive equations for the natural frequency and damping ratio of the equivalent SDF system; and that the SDF system response generally errs slightly on the conservative side.

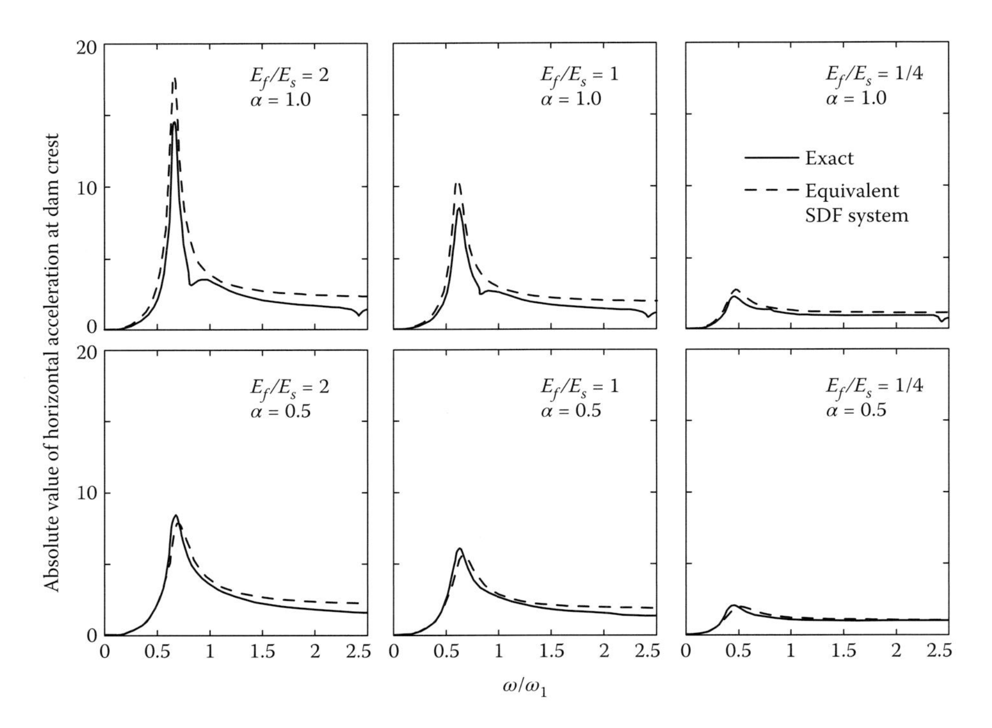

Figure 3.5.1 Comparison of exact and equivalent SDF system responses of dam–water–foundation system to horizontal ground motion for various values of E_f/E_s and α; $\zeta_1 = 5\%$. $\eta_f = 0.10$. Source: From Fenves and Chopra (1985d).

3.5.3 Peak Response†

Because the equivalent SDF system accurately predicts response of dams to harmonic ground motion over a complete range of excitation frequencies and for a wide range of parameters characterizing the properties of the dam, water, foundation, and reservoir-bottom materials, it can be used in the analysis of dam response to arbitrary ground motion. In particular, the peak displacements of the dam can be expressed as (Chopra 2017, Section 17.7)

$$r^k(x,y) = \left(\frac{\tilde{\omega}_f}{\omega_1}\right)^2 \tilde{\Gamma}_1 D\left(\tilde{T}_1, \tilde{\zeta}_1\right) \ \phi_1^k(x,y), \qquad k = x, y \tag{3.5.4}$$

in which $D\left(\tilde{T}_1, \tilde{\zeta}_1\right)$ is the ordinate of the deformation response (or design) spectrum for the earthquake ground motion evaluated at period $\tilde{T}_1 = 2\pi/\tilde{\omega}_1$ and damping ratio, $\tilde{\zeta}_1$. This equation is an integrated version of Eqs. (2.6.14) and (3.4.5) developed for the two special cases: dam–water system and dam–foundation system, respectively.

For the special case of rigid foundation, $\tilde{\omega}_f = \omega_1$ and $\zeta_f = 0$; Eq. (3.5.1) reduces to $\tilde{\omega}_1 = \tilde{\omega}_r$; Eq. (3.5.3) reduces to Eq. (2.6.12); and Eq. (3.5.4) reduces to Eq. (2.6.14). If the reservoir is empty, $\tilde{\Gamma}_1 = \Gamma_1$, $\tilde{\omega}_r = \omega_1$, and $\zeta_r = 0$; Eq. (3.5.1) reduces to $\tilde{\omega}_1 = \tilde{\omega}_f$; Eq. (3.5.3) reduces to Eq. (3.4.3), and Eq. (3.5.4) reduces to Eq. (3.4.5). If the reservoir impounds water, but the reservoir bottom is non-absorptive ($\alpha = 1$), then ζ_r is zero (Chopra 1978).

† A step-by-step summary to implement the analysis procedure developed in Section 3.5 is available in Fenves and Chopra (1985d).

The equivalent static forces – acting horizontally on the dam – associated with the peak displacements (Eq. (3.5.4)) of the dam are (see Chopra 1978 and Chopra 2017, Section 17.7):

$$f_1(x,y) = \tilde{\Gamma}_1 \frac{A\left(\tilde{T}_1, \tilde{\zeta}_1\right)}{g} \tilde{w}_x(x,y)\ \phi_1^x(x,y) \tag{3.5.5}$$

in which $A\left(\tilde{T}_1, \tilde{\zeta}_1\right)$ is the ordinate of the pseudo-acceleration response (or design) spectrum for the ground motion evaluated at period $\tilde{T}_1 = 2\pi/\tilde{\omega}_1$ and damping ratio $\tilde{\zeta}_1$; and $\tilde{w}_x(x,y) = g\tilde{m}_x(x,y)$ where $\tilde{m}_x(x,y)$ was first defined in Eq. (2.6.5a), repeated here for convenience:

$$\tilde{m}_x(x,y) = m_x(x,y) + m_a(y)\delta(x) \tag{3.5.6}$$

in which $m_x(x,y)$ is the mass density of the dam, and the "added hydrodynamic mass," $\tilde{m}_a(y)$, was first defined in Eq. (2.6.6), repeated here for convenience:

$$m_a(y) = \frac{\bar{p}_1\left(y, \tilde{\omega}_r\right)}{\phi_1^x(0,y)} \tag{3.5.7}$$

The substitution of Eqs. (3.5.6) and (3.5.7) into Eq. (3.5.5) gives the final expression for the lateral forces acting on the dam:

$$f_1(x,y) = \tilde{\Gamma}_1 \frac{A\left(\tilde{T}_1, \tilde{\zeta}_1\right)}{g} \left\{ w_x(x,y)\ \phi_1^x(x,y) + g\bar{p}_1\left(y, \tilde{\omega}_r\right)\delta(x) \right\} \tag{3.5.8}$$

in which $\tilde{w}_x(x,y) = g\tilde{m}_x(x,y)$ is the unit weight of the dam concrete. This equation is an integrated version of Eqs. (2.6.16) and (3.4.6) developed earlier for two special cases: dam–water system and dam–foundation system, respectively. It is essentially equivalent to Eq. (2.6.16) with the only change being the period and damping ratio at which the spectral ordinate is determined.

APPENDIX 3: EQUIVALENT SDF SYSTEM

The vibration properties and the frequency response function for the SDF system equivalent to the fundamental mode response of dams including dam–foundation interaction are derived in this appendix.

The frequency response function, $\bar{q}_1(\omega)$, for the fundamental modal coordinate of the dam–foundation system may be expressed from Eq. (3.2.7) as

$$\left\{ -\omega^2 M_1 + i\omega 2\zeta_1 M_1 \omega_1 + \omega_1^2 M_1 - \omega^2 M_1 \left(1 + \frac{2\zeta_1\omega}{\omega_1}\right) \mathcal{F}(\omega) \right\} \bar{q}_1(\omega) = -L_1 \tag{A3.1}$$

The frequency response function, $\bar{\bar{q}}_1(\omega)$, for the equivalent SDF system can be obtained from Eq. (A3.1) by evaluating the frequency-dependent foundation flexibility term, $\mathcal{F}(\omega)$, at the natural vibration frequency, $\tilde{\omega}_f$, of the equivalent SDF system, still to be determined. Separating the real- and imaginary-valued terms in Eq. (A3.1) and grouping them into the standard form for a frequency-domain equation for an SDF system gives

$$\begin{aligned} &\left\{ -\omega^2 M_1 \left(1 + \mathrm{Re}\left[\mathcal{F}\left(\tilde{\omega}_f\right)\right] - 2\zeta_1 \frac{\tilde{\omega}_f}{\omega_1} \mathrm{Im}\left[\mathcal{F}\left(\tilde{\omega}_f\right)\right] \right) \right. \\ &\left. + i\omega M_1 \left(2\zeta_1\omega_1 - \tilde{\omega}_f \mathrm{Im}\left[\mathcal{F}\left(\tilde{\omega}_f\right)\right] - 2\zeta_1 \frac{\tilde{\omega}_f^2}{\omega_1} \mathrm{Re}\left[\mathcal{F}\left(\tilde{\omega}_f\right)\right] \right) + \omega_1^2 M_1 \right\} \bar{\bar{q}}_1(\omega) = -L_1 \end{aligned} \tag{A3.2}$$

wherein the tilde symbol is intended to identify the equivalent SDF system. The real-valued term, $\left(2\zeta_1\tilde{\omega}_f/\omega_1\right)\mathrm{Im}\left[\mathcal{F}\tilde{\omega}_f\right]$, is a product of two small damping terms that can be neglected because its effect on the response of the equivalent SDF system is small.

The natural vibration frequency, $\tilde{\omega}_f$, is approximately given by the excitation frequency that makes the real-valued component of the left side of Eq. (A3.2) equal to zero:

$$-\tilde{\omega}_f^2 M_1 \left\{1 + \mathrm{Re}\left[\mathcal{F}\left(\tilde{\omega}_f\right)\right]\right\} + \omega_1^2 M_1 = 0 \tag{A3.3}$$

that directly leads to

$$\tilde{\omega}_f = \frac{\omega_1}{\sqrt{\left\{1 + \mathrm{Re}\left[\mathcal{F}\left(\tilde{\omega}_f\right)\right]\right\}}} \tag{A3.4}$$

which appeared as Eq. (3.4.1).

The frequency response function $\bar{\tilde{q}}_1(\omega)$ for the equivalent SDF system can be obtained from Eq. (A3.2) by using Eq. (A3.4) for the real-valued terms in the left-hand side and grouping the imaginary-valued terms:

$$\bar{\tilde{q}}_1(\omega) = \frac{-L_1}{-\omega^2 \left(\frac{\omega_1}{\tilde{\omega}_f}\right)^2 M_1 + i\omega\,\mathcal{D} + \omega_1^2 M_1} \tag{A3.5}$$

where a new symbol, $\mathcal{D}$, has been introduced temporarily for convenience. Multiplying both numerator and denominator by $\left(\tilde{\omega}_f/\omega_1\right)^2$ gives

$$\bar{\tilde{q}}_1(\omega) = \left(\frac{\tilde{\omega}_f}{\omega_1}\right)^2 \frac{-L_1}{-\omega^2 M_1 + i\omega\left(\tilde{\omega}_f/\omega_1\right)^2 \mathcal{D} + \tilde{\omega}_f^2 M_1} \tag{A3.6}$$

The imaginary-valued term can be expressed in terms of the damping ratio, $\tilde{\zeta}_f$, of the equivalent SDF system with mass M_1 and natural vibration frequency $\tilde{\omega}_f$:

$$i\omega\left(\tilde{\omega}_f/\omega_1\right)^2 \mathcal{D} = i\omega\left(2M_1\tilde{\omega}_f\right)\tilde{\zeta}_f \tag{A3.7}$$

which, when substituted in Eq. (A3.6), gives

$$\bar{\tilde{q}}_1(\omega) = \left(\frac{\tilde{\omega}_f^2}{\omega_1}\right)^2 \frac{-L_1}{-\omega^2 M_1 + i\omega\left(2\tilde{\zeta}_f M_1\tilde{\omega}_f\right) + \tilde{\omega}_f^2 M_1} \tag{A3.8}$$

The damping ratio, $\tilde{\zeta}_f$, is determined by matching the resonant response of the equivalent SDF system to that of the dam–foundation system. This requirement, $\bar{\tilde{q}}_1\left(\tilde{\omega}_f\right) = \bar{q}_1\left(\tilde{\omega}_f\right)$, upon use of Eq. (A3.4) leaves only the imaginary-valued components in the respective denominators that are equated to obtain

$$\tilde{\zeta}_f = \left(\frac{\tilde{\omega}_f}{\omega_1}\right)\zeta_1 - \left(\frac{\tilde{\omega}_f}{\omega_1}\right)^3 \zeta_1 \mathrm{Re}\left[\mathcal{F}\left(\tilde{\omega}_f\right)\right] - \frac{1}{2}\left(\frac{\tilde{\omega}_f}{\omega_1}\right)^2 \mathrm{Im}\left[\mathcal{F}\left(\tilde{\omega}_f\right)\right] \tag{A3.9}$$

Substituting Eq. (A3.3) into Eq. (A3.9) gives – after some manipulation – the final equations:

$$\tilde{\zeta}_f = \left(\frac{\tilde{\omega}_f}{\omega_1}\right)^3 \zeta_1 + \zeta_f \tag{A3.10}$$

where the added damping due to dam–foundation interaction is

$$\zeta_f = -\frac{1}{2}\left(\frac{\tilde{\omega}_f}{\omega_1}\right)^2 \mathrm{Im}\left[\mathcal{F}\left(\tilde{\omega}_f\right)\right] \tag{A3.11}$$

These equations first appeared as Eqs. (3.4.3) and (3.4.4).

4

Response Spectrum Analysis of Dams Including Dam–Water–Foundation Interaction

PREVIEW

The response spectrum analysis (RSA) procedure presented in this chapter to estimate the earthquake-induced stresses in concrete gravity dams considers only the more significant aspects of the response. Although the dynamics of the system including dam–water–foundation interaction is considered in estimating the response due to the fundamental vibration mode, the less significant part of the response due to higher modes is estimated by the static correction method. Only the horizontal component of ground motion is considered because the response due to the vertical component is known to be much smaller.

Dam–water–foundation interaction introduces frequency-dependent, complex-valued hydrodynamic and foundation terms in the governing equations. Based on a series of approximations, frequency-independent values of these terms were defined and an equivalent SDF system developed in Chapters 2 and 3 to estimate the fundamental mode response of dams. These concepts lead to the RSA procedure presented in this chapter. Recognizing that the cross-sectional geometry of concrete gravity dams does not vary widely, standard data for the vibrational properties of dams and for parameters that characterize dam–water interaction (including reservoir bottom absorption) and dam–foundation interaction are presented to facilitate implementation of the procedure. To enhance the accuracy of this RSA procedure, the possibility of calculating stresses by finite element analysis versus the traditional beam formulas is explored, and a correction factor for beam stresses on the downstream face of the dam is developed.

This chapter ends with a comprehensive evaluation of the accuracy of the RSA procedure by comparing its results with those obtained from response history analysis (RHA) of the dam modeled as a finite element system, including dam–water–foundation interaction; the RHA procedure is developed in Chapter 5. Comparison of the median of the peak responses of an actual dam to 58 ground motions determined by both procedures demonstrates that the RSA procedure estimates stresses to a degree of accuracy that is satisfactory for the preliminary phase in the design of new dams and in the safety evaluation of existing dams. This result is especially impressive when we consider that the RSA procedure models very complicated effects of dam–water–foundation interaction and reservoir bottom absorption on the dam response in a surprisingly simple but rational way.

4.1 EQUIVALENT STATIC LATERAL FORCES: FUNDAMENTAL MODE

The peak response of the dam in its fundamental vibration mode including dam–water–foundation interaction can be estimated by static analysis of the dam alone subjected to the lateral forces given by Eq. (3.5.8). These distributed forces vary over the cross section of the dam monolith.

4.1.1 One-Dimensional Representation

The variation of these forces across the breadth of the monolith is negligibly small, however, because the fundamental mode shape is essentially the same across the breadth. With this approximation, $f_1(x, y)$ is integrated over the breadth to obtain equivalent static lateral forces acting on the upstream face of the dam:

$$f_1(y) = \tilde{\Gamma}_1 \frac{A\left(\tilde{T}_1, \tilde{\xi}_1\right)}{g} \left[w_s(y)\, \phi_1(y) + g\bar{p}_1\left(y, \tilde{T}_r\right)\right] \tag{4.1.1}$$

in which $\phi_1(y) = \phi_1^x(0, y)$ is the horizontal component of displacement at the upstream face of the dam in the fundamental vibration mode shape of the dam supported on rigid foundation with empty reservoir; $w_s(y)$ is the weight per unit height of the dam (= integral of $w_s(x, y)$ over the monolith breadth at elevation y); $\tilde{\Gamma}_1 = \tilde{L}_1/\tilde{M}_1$, where $\tilde{M}_1$ and $\tilde{L}_1$ are given by Eqs. (2.6.9a) and (2.6.9c), repeated here for convenience:

$$\tilde{M}_1 = M_1 + \text{Re}\left[\int_0^H \bar{p}_1\left(y, \tilde{T}_r\right) \phi_1(y) dy\right] \tag{4.1.2a}$$

$$\tilde{L}_1 = L_1 + \int_0^H \bar{p}_1\left(y, \tilde{T}_r\right) dy \tag{4.1.2b}$$

in which H is the depth of impounded water. The generalized mass, M_1, and earthquake force coefficient, L_1, Eqs. (2.2.3) and (2.2.4), are now represented by one-dimensional integrals:

$$M_1 = \frac{1}{g} \int_0^{H_s} w_s(y)\, \phi_1^2(y) dy \tag{4.1.3a}$$

$$L_1 = \frac{1}{g} \int_0^{H_s} w_s(y)\, \phi_1(y) dy \tag{4.1.3b}$$

where H_s is the height of the dam; g is the acceleration due to gravity; and $A\left(\tilde{T}_1, \tilde{\xi}_1\right)$ is the pseudo-acceleration ordinate of the earthquake design spectrum evaluated at vibration period $\tilde{T}_1$ and damping ratio $\tilde{\xi}_1$ of the equivalent single-degree-of-freedom (SDF) system representing the dam–water–foundation system.

Recall from Section 2.3 that $\bar{p}_1\left(y, \tilde{T}_r\right)$ is the complex-valued function representing the hydrodynamic pressure on the upstream face due to harmonic acceleration, $\phi_1(y)e^{i\tilde{\omega}_r t}$, of the dam in its fundamental mode (the governing boundary value problem is shown in Figure 2.3.1c), where $\tilde{T}_r = 2\pi/\tilde{\omega}_r$ is the natural vibration period of the equivalent SDF system representing the fundamental mode response of the dam (on rigid foundation) with impounded water. Expressing Eq. (2.6.11) in terms of vibration periods leads to

$$\tilde{T}_r = R_r T_1 \tag{4.1.4a}$$

in which T_1 is the fundamental vibration period of the dam on rigid foundation with empty reservoir. Hydrodynamic effects lengthen the vibration period because of the frequency-dependent, added hydrodynamic mass arising from dam–water interaction (Eq. (2.6.6)). The period-lengthening ratio, R_r, which is greater than one, depends on the properties of the dam, the depth of the water, and the absorptiveness of the reservoir bottom.

The natural vibration period of the equivalent SDF system representing the fundamental mode response of the dam (with empty reservoir) on flexible foundation is obtained by expressing Eq. (3.4.1) in terms of periods:

$$\tilde{T}_f = R_f T_1 \tag{4.1.4b}$$

Dam–foundation interaction lengthens the vibration period because of the frequency-dependent foundation flexibility term, $\mathscr{F}(\omega)$ (Eq. (3.4.1)). The period-lengthening ratio, R_f, which is greater than one, depends on the properties of the dam and foundation, most importantly on the ratio E_f/E_s of the elastic moduli of the foundation rock and dam concrete.

The natural vibration period of the equivalent SDF system representing the fundamental mode response of the dam including dam–water–foundation interaction is obtained by expressing Eq. (3.5.1) in terms of vibration periods:

$$\tilde{T}_1 = R_r R_f T_1 \tag{4.1.4c}$$

The damping ratio of this equivalent SDF system, Eq. (3.5.3), can be expressed as

$$\tilde{\zeta}_1 = \frac{1}{R_r}\frac{1}{\left(R_f\right)^3}\zeta_1 + \zeta_r + \zeta_f \tag{4.1.5}$$

in which ζ_1 is the damping ratio of the dam on rigid foundation with empty reservoir; ζ_r is the added damping due to dam–water interaction and reservoir bottom absorption [Eq. (2.6.13)]; and ζ_f is the added damping due to dam–foundation interaction [Eq. (3.4.4)], which includes the combined effects of foundation radiation and material damping. Considering that $R_r > 1$ and $R_f > 1$, Eq. (4.1.5) implies that dam–water interaction and dam–foundation interaction reduce the effectiveness of structural damping. However, this reduction is usually more than compensated by the added damping due to (i) upstream propagation of hydrodynamic waves and their absorption at the reservoir bottom; and (ii) the energy dissipation due to material damping in the rock and radiation of elastic waves away from the vibrating dam into the half-space, thus leading to an increase in the overall damping of the system.

4.1.2 Approximation of Hydrodynamic Pressure

We will approximate – with justification – the lateral forces $f_1(y)$ that are, in general, complex-valued, by their real-valued components. The hydrodynamic pressure function $\tilde{p}_1\left(y, \tilde{T}_r\right)$ is real-valued at the period $\tilde{T}_r$ if the reservoir bottom is non-absorptive, i.e. $\alpha = 1$ (Chopra 1978), but complex-valued if the partial absorption of hydrodynamic waves at the reservoir bottom is considered, i.e. $\alpha < 1$ (Section 2.3). The real- and imaginary-valued components of lateral forces $f_1(y)$ at the upstream face of a typical concrete dam monolith with nearly full reservoir are shown in Figure 4.1.1, where it is apparent that the imaginary-valued component of lateral

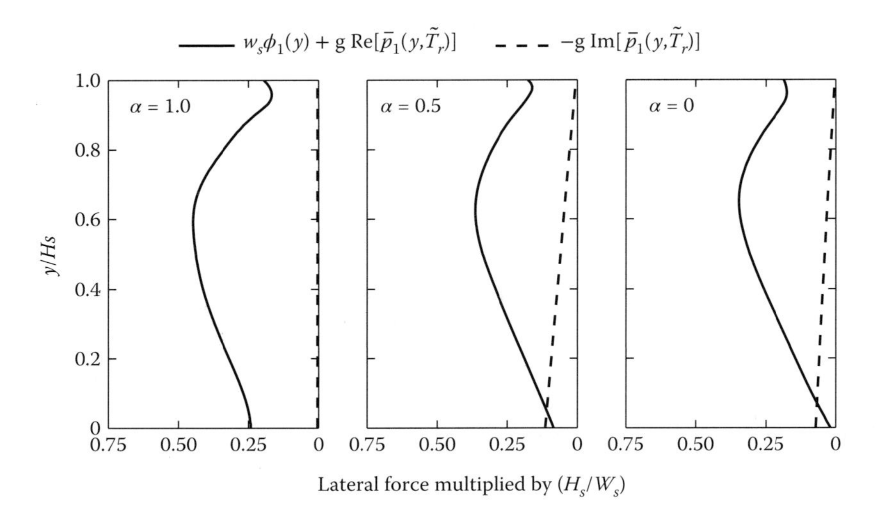

Figure 4.1.1 Real (Re) and imaginary (Im) valued components of equivalent lateral earthquake forces on the upstream face of a typical dam.

forces is small relative to the real-valued component, increasing near the base of the dam; as a result, it will have little influence on the stresses in the dam. Consequently, the imaginary-valued component of $\bar{p}_1\left(y, \tilde{T}_r\right)$ may be neglected in the evaluation of the lateral forces $f_1(y)$, and Eq. (4.1.1) becomes

$$f_1(y) = \tilde{\Gamma}_1 \frac{A\left(\tilde{T}_1, \tilde{\zeta}_1\right)}{g}\left[w_s(y)\,\phi_1(y) + gp\left(y, \tilde{T}_r\right)\right] \tag{4.1.6}$$

where $p\left(y, \tilde{T}_r\right) \equiv \mathrm{Re}\left[\bar{p}_1\left(y, \tilde{T}_r\right)\right]$. It is only in Eq. (4.1.6) that the imaginary-valued component of $\bar{p}_1\left(y, \tilde{T}_r\right)$ has been dropped; its more important effect – the contribution to added hydrodynamic damping ζ_r in Eq. (4.1.5) – is still included.

The generalized mass $\tilde{M}_1$ of the equivalent SDF system (Eq. (4.1.2a)) does not include the imaginary-valued component of hydrodynamic pressure, thus

$$\tilde{M}_1 = M_1 + \int_0^H p\left(y, \tilde{T}_r\right)\phi_1(y)dy \tag{4.1.7a}$$

where M_1 is defined in Eq. (4.1.3a). However, the generalized earthquake force coefficient $\tilde{L}_1$ (Eq. (4.1.2b)) contains both real-valued and imaginary-valued components of the hydrodynamic pressure. Again, dropping the small imaginary-valued component gives

$$\tilde{L}_1 = L_1 + \int_0^H p\left(y, \tilde{T}_r\right)dy \tag{4.1.7b}$$

where L_1 is defined in Eq. (4.1.3b).

4.2 EQUIVALENT STATIC LATERAL FORCES: HIGHER MODES

Although the fundamental vibration mode is dominant in the earthquake response of short vibration period structures, such as concrete gravity dams, the response contributions of the higher vibration modes may not be negligible because the effective mass (Chopra 1978; Chopra 2017, Chapter 13) in the fundamental vibration mode is typically only 30–50% of the total mass of the dam. Thus, the contributions of the higher vibration modes to the earthquake forces are included, but they are determined approximately using the "static correction" concept (Chopra 2017, Sections 12.12 and 13.1.5). This implies that the ordinates of the pseudo-acceleration design spectrum at higher-mode periods are approximated by the zero-period ordinate, i.e. the peak ground acceleration. The quality of this approximation depends on the dynamic amplification of the design spectrum at higher mode periods, as will be discussed in Section 4.7.

For dams (without impounded water) supported on rigid foundation, the maximum earthquake effects associated with the higher vibration modes can then be represented by the equivalent static lateral forces (Fenves and Chopra 1986; Appendix B).

$$f_{sc}(y) = \frac{a_g}{g} w_s(y) \left[1 - \frac{L_1}{M_1}\phi(y)\right] \tag{4.2.1}$$

in which a_g is the peak ground acceleration. Dam–foundation interaction effects may be neglected in a simplified procedure to compute the contributions of the higher vibration modes to the earthquake response of dams, just as soil-structure interaction could be ignored in the case of multistory buildings (Veletsos 1977). Thus, Eq. (4.2.1) is also valid for dams supported on flexible foundation.

Dam–water interaction introduces significant damping in the response of the higher vibration modes of concrete gravity dams, but it has little effect on the higher vibration periods (Fenves and Chopra 1985a). With increased damping, the approximation implied in the "static correction" method is even more appropriate. The equivalent static lateral forces associated with the higher vibration modes of dams, including the effects of the impounded water, are given by an extension of Eq. (4.2.1) (Fenves and Chopra 1986; Appendix B):

$$f_{sc}(y) = \frac{a_g}{g} \left\{ w_s(y) \left[1 - \frac{L_1}{M_1}\phi(y)\right] + \left[g p_0(y) - \frac{F_1}{M_1} w_s(y)\phi(y)\right] \right\} a_g \tag{4.2.2}$$

In this equation, $p_0(y)$ is a real-valued, frequency-independent function for hydrodynamic pressure on a rigid dam undergoing unit acceleration, with water compressibility neglected (shown in Figure 4.2.1 and tabulated in Fenves and Chopra (1986)) and F_1 provides a measure of the fundamental-mode component of $p_0(y)$:

$$F_1 = 0.20 \frac{F_{st}}{g} \left(\frac{H}{H_s}\right)^2 \tag{4.2.3}$$

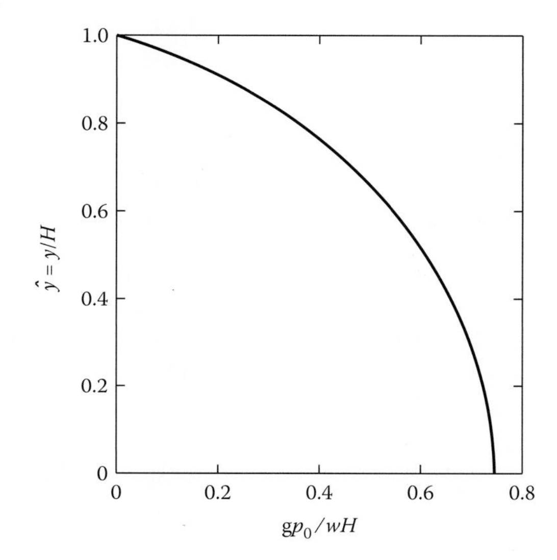

Figure 4.2.1 Standard values for hydrodynamic pressure function $p_0(\hat{y})$.

where F_{st} is the total hydrostatic force on the dam. The shape of only the fundamental vibration mode enters into Eq. (4.2.2), implying that the shapes of the higher modes are not required in estimating earthquake forces associated with these modes, thus simplifying the analysis considerably. In passing, we note that $p_0(y)$ is the same as added mass $m_a(y)$ of Eq. (2.3.26).

4.3 RESPONSE ANALYSIS

4.3.1 Dynamic Response

As shown in the preceding two sections, the maximum effects of earthquake ground motion in the fundamental vibration mode of the dam have been represented by equivalent static lateral forces $f_1(y)$ and those due to all the higher modes by $f_{sc}(y)$, both determined directly from the response (or design) spectrum – without any response history analysis (RHA). Static analysis of the dam alone for these two sets of forces provide the values r_1 and r_{sc} for any response quantity, r, e.g. the shear force or bending moment at any horizontal section, or the shear stress or vertical bending stress at any point. According to the square-root-of-the-sum-of-squares (SRSS) rule, the peak value of dynamic response may be estimated by

$$r_o \simeq \sqrt{(r_1)^2 + (r_{sc})^2} \tag{4.3.1}$$

Because the natural frequencies of *lateral* vibration of a concrete dam are well separated, it is not necessary to include the correlation of modal responses in the combination rule (Chopra 2017, Section 13.8).

The SRSS combination rule is applicable to the computation of any response quantity that is proportional to the modal coordinates (Chopra 2017, Section 13.8). Thus it is valid for a single stress component, such as vertical stress, but generally not for principal stresses. A response-spectrum-based procedure for estimation of principal stresses is available (Chopra 2017, Section 13.11). However, it is not essential to call upon this general procedure because the much simpler SRSS rule can be justified to estimate principal stresses at the two faces of a dam monolith if the upstream face is nearly vertical and the effects of tail water at the downstream face are small. Under these restrictive conditions, the principal stresses are proportional to the vertical bending stress, which in turn is proportional to the modal coordinate (Fenves and Chopra 1986; Appendix C). Thus, the principal stresses at the two faces of the dam (*but not in its interior*) due to the first mode and higher modes may be combined by the SRSS rule.

4.3.2 Total Response

In order to obtain the total value of any response quantity r, the SRSS estimate of dynamic response r_d should be combined with the static effects r_{st}. The latter may be determined by standard analysis procedures to compute the initial value of any response quantity – prior to the earthquake – due to the self-weight of the dam (with the construction sequence modeled), hydrostatic pressures (with the filling of reservoir modeled), ice loads, and thermal effects. In order to recognize that the direction of lateral earthquake forces is reversible, combinations of static and dynamic stresses should allow for the worse case, leading to the peak value of total response:

$$r_{o,total} = r_{st} \pm \sqrt{(r_1)^2 + (r_{sc})^2} \tag{4.3.2}$$

This combination of static and dynamic responses is appropriate if r_{st}, $r_1 r_1$, and r_{sc} are oriented similarly. Such is the case for the shear and bending stresses at any point, but generally not for principal stresses except under the restricted conditions previously mentioned.

4.4 STANDARD PROPERTIES FOR FUNDAMENTAL MODE RESPONSE

Direct evaluation of Eq. (4.1.6) would require a complicated computation of several quantities: $p(y, \tilde{T}_r)$ from the infinite series expression in Eq. (2.3.12c); the period lengthening ratios R_r and R_f due to dam–water and dam–foundation rock interactions by iterative solution of Eqs. (2.6.11) and (3.4.1) involving frequency-dependent hydrodynamic and foundation-flexibility terms; damping ratios ζ_r and ζ_f from Eqs. (2.6.13) and (3.4.4), respectively, involving the same terms at a specific frequency; the integrals in Eq. (4.1.7); and the fundamental vibration period and mode shape of a finite element model of the dam. Such computations can be avoided by recognizing that the cross-sectional geometry of non-overflow monoliths of concrete gravity dams does not vary widely. Consequently, it was possible to develop standard values for fundamental vibration period and mode of dams, and quantities that depend on them (Fenves and Chopra 1986; Løkke and Chopra 2013) that enter into Eq. (4.1.6). Similar standard data for overflow monoliths is available in Chopra and Tan (1989).

4.4.1 Vibration Period and Mode Shape

Computed by the finite element method, the fundamental vibration periods and mode shapes are presented in Figure 4.4.1a for non-overflow monoliths of two actual dams and four idealized dams; the latter cover the range of slopes for the downstream and upstream faces encountered in actual dams. Because the fundamental vibration periods and mode shapes for these cross sections are similar, a standard vibration period and mode shape may be used in computing the equivalent static lateral forces. These are selected as the vibration properties of the "standard" cross section (Figure 4.4.1b) on rigid foundation with empty reservoir. The standard vibration mode shape is

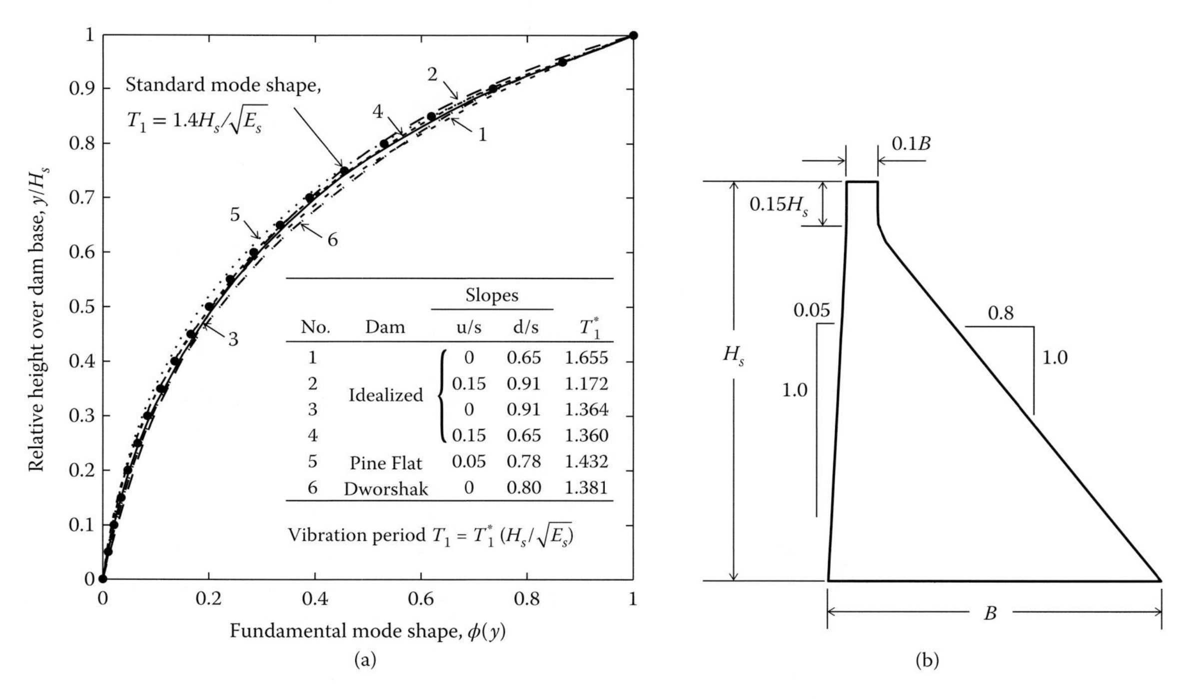

Figure 4.4.1 (a) Comparison of standard vibration period and mode shape with vibration properties of six dams; and (b) "standard" cross section.

included in Figure 4.4.1a, and the fundamental vibration period, in seconds, of a "standard" cross section is (Chopra 1978)

$$T_1 = 1.4\frac{H_s}{\sqrt{E_s}} \tag{4.4.1}$$

where H_s is the height of the dam in feet, and E_s is the modulus of elasticity of the concrete in psi.

4.4.2 Modification of Period and Damping: Dam–Water Interaction

Dam–water interaction and reservoir bottom absorption lengthen the natural vibration period [Eq. (4.1.4a)] and modify the damping ratio [Eq. (4.1.5)] of the equivalent SDF system representing the fundamental vibration mode response of the dam. For the standard dam cross section, the period-lengthening ratio R_r and added damping ζ_r depend on several parameters, the more significant of which are: Young's modulus E_s of the dam concrete, ratio H/H_s of water depth to dam height, and wave reflection coefficient α. The results of many analyses of the "standard" dam cross section, using the procedures developed in Chapter 2 and modified for dams with unusually large elastic modulus E_s, are available in Fenves and Chopra (1986). The period-lengthening ratio R_r and added damping ζ_r are presented as functions of H/H_s for the full range of values of E_s and α; the data for one selected value of E_s are shown in Figure 4.4.2. These and additional results (not included here) demonstrate that the effects of dam–water interaction and reservoir bottom absorption may be neglected if the reservoir depth is less than half full, $H/H_s < 0.5$. For these conditions the dam may be analyzed as if there is no impounded water, i.e. $R_r = 1$, $\zeta_r = 0$, $\tilde{L}_1 = L_1$, $\tilde{M}_1 = M_1$, and $\tilde{\Gamma}_1 = \Gamma_1$.

4.4.3 Modification of Period and Damping: Dam–Foundation Interaction

Dam–foundation interaction lengthens the natural vibration period (Eq. (4.1.4b)) and increases the damping ratio (Eq. (4.1.5)) of the equivalent SDF system representing the fundamental vibration mode response of the dam. For the "standard" dam cross section, the period-lengthening

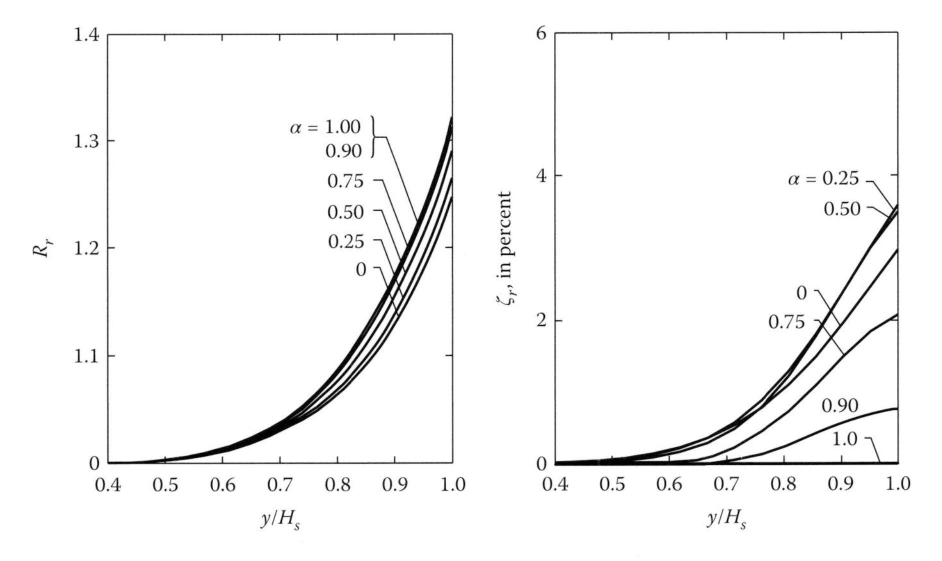

Figure 4.4.2 Standard values for period-lengthening ratio R_r, and added damping ratio ζ_r, due to hydrodynamics effects; $E_s = 3.0$ million psi.

ratio R_f and the added damping ζ_f due to dam–foundation interaction depend on several parameters, the more significant of which are: moduli ratio E_f/E_s, where E_s and E_f are the Young's moduli of the dam concrete and foundation rock, respectively; and the constant hysteretic damping factor η_f for the foundation rock. The period-lengthening ratio R_f is, however, relatively insensitive to η_f.

By performing many analyses of the "standard" dam cross section, using the procedures developed in Chapter 3, the period-lengthening ratio R_f and added damping ζ_f were initially computed for a range of values of E_f/E_s and $\eta_f = 0.01$, 0.10, 0.25, and 0.50 (Fenves and Chopra 1986), which in retrospect turned out to be too coarse. The added damping, ζ_f, has now been recomputed for a closely-spaced set of η_f values; selected results are presented in Figure 4.4.3 and the complete data are available in Løkke and Chopra (2013). These results suggest that the foundation may be considered rigid in the simplified analysis if $E_f/E_s > 5$, for these conditions the dam may be analyzed as if the foundation rock is rigid, i.e. $R_f = 1$ and $\zeta_f = 0$.

4.4.4 Hydrodynamic Pressure

In order to provide a convenient means for determining $p\left(y, \tilde{T}_r\right)$, required to evaluate Eqs. (4.1.6) and (4.1.7), a non-dimensional form of this function, $gp(\hat{y})/wH$, where $\hat{y} = y/H$ and w is the unit weight of the water, has been computed from Eq. (2.3.12c), for several values of the wave reflection coefficient, α, and the necessary range of values of

$$R_w = \frac{T_1^r}{\tilde{T}_r} \tag{4.4.2}$$

in which the fundamental vibration period of the impounded water $T_1^r = 4H/C$, where C is the speed of pressure waves in water. Results are presented in Fenves and Chopra (1986) for several values of α in the form of plots similar to Figure 4.4.4 and also in tables. All these data are for full reservoir, $H/H_s = 1$. The function $gp(\hat{y})/wH$ for a partially filled reservoir is approximately equal to $(H/H_s)^2$ times the function for $H/H_s = 1$ (Chopra 1978).

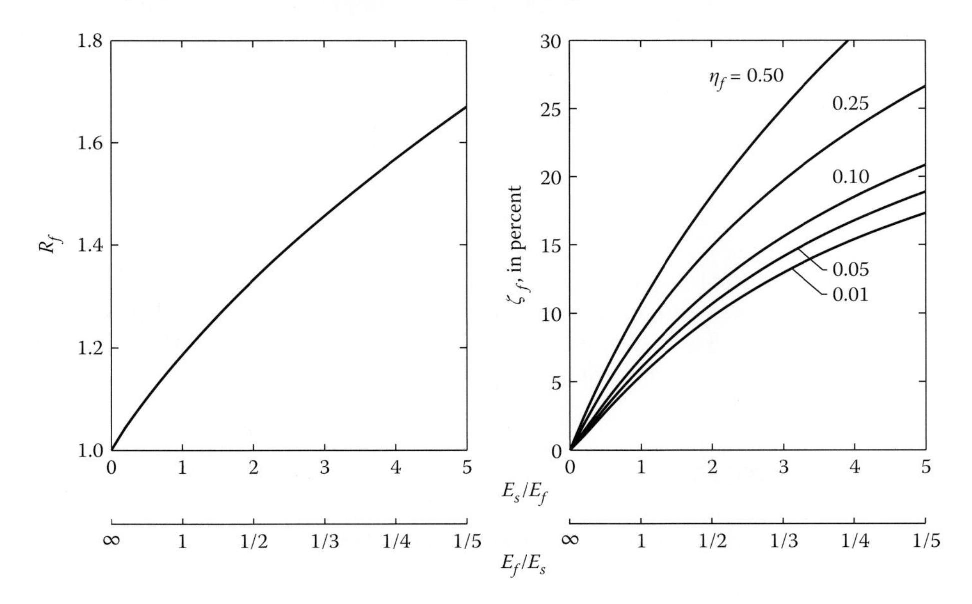

Figure 4.4.3 Standard values for period-lengthening ratio R_f and added damping ratio ζ_f due to dam–foundation interaction.

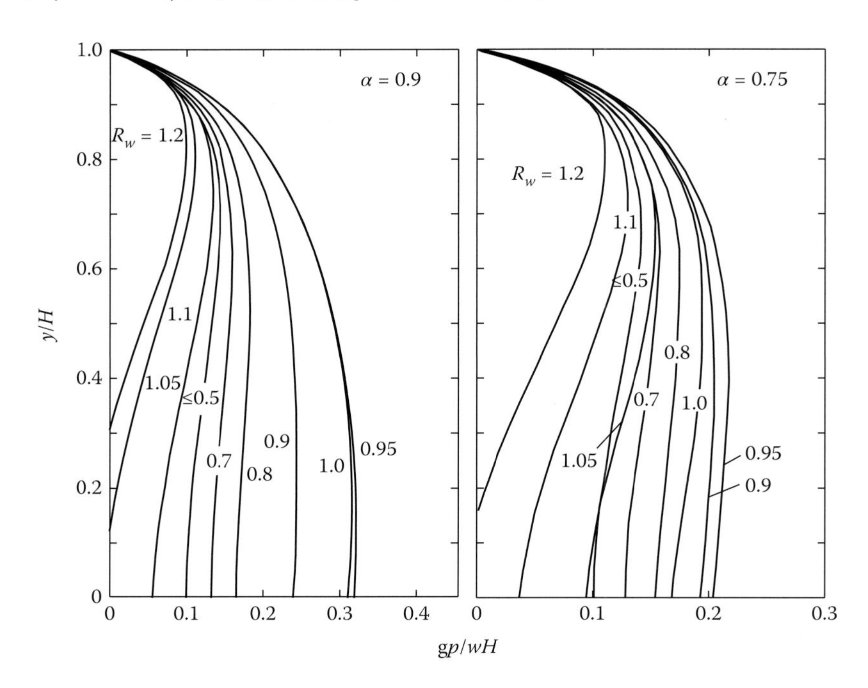

Figure 4.4.4 Standard values for the hydrodynamic pressure function $p(\hat{y})$ for full reservoir, i.e. $H/H_s = 1$; $\alpha = 0.90$ and 0.75.

4.4.5 Generalized Mass and Earthquake Force Coefficient

The generalized mass $\tilde{M}_1$ (Eq. (4.1.7a)) of the equivalent SDF system representing the dam, including hydrodynamic effects, can be conveniently computed from Eq. (2.6.11), which can be rewritten as

$$\tilde{M}_1 = (R_r)^2 M_1 \tag{4.4.3a}$$

in which M_1 is given by Eq. (4.1.3a). In order to provide a convenient means to compute the generalized earthquake force coefficient L_1, Eq. (4.1.7b) is expressed as

$$\tilde{L}_1 = L_1 + \frac{1}{g} F_{\text{st}} \left(\frac{H}{H_s} \right)^2 A_p \tag{4.4.3b}$$

where $F_{\text{st}} = wH^2/2$ is the total hydrostatic force on the dam; and A_p is the integral of the function $2gp(\hat{y})/wH$ over the depth of the impounded water, for $H/H_s = 1$. The hydrodynamic force coefficient A_p is tabulated in Fenves and Chopra (1986) for a range of values for the period ratio R_w and the wave reflection coefficient α.

4.5 COMPUTATIONAL STEPS

The computation of earthquake-induced stresses in the dam is organized in three parts.

Part I Utilizing the standard data available in Løkke and Chopra (2013), the earthquake forces and stresses due to the fundamental vibration mode can be determined by the following computational steps:

1. Compute T_1, the fundamental vibration period of the dam, in seconds, on rigid foundation with an empty reservoir from Eq. (4.4.1) in which H_s is the height of the dam in feet, and E_s is the design value for Young's modulus of elasticity of concrete in psi.
2. Compute $\tilde{T}_r$, the fundamental vibration period of the dam, in seconds, including the influence of impounded water from Eq. (4.1.4a) in which T_1 was computed in step 1; R_r is the period ratio determined from data,† such as Figure 4.4.2, for the design values of E_s, the wave reflection coefficient α, and the depth ratio H/H_s, where H is the depth of the impounded water, in feet. If $H/H_s < 0.5$, computation of R_r may be avoided; use $R_r = 1$.
3. Compute the period ratio R_w from Eq. (4.4.2) in which $\tilde{T}_r$ was computed in step 2, and $T_1^r = 4H/C$ where $C = 4720$ ft/sec.
4. Compute $\tilde{T}_1$, the fundamental vibration period of the dam in sec, including the effects of dam–water–foundation interaction from Eq. (4.1.4c) in which R_r was determined in step 2; R_f is the period ratio determined from data such as Figure 4.4.3, or from standard data,† for the design value of E_f/E_s. If $E_f/E_s > 5$, use $R_f = 1$.
5. Compute the damping ratio $\tilde{\zeta}_1$ of the dam from Eq. (4.1.5) using the period ratios R_r and R_f determined in steps 2 and 4, respectively; ζ_1 is the viscous damping ratio for the dam on rigid foundation with empty reservoir; ζ_r is the added damping ratio due to dam–water interaction and reservoir bottom absorption, obtained from standard data†, such as Figure 4.4.2, for the selected values of E_s, α, and H/H_s; and ζ_f is the added damping ratio due to dam–foundation interaction, obtained from Figure 4.4.3 presented here or standard data,† for the design values of E_f/E_s and η_f. If $H/H_s < 0.5$, use $\zeta_r = 0$; if $E_f/E_s > 5$, use $\zeta_f = 0$; and if the computed value of $\tilde{\zeta}_1 < \zeta_1$, use $\tilde{\zeta}_1 = \zeta_1$.
6. Determine $gp\left(\hat{y}, \tilde{T}_r\right)$ from standard data,† such as Figure 4.4.4, corresponding to the value of R_w computed in step 3 (by interpolating, if necessary, between data for the nearest values of R_w), the design value of α, and for $H/H_s = 1$; the result is multiplied by $(H/H_s)^2$. If $H/H_s < 0.5$, $p\left(y, \tilde{T}_r\right) \approx 0$ and computation of $p\left(y, \tilde{T}_r\right)$ may be avoided.
7. Compute the generalized mass $\tilde{M}_1$ from Eq. (4.4.3a) in which R_r was computed in step 2, and M_1 is computed from Eq. (4.1.3a), in which $w_s(y)$ is the weight of the dam/unit height; the fundamental vibration mode shape $\phi(y)$ is given by Figure 4.4.1a or standard data;† and $g = 32.2$ ft/sec^2.
8. Compute the effective earthquake force coefficient $\tilde{L}_1$ from Eq. (4.4.3b) in which L_1 is computed from Eq. (4.1.3b), $F_{st} = wH^2/2$, and A_p is given in standard data† for the values of R_w and α used in step 6. If $H/H_s < 5$, $\tilde{L}_1 \simeq L_1$ and computation of $\tilde{L}_1$ may be avoided.
9. Compute $f_1(y)$, the equivalent lateral earthquake forces associated with the fundamental vibration mode from Eq. (4.1.6) in which $A\left(\tilde{T}_1, \tilde{\zeta}_1\right)$ is the pseudo-acceleration ordinate of the earthquake design spectrum at period $\tilde{T}_1$ determined in step 4 and damping ratio $\tilde{\zeta}_1$ determined in step 5; $w_s(y)$ is the weight/unit height of the dam; $\phi_1(y)$ is the fundamental vibration mode shape of the dam from Figure 4.4.1 or standard data;† $\tilde{M}_1$ and $\tilde{L}_1$ are the generalized mass and earthquake force coefficient, respectively, determined in steps 7 and 8; and the hydrodynamic pressure term $gp\left(\hat{y}, \tilde{T}_r\right)$ was determined in step 6.
10. Determine all response quantities of interest by static analysis of the dam fixed (or clamped) at the base subjected to equivalent lateral forces $f_1(y)$, from step 9, applied to the upstream face of the dam. Traditional procedures for design calculations may be used wherein the normal bending stresses across a horizontal section are computed by elementary formulas for stresses in beams. However, beam theory overestimates the stresses near the sloped downstream face by a factor that depends on this slope and the height-wise distribution of equivalent lateral forces. For most dams, a correction factor of 0.75 is appropriate for the

† Numerical values for the standard data are available in Løkke and Chopra (2013).

sloping part of the downstream face (Løkke and Chopra 2013). The maximum principal stresses at the upstream and downstream faces can be computed from the normal bending stresses σ_{y1} by an appropriate transformation. Alternatively, the finite element method may be used for a more accurate static stress analysis.

Part II The earthquake forces and stresses due to the higher vibration modes can be estimated for purposes of preliminary design by the following computational steps:

11. Compute $f_{sc}(y)$, the lateral forces associated with the higher vibration modes from Eq. (4.2.2) in which M_1 and L_1 were determined in steps 7 and 8, respectively; $gp_0(y)$ is determined from Figure 4.2.1 or standard data;[†] F_1 is computed from Eq. (4.2.3); and a_g is the peak ground acceleration for the design earthquake. If $H/H_s < 0.5, p_0(y) \simeq 0$ and computation of $p_0(y)$ may be avoided, thus $B_1 = 0$.
12. Determine all response quantities of interest by static analysis of the dam subjected to the equivalent lateral forces $f_{sc}(y)$, from Step 11, applied to the upstream face of the dam. In particular, stresses may be determined by the procedures and correction factor mentioned in Step 10. Alternatively, the finite element method may be used for a more accurate static stress analysis.

Part III The total value of any response quantity is determined as follows:

13. Compute the total value of any response quantity by Eq. (4.3.2) in which r_1 and r_{sc} are values of the response quantity determined in steps 10 and 12 associated with the fundamental and higher vibration modes, respectively, and r_{st} is its initial value prior to the earthquake due to various effects, including the self-weight of the dam, hydrostatic pressure, ice loads, creep, construction sequence, and thermal effects.

Use of *S.I.* Units. Because the standard values for most quantities required in the simplified analysis procedure are presented in non-dimensional form, implementation of the procedure in *S.I.* units is straightforward. The expressions and data requiring conversion to metric units are noted here:

1. The fundamental vibration period T_1 of the dam on rigid foundation rock with empty reservoir (step 1), in sec, is given by:

$$T_1 = 0.38\frac{H_s}{\sqrt{E_s}} \tag{4.5.1}$$

   where H_s is the height of the dam in meters; and E_s is the Young's modulus of elasticity of the dam concrete in mega-Pascals (MPa).
2. The period ratio R_r and added damping ratio ζ_r due to dam–water interaction, presented in Figure 4.4.2 and "standard" data is for specified values of E_s in pounds per square inch (psi), which should be converted to S.I. units: 1 million psi $\approx$ 6895 MPa.
3. Where required in the calculation, the unit weight of water w = 9.81 k-Newtons/m^3; the acceleration due to gravity g = 9.81 m/sec^2; and the velocity of pressures waves in water C = 1440 m/sec.

4.6 CADAM COMPUTER PROGRAM

CADAM – computer-aided stability analysis of gravity dams – a program developed at the École Polytechnique de Montreal, Canada (Leclerc et al. 2003), implements the response spectrum analysis (RSA) procedure described in the preceding sections of this chapter. This procedure is referred to as the "pseudo-dynamic method" in CADAM, which also includes several other analysis options. An object-oriented program that offers a versatile computing environment, CADAM is convenient to use. It can be downloaded from http://www.polymtl.ca/structures/en/telecharg/cadam/telechargement.php, where a Users Manual is also available.

4.7 ACCURACY OF RESPONSE SPECTRUM ANALYSIS PROCEDURE

As mentioned in Chapters 2 and 3 and in the preceding sections of this chapter, various approximations were introduced to develop the RSA procedure, and their implications were individually evaluated to ensure that they would lead to satisfactory results (Fenves and Chopra 1985c, 1985d). Presented in this section is an overall evaluation of the accuracy of the RSA procedure by comparing its results with those obtained from RHA of the dam, modeled as a finite element system, including dam–water–foundation interaction and reservoir bottom absorption effects; the latter set of results were computed by a newer version of the program EAGD-84 (Fenves and Chopra 1984c) that implements the RHA procedure developed in Chapter 5.

4.7.1 System Considered

The system considered is the tallest, non-overflow monolith of the Pine Flat Dam shown in Figure 4.7.1; height of the dam, H_s = 400 ft, modulus of elasticity of concrete, E_s = 3.25 million psi; unit weight of concrete w_s = 155 pcf; viscous damping ratio of the dam alone, ζ_1 = 2% in the RSA procedure, which correspond to constant hysteric damping factor η_s = 4% in the RHA procedure; modulus of elasticity of foundation rock E_f = 3.25 million psi; constant hysteretic damping factor for foundation rock, η_f = 4%; depth of water, H = 381 ft; and the wave reflection coefficient at the reservoir bottom, α = 0.75.

4.7.2 Ground Motions

Based on a probabilistic seismic hazard analysis (PSHA) for the Pine Flat Dam site at the 1% in 100 years hazard level, a Conditional Mean Spectrum (Baker 2011) was developed. A total of 29 ground motion records on rock or NEHRP soil class D or stiffer sites, at a distance

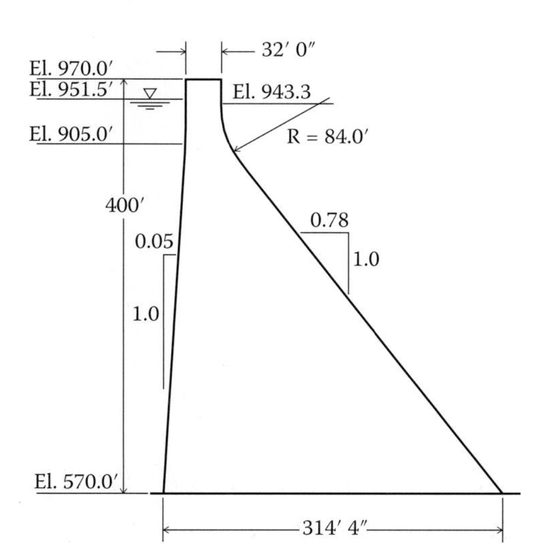

Figure 4.7.1 Tallest, non-overflow monolith of Pine Flat Dam.

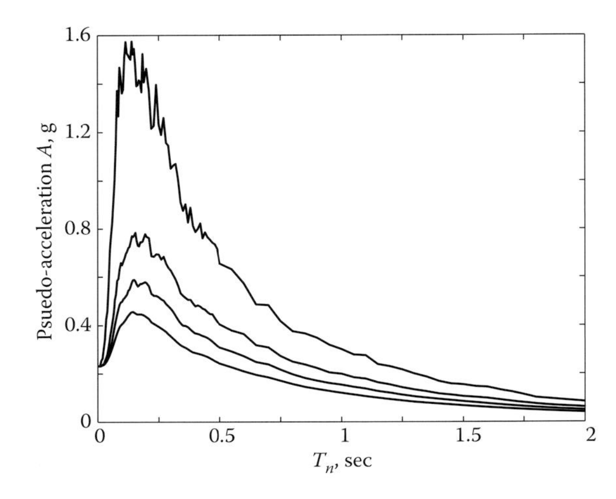

Figure 4.7.2 Median response spectrum for 58 ground motions; $\zeta = 0$, 2, 5, and 10%.

$R = 0$–50 km from earthquakes of magnitude $M_w = 5.0$–7.5 were selected; the selected range of M_w and R is consistent with the disaggregation of the seismic hazard at the site. Each of the resulting 58 ground motions (two horizontal components of 29 records) was amplitude-scaled to minimize the mean square difference between the response spectrum and the target spectrum over the period range of interest $0.3 \leq T \leq 0.5$ sec. Results of the PSHA, as well as the selection and scaling of records is presented in Løkke and Chopra (2013). The median (or geometric mean) of the response spectra for the 58 ground motions is presented in Figure 4.7.2.

4.7.3 Response Spectrum Analysis

With the earthquake excitation defined by the median response spectrum of Figure 4.7.2, the dam is analyzed by the RSA procedure for the four cases listed in Table 4.7.1. The vibration period and damping ratio of the equivalent SDF system with the corresponding spectral ordinates are presented in Table 4.7.1, and the equivalent static lateral forces $f_1(y)$ and $f_{sc}(y)$, representing the maximum earthquake effects of the fundamental and higher modes of vibration, respectively, are presented in Figure 4.7.3. Details of the computations are available in Løkke and Chopra (2013).

The vertical stresses $\sigma_{y,1}$ and $\sigma_{y,sc}$ due to the two sets of forces f_1 and f_{sc} are computed by static stress analysis of the dam alone by two methods: (i) elementary formulas for stresses in beams; and (ii) finite element analysis of the dam. Combining $\sigma_{y,1}$ and $\sigma_{y,sc}$ according to Eq. (4.3.1) leads to the earthquake induced stresses, $\sigma_{y,d}$, presented in Figure 4.7.4; note that stresses due to initial static loads are not included. The stress values presented occur as tensile stresses at the upstream face when the earthquake forces act in the direction shown in Figure 4.7.3, and at the

Table 4.7.1 Pine Flat Dam analysis cases, fundamental mode properties, and corresponding pseudo-acceleration ordinates.

Analysis case	Foundation	Reservoir	$\tilde{T}_1$, sec	$\tilde{\zeta}_1$, percent	$A\left(\tilde{T}_1, \tilde{\zeta}_1\right) \div g$
1	Rigid	Empty	0.311	2.0	0.606
2	Rigid	Full	0.387	3.9	0.409
3	Flexible	Empty	0.369	7.1	0.347
4	Flexible	Full	0.459	9.2	0.274

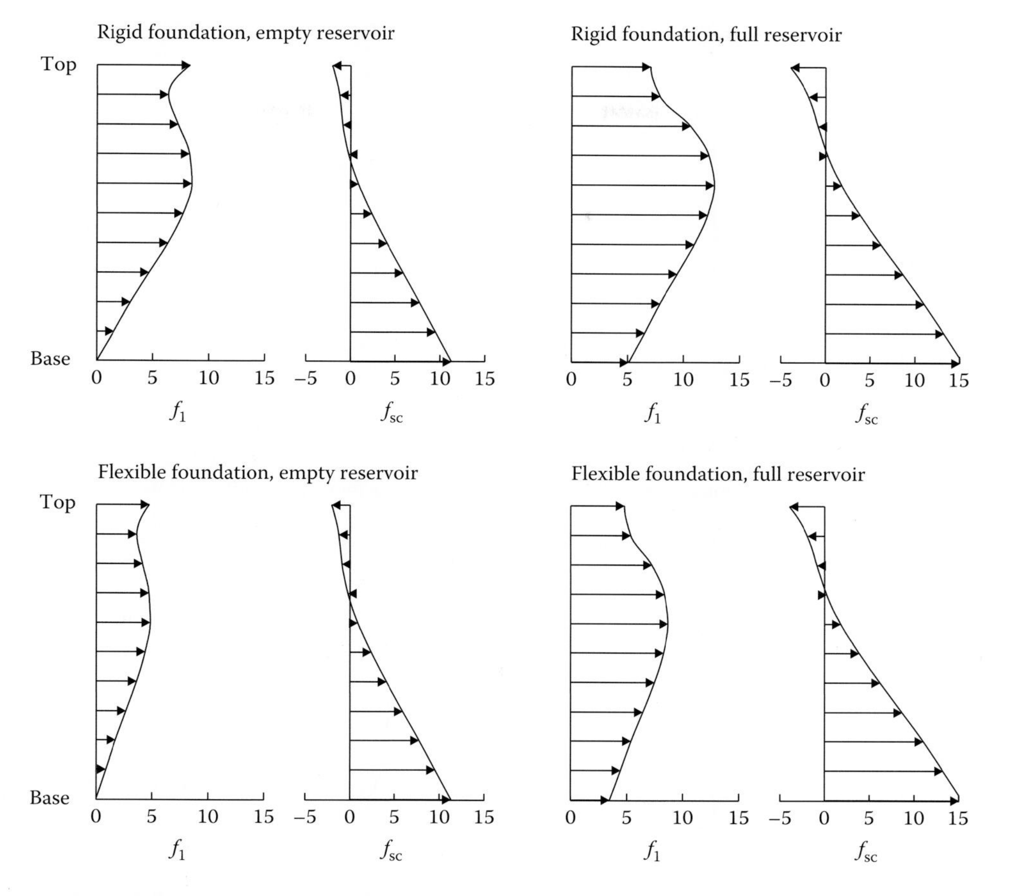

Figure 4.7.3 Equivalent static lateral forces, f_1 and f_{sc} in kips per foot height of the dam, computed by the RSA procedure for four cases.

downstream face when the earthquake forces act in the opposite direction. A detailed description of the computations is available in Løkke and Chopra (2013).

The results presented in Figure 4.7.4 confirm that the correction factor of 0.75 for stresses computed by beam theory at the sloping part of the downstream face is satisfactory for all four cases. The stresses determined by beam theory with the correction factor are very close to those determined by finite element analysis except near the heel and toe of the dam and near the location where the downstream face changes slope. Therefore, only the stresses determined by beam theory in RSA are compared with the results from RHA in Section 4.7.4.

4.7.4 Comparison with Response History Analysis

Response history analysis of the dam monolith modeled as a finite element system, considering rigorously the effects of dam–water–foundation interaction and reservoir bottom absorption, is implemented by a newer version of the computer program EAGD-84† for each of the 58 ground motions; the finite-element model is shown in Figure 6.1.2. Results computed by RSA and RHA procedures are compared in the following sections.

† Developed by A. Lökke, the new EAGD-84 Matlab modules, updated source code, and foundation compliance data may be downloaded from: http://nisee.berkeley.edu/elibrary/getpkg?id=EAGD84.

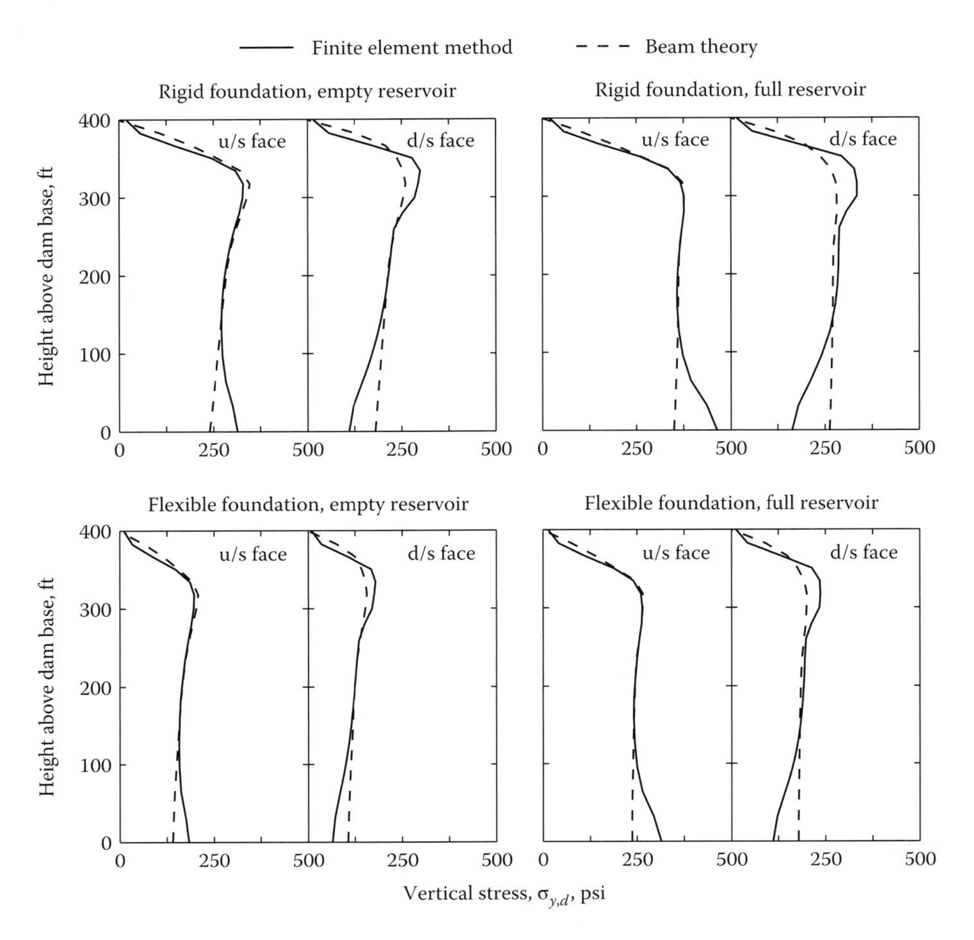

Figure 4.7.4 Comparison of vertical stresses computed in the RSA procedure by two methods: beam theory and the finite element method for four cases.

Fundamental Mode Properties. The fundamental vibration period and the effective damping ratio at this period are estimated using Eqs. (4.1.4a) and (4.1.5) in the RSA procedure. These vibration properties are not needed in the RHA procedure; however, for the purposes of evaluating the accuracy of the approximate results, they are determined – by the half-power bandwidth method – from the frequency response function for the fundamental mode response of the dam–water–foundation system computed in the EAGD-84 program. These are referred to as the "exact" results in Table 4.7.2.

It is apparent that the approximate procedure provides excellent estimates for the resonant period and effective damping ratio of the system in its fundamental mode, confirming that the equivalent SDF model for the dam–water–foundation system is able to represent the important effects of dam–water interaction, reservoir bottom absorption, and dam–foundation interaction.

Stresses. The peak value of the maximum principal stress at a location over the duration of each ground motion is determined from the response history computed by the EAGD-84 program (Løkke and Chopra 2013; Appendix C). At the two faces of the dam, the principal stresses are essentially parallel to the faces if the upstream face is nearly vertical and the stresses due to tail water at the downstream face are negligible (Fenves and Chopra 1986); these conditions are usually satisfied in practical problems. This implies that the direction of the peak value of maximum principal stress at locations on a dam face is essentially invariant among ground motions,

Table 4.7.2 "Exact" and approximate fundamental mode properties.

			Vibration period, $\tilde{T}_1$, sec		Damping ratio, $\tilde{\zeta}_1$, percent	
Case	Foundation	Reservoir	Approx.	Exact	Approx.	Exact
1	Rigid	Empty	0.311	0.318	2.0	2.0
2	Rigid	Full	0.387	0.395	3.9	3.2
3	Flexible	Empty	0.369	0.390	7.1	8.7
4	Flexible	Full	0.459	0.491	9.2	9.8

therefore the peak stress values due to the 58 ground motions lend themselves to statistical analysis.

At each location on the two faces of the dam the median value is computed as the geometric mean of the data set; results are presented in Figure 4.7.5 where they are compared with the RSA results. The maximum principal stresses in the RSA procedure are obtained by a transformation of the vertical stresses determined by beam theory (Fenves and Chopra 1986).

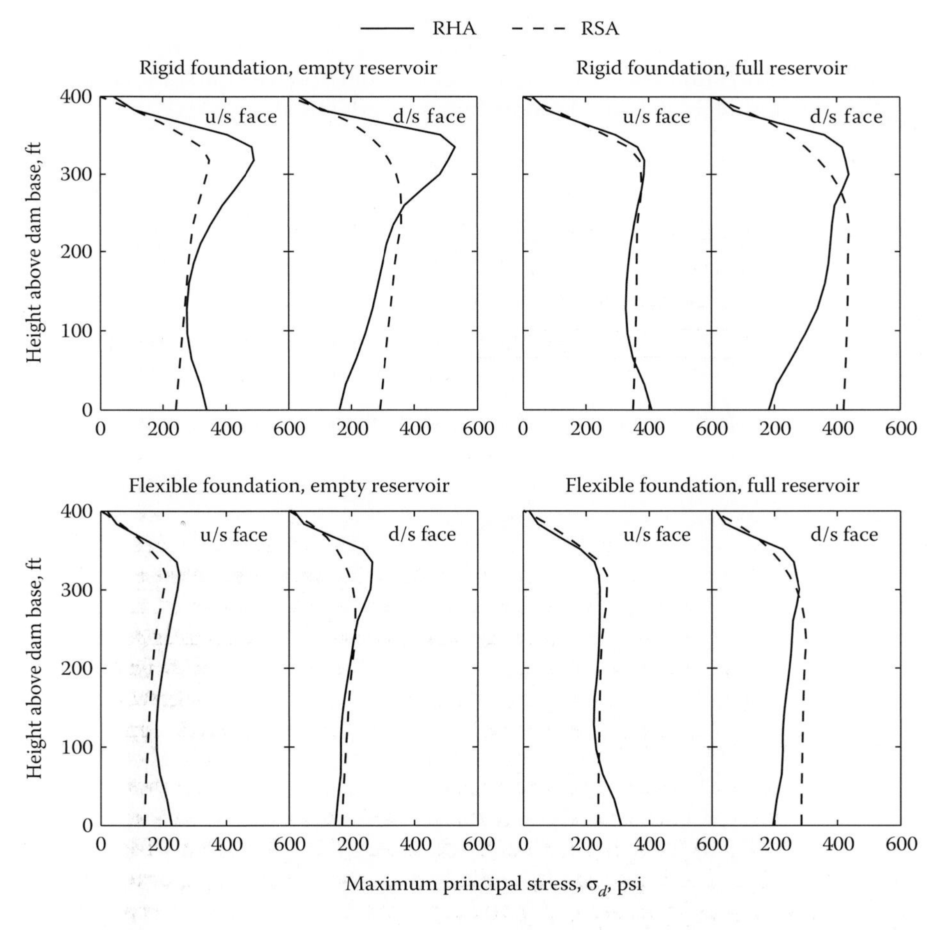

Figure 4.7.5 Comparison of peak values of maximum principal stresses computed by RSA and RHA procedures for four cases; initial static stresses are excluded.

Case 1 (rigid foundation, empty reservoir) is an example where higher mode contributions are considerable, primarily in the upper part of the dam, as expected, where the steep stress gradients are evident in the RHA results (Figure 4.7.5). The RSA procedure underestimates these higher-mode contributions because the vibration periods are not short enough for the static correction approximation to be valid. As shown in Figure 4.7.6, the spectral accelerations at the second- and third-mode periods are more than three times the peak ground acceleration that is used instead in the static correction method. Thus, the static correction method grossly underestimates the higher-mode stresses. For the median response spectrum considered, such discrepancy would be much smaller in the case of a dam of lower height with shorter periods. For Cases 2–4, the RSA procedure provides very good estimates of the maximum principal stresses.

The RSA procedure tends to be more conservative–relative to the RHA result – at the downstream face of the dam than at the upstream face (Figure 4.7.5). An investigation revealed that the underlying reason is the one-dimensional approximation of the equivalent static lateral forces in Eq. (4.1.1), wherein any variation of the fundamental mode shape over the breadth of the dam was neglected, thus ignoring the horizontal variation of the lateral forces.

The preceding results demonstrate that the RSA procedure estimates stresses to a degree of accuracy that is satisfactory for the preliminary phase in the design of new dams and in the safety evaluation of existing dams. This conclusion is impressive in light of the complicated effects of dam–water–foundation interaction and reservoir bottom absorption on the dam response, and the number of approximations necessary to develop the procedure. The accuracy of the computed results depends on several factors, including how well the fundamental resonant period and damping ratio are estimated in the RSA procedure, and how well the static correction method is able to account for the higher-mode contributions to the total response.

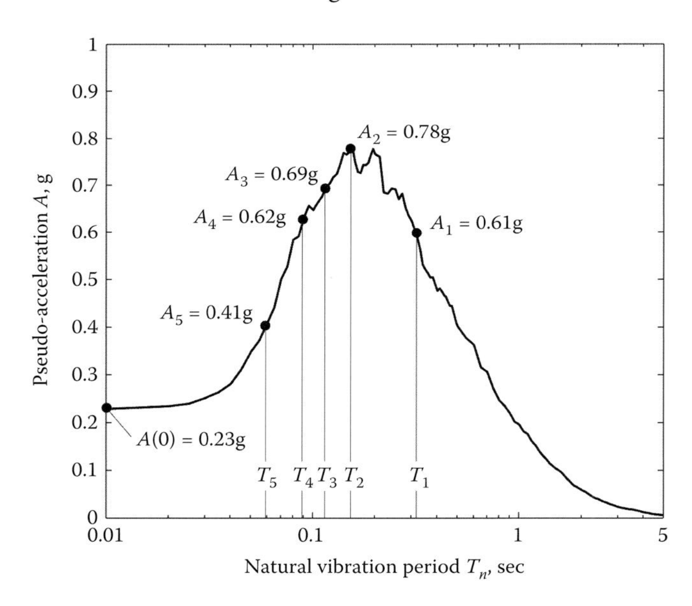

Figure 4.7.6 Spectral accelerations at the first five natural vibration periods of Pine Flat Dam on rigid foundation with empty reservoir; damping, $\zeta = 2\%$.

5

Response History Analysis of Dams Including Dam–Water–Foundation Interaction

PREVIEW

A substructure method for response history analysis (RHA) of concrete gravity dams is presented in this chapter. Included are all the significant effects of dam–water–foundation interaction and sedimentary deposits at the reservoir bottom, as identified in Chapters 2 and 3. By representing the dam, impounded water, and supporting foundation as three substructures of the complete system, it is possible to select the idealization most appropriate for each substructure: finite-element model for the dam; continuum model for the fluid domain, unbounded in the upstream direction; and a viscoelastic half-space (semi-unbounded) continuum model for the foundation. The substructure method is formulated in the frequency domain to determine the complex-valued frequency response functions, followed by Fourier synthesis of the responses to individual harmonic components to determine the responses – displacements and stresses – of the dam to free-field ground motion specified at the dam–foundation interface. This chapter ends with a brief mention of EAGD-84, the computer program that implements the analysis procedure.

5.1 DAM–WATER–FOUNDATION SYSTEM

5.1.1 Two-Dimensional Idealization

Although concrete gravity dams are three-dimensional structures (Figure 5.1.1) and strictly speaking they should be analyzed as such, useful results can often be obtained by analyzing two-dimensional idealizations. At small amplitudes of vibration, a mass-concrete dam will behave as a three-dimensional solid even though the contraction joints between the monoliths may slip, as demonstrated by forced vibration tests of Pine Flat Dam (Rea et al. 1975). At larger-amplitude motions, the behavior of a dam depends on the extent to which inertia forces

Earthquake Engineering for Concrete Dams: Analysis, Design, and Evaluation, First Edition. Anil K. Chopra.

Figure 5.1.1 Pine Flat Dam, near Fresno, California.

Figure 5.1.2 Olivenhain Dam, Escondido, California.

can be transmitted across joints. For dams with straight joints, grouted or ungrouted, the inertia forces that develop at larger amplitudes of motion are much larger than the shear forces that the joints can transmit. Consequently, the joints would slip and monoliths vibrate independently as evidenced by the spalled concrete and water leakage at the joints of Koyna Dam during the Koyna earthquake of 11 December 1967 (Chopra and Chakrabarti 1973). Although keyed joints offer larger shear resistance compared to straight joints, at large-enough amplitude of motions the shear keys are expected to break and the monoliths would tend to vibrate independently. Thus, a two-dimensional plane stress model of the individual monoliths is appropriate for estimating the earthquake response of dams to intense ground motions.

On the other hand, roller-compacted concrete gravity dams that are built without contraction joints are better idealized as plane strain systems, a two-dimensional model that is especially appropriate if the dam is located in a wide canyon, e.g. Olivenhain Dam (Figure 5.1.2). However, this idealization may not be appropriate for a dam in a narrow canyon.

For all the aforementioned reasons, we will limit the scope of this chapter to two-dimensional – plane stress or plane strain – idealizations of dams. For dams not amenable to such idealizations, the three-dimensional analysis procedures developed in Chapter 8 may be employed.

5.1.2 System Considered

The two-dimensional system considered consists of a monolith (or cross section) of a concrete gravity dam supported on the horizontal surface of underlying flexible foundation rock and impounding a reservoir of water with alluvium and sediments deposited at the bottom of the

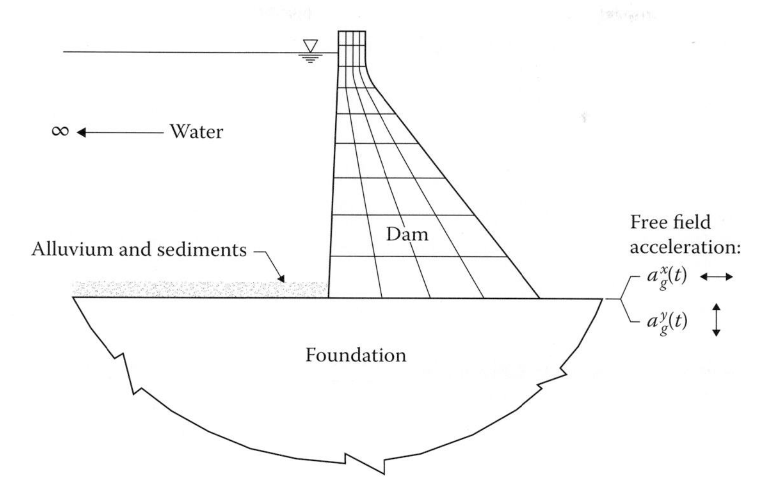

Figure 5.1.3 Dam–water–foundation system.

reservoir (Figure 5.1.3); in contrast to Chapter 3, the base of the dam is no longer restricted to rigid body motions. The system is analyzed under the assumption of linear behavior for the concrete dam, the impounded water, and the foundation rock. Thus, the possibility of cracking in concrete, sliding or opening at lift joints or at the dam–foundation interface, or any other source of nonlinearity are not considered.

The selected monolith of a dam is idealized as a two-dimensional – generalized plane stress or plane strain – finite element system in order to model arbitrary geometry and variation of material properties. However, certain restrictions on the geometry are imposed to permit a continuum solution for the hydrodynamic pressure. For the purpose of determining hydrodynamic effects, and only for this purpose, the upstream face of the dam is assumed to be vertical, an assumption justified in Section 2.1. The impounded water in the reservoir is idealized by a fluid domain of constant depth and infinite length in the upstream direction. Although this idealization is often appropriate, discrete modeling procedures have been developed for fluid domains of arbitrary geometry extending very long or relatively short distances in the upstream direction (Hall and Chopra 1982).

The foundation is idealized as a homogeneous, isotropic, viscoelastic half-plane. This idealization is selected to recognize that, at many dam sites, similar rocks extend to great depths and no obvious boundary such as a soil-rock interface exists to justify a finite-sized model for the foundation. However, the half-space model is not appropriate to represent the effect of sedimentary materials deposited at the reservoir bottom. They are approximately modeled, as described in Section 2.3, by a reservoir bottom that partially absorbs incident hydrodynamic pressure waves.

5.1.3 Ground Motion

The earthquake excitation for the dam–water–foundation system is defined by the two components of free-field ground acceleration in the plane of the monolith (or a cross section) of the dam at its base: the horizontal component $a_g^x(t)$ transverse to the dam axis, and the vertical component $a_g^y(t)$. The free-field ground motion is assumed identical at all nodal points on the horizontal base of the dam. This assumption implies that spatial variations in ground motion across the base are not included in the analysis procedure; however, it can be extended, as demonstrated in Chapter 8 for arch dams.

5.2 FREQUENCY-DOMAIN EQUATIONS: DAM SUBSTRUCTURE

The equations of motion for the dam idealized as a planar, two-dimensional finite-element system (Figure 5.2.1) are

$$\mathbf{m}_c\ddot{\mathbf{r}}_c + \mathbf{c}_c\dot{\mathbf{r}}_c + \mathbf{k}_c\mathbf{r}_c = -\mathbf{m}_c\boldsymbol{\iota}_c^x a_g^x(t) - \mathbf{m}_c\boldsymbol{\iota}_c^y a_g^y(t) + \mathbf{R}_c(t) \tag{5.2.1}$$

in which $\mathbf{m}_c$, $\mathbf{c}_c$, and $\mathbf{k}_c$ are the mass, damping, and stiffness matrices for the finite element system; $\mathbf{r}_c$ is the vector of nodal displacements relative to the free-field ground displacement (Figure 5.2.1):

$$\mathbf{r}_c^T = \left\langle r_1^x \;\; r_1^y \;\; r_2^x \;\; r_2^y \ldots r_n^x \;\; r_n^y \ldots r_{N+N_b}^x \;\; r_{N+N_b}^y \right\rangle \tag{5.2.2}$$

where r_n^x and r_n^y are the x and y components of displacements of node n; N is the number of nodes above the base, N_b is the number of nodes on the base; and

$$\begin{aligned} \boldsymbol{\iota}_c^x &= \langle 1 \quad 0 \quad 1 \quad 0 \ldots 1 \quad 0 \ldots 1 \quad 0 \rangle^T \\ \boldsymbol{\iota}_c^y &= \langle 0 \quad 1 \quad 0 \quad 1 \ldots 0 \quad 1 \ldots 0 \quad 1 \rangle^T \end{aligned} \tag{5.2.3}$$

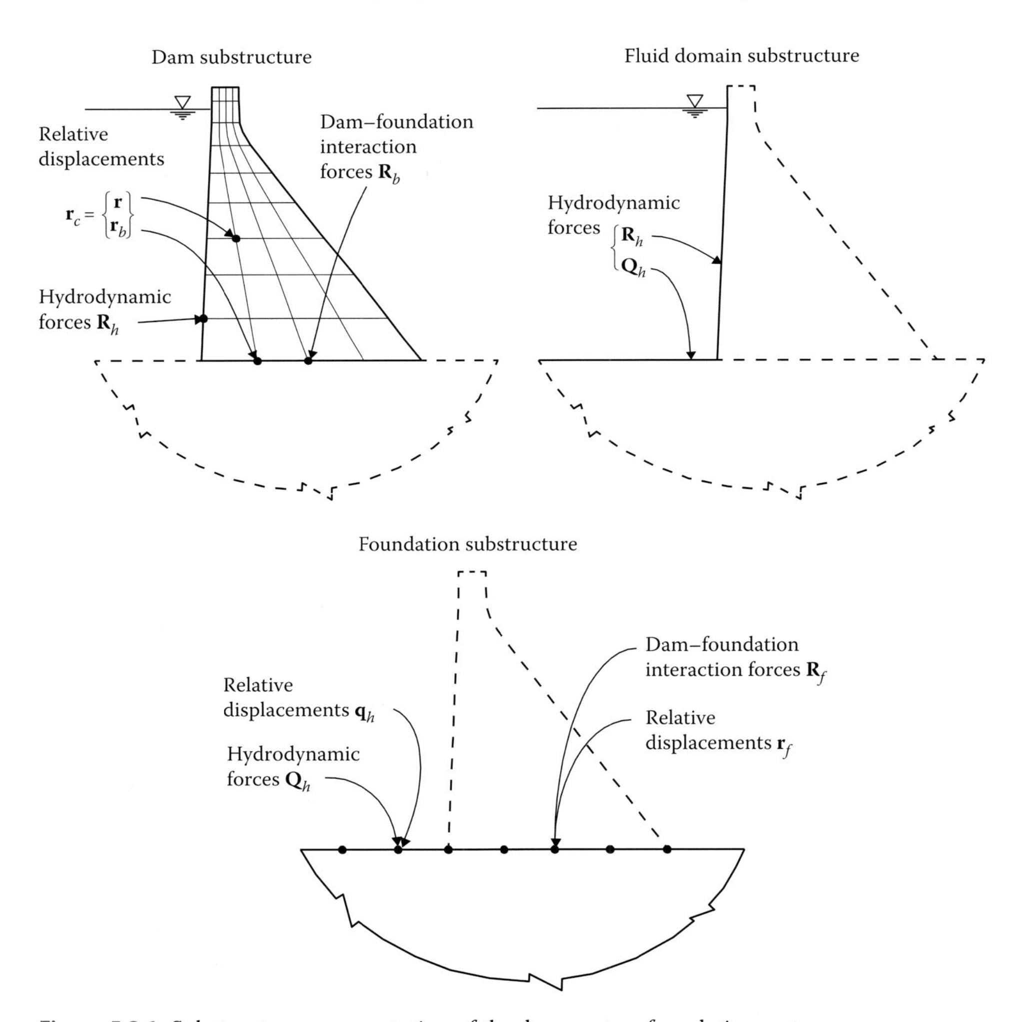

Figure 5.2.1 Substructure representation of the dam–water–foundation system.

The force vector $\mathbf{R}_c(t)$ includes hydrodynamic forces $\mathbf{R}_h(t)$ at the upstream face of the dam due to dam–water interaction and forces $\mathbf{R}_b(t)$ at the base of the dam due to dam–foundation interaction.

We will first determine the response to harmonic ground acceleration: $a_g^l(t) = e^{i\omega t}$ in the $l = x$ (horizontal) or $l = y$ (vertical) direction. The resulting displacements and forces can be expressed in terms of their complex-valued frequency response functions: $\mathbf{r}_c(t) = \bar{\mathbf{r}}_c^l(\omega)\, e^{i\omega t}$, $\mathbf{R}_c(t) = \bar{\mathbf{R}}_c^l(\omega)\, e^{i\omega t}$, $\mathbf{R}_h(t) = \bar{\mathbf{R}}_h^l(\omega)\, e^{i\omega t}$, and $\mathbf{R}_b(t) = \bar{\mathbf{R}}_b^l(\omega)\, e^{i\omega t}$. Substituting these expressions in Eq. (5.2.1) and canceling the $e^{i\omega t}$ term, leads to

$$\left[-\omega^2 \mathbf{m}_c + i\omega \mathbf{c}_c + \mathbf{k}_c\right] \bar{\mathbf{r}}_c^l(\omega) = -\mathbf{m}_c \boldsymbol{\iota}_c^l + \bar{\mathbf{R}}_c^l(\omega); \quad l = x \text{ or } y \tag{5.2.4}$$

If the energy dissipation in the dam is modeled by constant hysteretic damping, Eq. (5.2.4) becomes

$$\left[-\omega^2 \mathbf{m}_c + \left(1 + i\eta_s\right) \mathbf{k}_c\right] \bar{\mathbf{r}}_c^l(\omega) = -\mathbf{m}_c \boldsymbol{\iota}_c^l + \bar{\mathbf{R}}_c^l(\omega); \quad l = x \text{ or } y \tag{5.2.5}$$

where η_s is the constant hysteretic damping factor. Partitioning $\mathbf{r}_c$ into $\mathbf{r}$ for nodal points above the base and $\mathbf{r}_b$ for nodal points on the base, as shown in Figure 5.2.1a, the preceding equation can be expressed as

$$\left(-\omega^2 \begin{bmatrix} \mathbf{m} & \mathbf{0} \\ \mathbf{0} & \mathbf{m}_b \end{bmatrix} + \left(1 + i\eta_s\right) \begin{bmatrix} \mathbf{k} & \mathbf{k}_b \\ \mathbf{k}_b^T & \mathbf{k}_{bb} \end{bmatrix}\right) \begin{Bmatrix} \bar{\mathbf{r}}^l(\omega) \\ \bar{\mathbf{r}}_b^l(\omega) \end{Bmatrix} = -\begin{Bmatrix} \mathbf{m}\boldsymbol{\iota}^l \\ \mathbf{m}_b \boldsymbol{\iota}_b^l \end{Bmatrix} + \begin{Bmatrix} \bar{\mathbf{R}}_h^l(\omega) \\ \bar{\mathbf{R}}_b^l(\omega) \end{Bmatrix} \tag{5.2.6}$$

Dam–foundation interaction forces, $\mathbf{R}_b$, will be expressed in terms of the displacements at the base by analysis of the foundation substructure (Section 5.3), and the hydrodynamic forces $\mathbf{R}_h$ will be expressed in terms of the acceleration of the upstream face of the dam by analysis of the fluid domain substructure (Section 5.5).

5.3 FREQUENCY-DOMAIN EQUATIONS: FOUNDATION SUBSTRUCTURE

The forces acting at the foundation surface include the forces $\mathbf{R}_f$ at the base of the dam due to dam–foundation interaction and the hydrodynamic forces $\mathbf{Q}_h$ at the bottom of the reservoir (Figure 5.2.1). Beyond a certain distance upstream of the dam, the hydrodynamic pressures will become small enough in the sense that their effects on the dam would be negligible; therefore, only the significant forces are included in $\mathbf{Q}_h$. For unit harmonic ground acceleration these forces can be expressed in terms of their complex-valued frequency response functions; thus $\mathbf{R}_f(t) = \bar{\mathbf{R}}_f(\omega)\, e^{i\omega t}$ and $\mathbf{Q}_h(t) = \bar{\mathbf{Q}}_h(\omega)\, e^{i\omega t}$. The corresponding displacements relative to the free-field ground displacement are $\mathbf{r}_f(t) = \bar{\mathbf{r}}_f(\omega)\, e^{i\omega t}$ and $\mathbf{q}_h(t) = \bar{\mathbf{q}}_h(\omega)\, e^{i\omega t}$. The forces and displacements can be related as follows:

$$\begin{bmatrix} \mathcal{S}_{rr}(\omega) & \mathcal{S}_{rq}(\omega) \\ \mathcal{S}_{rq}^T(\omega) & \mathcal{S}_{qq}(\omega) \end{bmatrix} \begin{Bmatrix} \bar{\mathbf{r}}_f(\omega) \\ \bar{\mathbf{q}}_h(\omega) \end{Bmatrix} = \begin{Bmatrix} \bar{\mathbf{R}}_f(\omega) \\ \bar{\mathbf{Q}}_h(\omega) \end{Bmatrix} \tag{5.3.1}$$

where the elements of the complex-valued frequency-dependent (dynamic) stiffness matrix, $\mathcal{S}(\omega)$, for the foundation are determined by solving boundary value problems with unit displacements imposed at nodal points contained in $\mathbf{r}_f$ and $\mathbf{q}_h$, and zero tractions outside these nodal points. In particular, $\mathcal{S}_{ij}(\omega)$, the ijth-element of this matrix, is determined by imposing a unit harmonic displacement in degrees of freedom (DOF) j, as shown in Figure 5.3.1.

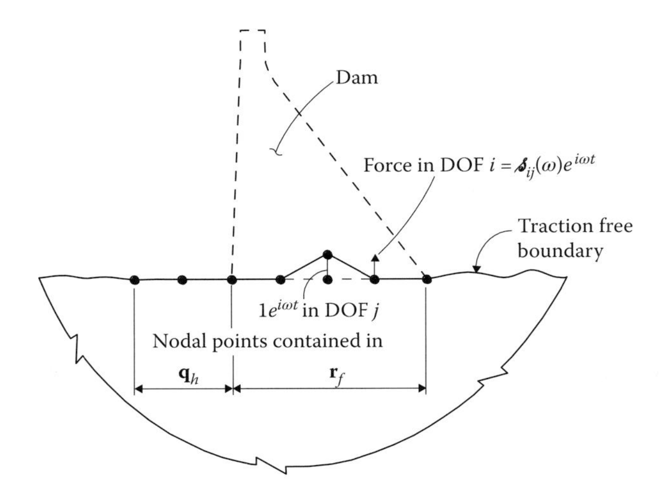

Figure 5.3.1 Definition of $\mathcal{S}_{ij}(\omega)$, the ijth element of the dynamic stiffness matrix $\mathcal{S}(\omega)$ of the foundation.

The second matrix equation from Eq. (5.3.1) can be expressed as

$$\bar{\mathbf{q}}_h(\omega) = \mathcal{S}_{qq}^{-1}(\omega)\left[\bar{\mathbf{Q}}_h(\omega) - \mathcal{S}_{rq}^{T}(\omega)\bar{\mathbf{r}}_f(\omega)\right] \tag{5.3.2}$$

which is substituted into the first matrix equation, resulting in

$$\mathcal{S}_f(\omega)\bar{\mathbf{r}}_f(\omega) = \bar{\mathbf{R}}_f(\omega) - \mathcal{S}_{rq}(\omega)\,\mathcal{S}_{qq}^{-1}(\omega)\,\bar{\mathbf{Q}}_h(\omega) \tag{5.3.3}$$

where

$$\mathcal{S}_f(\omega) = \mathcal{S}_{rr}(\omega) - \mathcal{S}_{rq}(\omega)\,\mathcal{S}_{qq}^{-1}\,\mathcal{S}_{rq}^{T}(\omega) \tag{5.3.4}$$

The dynamic stiffness matrix $\mathcal{S}_f(\omega)$ in Eq. (5.3.4) can be determined by a separate analysis of the foundation substructure, idealized as a homogeneous, viscoelastic half-space in generalized plane stress or plane strain, utilizing available methods and data (Dasgupta and Chopra 1979).

5.4 DAM–FOUNDATION SYSTEM

5.4.1 Frequency-Domain Equations

Equilibrium of the interaction forces between the dam and foundation substructures at the dam–foundation interface requires that

$$\bar{\mathbf{R}}_b^l(\omega) + \bar{\mathbf{R}}_f^l(\omega) = 0 \tag{5.4.1}$$

and compatibility of interaction displacements at the interface requires that

$$\bar{\mathbf{r}}_b^l(\omega) = \bar{\mathbf{r}}_f^l(\omega) \tag{5.4.2}$$

Upon use of Eqs. (5.4.1) and (5.4.2), (5.3.3) becomes

$$\bar{\mathbf{R}}_b^l(\omega) = -\mathcal{S}_f(\omega)\bar{\mathbf{r}}_b^l(\omega) - \mathcal{S}_{rq}(\omega)\,\mathcal{S}_{qq}^{-1}(\omega)\,\bar{\mathbf{Q}}_h^l(\omega) \tag{5.4.3}$$

which upon substitution into Eq. (5.4.6) gives

$$\left(-\omega^2\begin{bmatrix}\mathbf{m} & \mathbf{0}\\ \mathbf{0} & \mathbf{m}_b\end{bmatrix} + (1+i\eta_s)\begin{bmatrix}\mathbf{k} & \mathbf{k}_b\\ \mathbf{k}_b^T & \mathbf{k}_{bb}\end{bmatrix} + \begin{bmatrix}\mathbf{0} & \mathbf{0}\\ \mathbf{0} & \mathcal{S}_f(\omega)\end{bmatrix}\right)\begin{Bmatrix}\bar{\mathbf{r}}^l(\omega)\\ \bar{\mathbf{r}}_b^l(\omega)\end{Bmatrix} =$$

$$-\begin{Bmatrix}\mathbf{m}\iota^l\\ \mathbf{m}_b\iota_b^l\end{Bmatrix} + \begin{Bmatrix}\bar{\mathbf{R}}_h^l(\omega)\\ -\mathcal{S}_{rq}\,\mathcal{S}_{qq}^{-1}\bar{\mathbf{Q}}_h^l(\omega)\end{Bmatrix} \tag{5.4.4}$$

The vector $\bar{\mathbf{R}}_h^l(\omega)$ of frequency response functions for hydrodynamic forces contains non-zero terms corresponding only to the x-DOFs on the upstream face of the dam. The vector $\bar{\mathbf{Q}}_h^l(\omega)$ of frequency response functions for hydrodynamic forces at the reservoir bottom contains non-zero terms corresponding only to the y-DOFs. These hydrodynamic forces will later (Section 5.5) be expressed in terms of the accelerations of the upstream face of the dam and of the reservoir bottom, respectively, by analysis of the fluid domain substructure.

5.4.2 Reduction of Degrees of Freedom

The set of $2(N + N_b)$ frequency-dependent and complex-valued Eq. (5.4.4) are to be solved for many values of the excitation frequency. To reduce the computational effort, the displacements $\mathbf{r}_c$ relative to the free-field ground motion are expressed as a linear combination of J Ritz vectors, selected as the eigenvectors of an associated dam–foundation system that remains to be defined:

$$\mathbf{r}_c(t) = \sum_{j=1}^{J} Z_j(t)\,\boldsymbol{\psi}_j \tag{5.4.5}$$

where $Z_j(t)$ is the generalized co-ordinate that corresponds to the jth Ritz vector $\boldsymbol{\psi}_j$. For response to the l-component of ground motion, Eq. (5.4.5) can be expressed in terms of the complex-valued frequency response functions for the generalized co-ordinates:

$$\mathbf{r}_c^l(\omega) = \sum_{j=1}^{J} \bar{Z}_j^l(\omega)\,\boldsymbol{\psi}_j \tag{5.4.6}$$

Equation (5.4.6) contains the following equation for the interaction displacements at the base of the dam:

$$\mathbf{r}_b^l(\omega) = \sum_{j=1}^{J} \bar{Z}_j^l(\omega)\,\boldsymbol{\psi}_{bj} \tag{5.4.7}$$

where $\boldsymbol{\psi}_{bj}$ is the subvector of $\boldsymbol{\psi}_j$ corresponding to the $2N_b$ DOFs at the base.

The eigenvalues λ_j and Ritz vectors $\boldsymbol{\psi}_j$ are the solutions of the eigenvalue problem for an associated dam–foundation system, defined by replacing $\mathcal{S}_f(\omega)$ by its static value $\mathcal{S}_f(0)$:

$$\left[\mathbf{k}_c + \tilde{\mathcal{S}}_f(0)\right]\boldsymbol{\psi}_j = \lambda_j^2\ \mathbf{m}_c\ \boldsymbol{\psi}_j \tag{5.4.8}$$

where

$$\tilde{\mathcal{S}}_f(\omega) = \begin{bmatrix} \mathbf{0} & \mathbf{0} \\ \mathbf{0} & \mathcal{S}_f(\omega) \end{bmatrix} \tag{5.4.9}$$

The Ritz vectors are normalized, such that $\boldsymbol{\psi}_j^T \mathbf{m}_c\ \boldsymbol{\psi}_j = 1$.

The vector $\bar{\mathbf{q}}_h(\omega)$ of frequency response functions for the displacements at the reservoir bottom, relative to the free-field ground motion, may be expressed in terms of the generalized coordinates Z_j by substituting Eqs. (5.4.2) and (5.4.7) into Eq. (5.3.2):

$$\bar{\mathbf{q}}_h^l(\omega) = \mathcal{S}_{qq}^{-1}(\omega)\,\bar{\mathbf{Q}}_h^l(\omega) + \sum_{j=1}^{J} \bar{Z}_j^l(\omega)\,\boldsymbol{\chi}_j \tag{5.4.10}$$

where $\boldsymbol{\chi}_j$ is given by

$$\boldsymbol{\chi}_j = -\,\mathcal{S}_{qq}^{-1}(0)\ \mathcal{S}_{rq}^{T}(0)\,\boldsymbol{\psi}_{bj} \tag{5.4.11}$$

Because the Ritz vectors $\boldsymbol{\psi}_j$ were determined from Eq. (5.4.8) using $\mathcal{S}_f(0)$, for consistency the zero frequency values of $\mathcal{S}_{qq}$ and $\mathcal{S}_{rq}^T$ are used to express $\boldsymbol{\chi}_j$ in terms of $\boldsymbol{\psi}_{bj}$. The first term on the right side of Eq. (5.4.10) relates to the hydrodynamic forces at the reservoir bottom to the relative displacements, i.e. it represents the interaction between the fluid domain and the foundation region under the reservoir.

Introducing the transformation Eq. (5.4.6) into Eq. (5.4.4), pre-multiplying by $\boldsymbol{\psi}_n^T$, and introducing the orthogonality properties of the eigenvectors of the associated dam–foundation system with respect to the "stiffness" and mass matrices of Eq. (5.4.8), leads to

$$\mathbf{S}(\omega)\,\bar{\mathbf{Z}}^l(\omega) = \mathbf{L}^l(\omega) \tag{5.4.12}$$

where the elements of the matrix $\mathbf{S}(\omega)$ and the vector $\mathbf{L}^l(\omega)$ are

$$S_{nj}(\omega) = \left[-\omega^2 + \left(1 + i\eta_s\right)\lambda_n^2\right]\delta_{nj} + \boldsymbol{\psi}_n^T\left[\tilde{\mathcal{S}}_f(\omega) - \left(1 + \eta_s\right)\tilde{\mathcal{S}}_f(0)\right]\boldsymbol{\psi}_j \tag{5.4.13a}$$

$$L_n^l(\omega) = -\boldsymbol{\psi}_n^T\mathbf{m}_c\boldsymbol{\iota}_c^l + \boldsymbol{\psi}_{fn}^T\bar{\mathbf{R}}_h^l(\omega) - \boldsymbol{\psi}_{bn}^T\,\mathcal{S}_{rq}(\omega)\,\mathcal{S}_{qq}^{-1}(\omega)\,\bar{\mathbf{Q}}_h^l(\omega) \tag{5.4.13b}$$

for $n, j = 1, 2, 3, \ldots J$; $\bar{\mathbf{Z}}^l(\omega)$ is the vector of frequency response functions $\bar{Z}_j^l(\omega)$ for the generalized coordinates $Z_j(t)$; δ_{nj} is the Kronecker delta function; and $\boldsymbol{\psi}_{fn}$ is the subvector of $\boldsymbol{\psi}_n$ that contains only the elements corresponding to the nodal points at the upstream face of the dam.

Equations (5.4.12) and (5.4.13b) represent J complex-valued, frequency-dependent, algebraic equations in the unknown generalized coordinates. These equations need to be solved for several hundred to a few thousand values of the excitation frequency to determine the frequency response functions. Fortunately, accurate solutions can be obtained by including only a small number of Ritz vectors, typically less than 10, which profoundly reduces the computational effort (Chopra and Chakrabarti 1981).

5.5 FREQUENCY–DOMAIN EQUATIONS: FLUID DOMAIN SUBSTRUCTURE

5.5.1 Boundary Value Problems

The unknown hydrodynamic forces $\mathbf{R}_h(t)$ and $\mathbf{Q}_h(t)$, whose frequency response functions appear in Eq. (5.4.13b), can be expressed in terms of accelerations of the upstream face of the dam and of the reservoir bottom by analyzing the fluid domain. The motion of water is governed by the two-dimensional wave equation (Eq. 2.3.1). For harmonic ground acceleration $a_g^l(t) = e^{i\omega t}$, the hydrodynamic pressure can be expressed as $p\left(x, y, t\right) = \bar{p}^l\left(x, y, \omega\right)e^{i\omega t}$, where $\bar{p}^l\left(x, y, \omega\right)$ is the complex-valued frequency response function governed by the Helmholtz equation (2.3.8).

The hydrodynamic pressure is generated by horizontal acceleration of the vertical upstream face of the dam,

$$\ddot{\mathbf{u}}(t) = \left[\delta_{xl}\boldsymbol{\iota}^x + \sum_{j=1}^{J}\boldsymbol{\psi}_{fj}^x\ddot{\bar{Z}}_j^l(\omega)\right]e^{i\omega t} \qquad l = x, y \tag{5.5.1}$$

and by vertical acceleration of the horizontal reservoir bottom.

$$\ddot{\mathbf{v}}(t) = \left[\delta_{yl}\boldsymbol{\iota}^y + \sum_{j=1}^{J}\boldsymbol{\chi}_j^y\ddot{\bar{Z}}_j^l(\omega)\right]e^{i\omega t} \qquad l = x, y \tag{5.5.2}$$

In Eqs. (5.5.1) and (5.5.2), $\boldsymbol{\iota}^x$ = <1 0 1 0......1 0> with unit values corresponding to x-DOFs, $\boldsymbol{\iota}^y$ = <0 1 0 1 ......0 1> with unit values corresponding to y-DOFs; the vector $\boldsymbol{\psi}_{fj}^x$ is a subvector of $\boldsymbol{\psi}_j$ containing only the elements corresponding to the x-DOFs of the nodal points on the upstream face of the dam, and $\boldsymbol{\chi}_j^y$ is a subvector of $\boldsymbol{\chi}_j$ containing only elements corresponding to the y-DOFs of the nodal points at the bottom of the reservoir. Strictly for purposes of notational convenience, the vectors $\boldsymbol{\psi}_{fj}^x$ and $\boldsymbol{\chi}_j^y$ are replaced by their continuous analog functions $\boldsymbol{\psi}_j(y)$ and $\boldsymbol{\chi}_j(x)$, respectively, to obtain

$$\ddot{u}(0, y, t) = \left[\delta_{xl} + \sum_{j=1}^{J} \boldsymbol{\psi}_j(y)\,\bar{\ddot{Z}}_j^l(\omega)\right] e^{i\omega t} \tag{5.5.3}$$

$$\ddot{v}(x, 0, t) = \left[\delta_{yl} + \sum_{j=1}^{J} \boldsymbol{\chi}_j(x)\,\bar{\ddot{Z}}_j^l(\omega)\right] e^{i\omega t} \tag{5.5.4}$$

These motions of the two boundaries are related to the hydrodynamic pressure by the boundary conditions in Eqs. (5.5.5) and (5.5.6). The boundary condition at the upstream face of the dam is a generalization of Eq. (2.4.8a), wherein the acceleration term in brackets is replaced by the right side of Eq. (5.5.3) excluding $e^{i\omega t}$ to obtain

$$\frac{\partial}{\partial x}\bar{p}(0, y, \omega) = -\rho\left[\delta_{xl} + \sum_{j=1}^{J} \boldsymbol{\psi}_j(y)\,\bar{\ddot{Z}}_j^l(\omega)\right] \qquad l = x, y \tag{5.5.5}$$

The boundary condition at the reservoir bottom is a generalization of Eq. (2.4.8b), wherein the acceleration term δ_{yl} is replaced by the right side of Eq. (5.5.4) excluding $e^{i\omega t}$ to obtain

$$\left(\frac{\partial}{\partial y} - i\omega\xi\right)\bar{p}(x, 0, \omega) = -\rho\left[\delta_{yl} + \sum_{j=1}^{J} \chi_j(x)\,\bar{\ddot{Z}}_j^l(\omega)\right] \qquad l = x, y \tag{5.5.6}$$

where δ_{kl} is the Kronecker delta function. This boundary condition at the reservoir bottom is equivalent to replacing the compliance matrix $\mathcal{S}_{qq}^{-1}(\omega)$ in Eq. (5.4.10) by the approximate compliance function, $\mathcal{C}(\omega)$ for the foundation, derived in Sections 2.A.1 and 2.A.2.

Neglecting the effects of surface waves, which are known to be small in the earthquake response of dams (Chopra 1967), the boundary condition at the free surface is

$$\bar{p}(x, H, \omega) = 0 \tag{5.5.7}$$

where H is the depth of the impounded water.

5.5.2 Solutions for Hydrodynamic Pressure Terms

The frequency response function $\bar{p}^l(x, y, \omega)$ for the hydrodynamic pressure is the solution of Eq. (2.3.8) subject to the boundary conditions in Eqs. (5.5.5)–(5.5.7) and the radiation condition in the upstream direction (negative x-direction). The linear form of the governing equation and boundary conditions allow $\bar{p}^l(x, y, \omega)$ to be expressed as

$$\bar{p}^l(x, y, \omega) = \bar{p}_0^l(x, y, \omega) + \sum_{j=1}^{J} \bar{\ddot{Z}}_j^l(\omega)\left[\bar{p}_j^f(x, y, \omega) + \bar{p}_j^b(x, y, \omega)\right] \tag{5.5.8}$$

In Eq. (5.5.8), the frequency response function $\bar{p}_0^x(x, y, \omega)$ for the hydrodynamic pressure due to the horizontal (x) component of ground acceleration with the dam rigid is the solution of Eq. (2.3.8) subject to the boundary conditions of Eq. (2.3.9). The frequency response function $\bar{p}_0^y(x, y, \omega)$ for the hydrodynamic pressure due to the vertical (y) component of ground acceleration with the dam rigid is the solution of Eq. (2.3.8) subject to the boundary conditions of Eq. (2.3.10). The frequency response function $\bar{p}_j^f(x, y, \omega)$ for the hydrodynamic pressure due to horizontal acceleration of the upstream face of the dam corresponding to the jth Ritz vector with no motion at the reservoir bottom, is the solution of Eq. (2.3.8) subject to the following boundary conditions:

$$\frac{\partial \bar{p}}{\partial x}(0, y, \omega) = -\rho \psi_j(y) \qquad \left[\frac{\partial}{\partial_y} - i\omega\xi\right] \bar{p}(x, 0, \omega) = 0 \qquad \bar{p}(x, H, \omega) = 0 \tag{5.5.9}$$

which are the same as Eq. (2.3.11) with $\phi_1^x(0, y)$ replaced by $\psi_j(y)$. The frequency response function $\bar{p}_j^b(x, y, \omega)$ for the hydrodynamic pressure due to vertical acceleration $\chi_j(x)$ of the reservoir bottom that corresponds to the jth Ritz vector, with no motion of the dam, is the solution of Eq. (2.3.8) subject to the following boundary conditions:

$$\frac{\partial \bar{p}}{\partial_x}(0, y, \omega) = 0 \qquad \left[\frac{\partial}{\partial_y} - i\omega\xi\right] \bar{p}(x, 0, \omega) = -\rho \chi_j(x) \qquad \bar{p}(x, H, \omega) = 0 \tag{5.5.10}$$

The complex-valued frequency response functions $\bar{p}_0^l(x, y, \omega)$ and $\bar{p}_j^f(x, y, \omega)$ were obtained using standard solution methods for boundary value problems. Specialized for the upstream face of the dam ($x = 0$) they are given by Eq. (2.3.12), repeated here for convenience

$$\bar{p}_0^x(0, y, \omega) = -2\rho H \sum_{n=1}^{\infty} \frac{\mu_n^2(\omega)}{H\left[\mu_n^2(\omega) - (\omega\xi)^2\right] + i(\omega\xi)} \frac{I_{0n}(\omega)}{\sqrt{\mu_n^2(\omega) - (\omega^2/C^2)}} Y_n(y, \omega) \tag{5.5.11a}$$

$$\bar{p}_0^y(0, y, \omega) = \frac{\rho}{\omega/C} \frac{1}{\cos(\omega H/C) + i\xi C \sin(\omega H/C)} \sin\frac{\omega(H - y)}{C} \tag{5.5.11b}$$

$$\bar{p}_j^f(0, y, \omega) = -2\rho H \sum_{n=1}^{\infty} \frac{\mu_n^2(\omega)}{H\left[\mu_n^2(\omega) - (\omega\xi)^2\right] + i(\omega\xi)} \frac{I_{jn}(\omega)}{\sqrt{\mu_n^2(\omega) - (\omega^2/C^2)}} Y_n(y, \omega) \tag{5.5.11c}$$

where

$$I_{0n}(\omega) = \frac{1}{H}\int_0^H Y_n(y, \omega)\, dy \qquad I_{jn}(\omega) = \frac{1}{H}\int_0^H \boldsymbol{\psi}_j(y)\, Y_n(y, \omega)\, dy \tag{5.5.12}$$

where the eigenvalues $\mu_n(\omega)$, which are complex-valued and depend on the excitation frequency, satisfy Eq. (2.3.14), and the eigenfunctions $Y_n(y, \omega)$ are defined by Eq. (2.3.15).

The solution for the frequency response function $\bar{p}_j^b(x, y, \omega)$ may be obtained by employing a Fourier transform with respect to the spatial x-coordinate. Such a general solution is not necessary, however, because the resulting hydrodynamic pressure has little effect on the response of the dam, as will be demonstrated in Appendix 5.

5.5.3 Hydrodynamic Force Vectors

The frequency response functions for $\mathbf{R}_h^l(t)$, the vector of hydrodynamic forces at the upstream face of the dam, and $\mathbf{Q}_h^l(t)$, the vector of hydrodynamic forces at the reservoir bottom, are from Eq. (5.5.8):

$$\bar{\mathbf{R}}_h^l(\omega) = \bar{\mathbf{R}}_0^l(\omega) + \sum_{j=1}^{J} \bar{\bar{Z}}_j^l(\omega)\left[\bar{\mathbf{R}}_j^f(\omega) + \bar{\mathbf{R}}_j^b(\omega)\right] \tag{5.5.13a}$$

$$\bar{\mathbf{Q}}_h^l(\omega) = \bar{\mathbf{Q}}_0^l(\omega) + \sum_{j=1}^{J} \bar{\bar{Z}}_j^l(\omega)\left[\bar{\mathbf{Q}}_j^f(\omega) + \bar{\mathbf{Q}}_j^b(\omega)\right] \tag{5.5.13b}$$

in which the x-DOF elements of the vectors $\bar{\mathbf{R}}_0^l(\omega)$, $\bar{\mathbf{R}}_j^f(\omega)$, and $\bar{\mathbf{R}}_j^b(\omega)$ are the nodal forces statically equivalent to the corresponding pressure functions at the upstream face of the dam:

$\bar{p}_0^l(0,y,\omega)$, $\bar{p}_j^f(0,y,\omega)$, and $\bar{p}_j^b(0,y,\omega)$, respectively; the y-DOF elements of the vectors $-\bar{\mathbf{Q}}_0^l(\omega)$, $-\bar{\mathbf{Q}}_j^f(\omega)$, and $-\bar{\mathbf{Q}}_j^b(\omega)$ are the nodal forces statically equivalent to the corresponding pressure functions at the reservoir bottom: $\bar{p}_0^l(x,0,\omega)$, $\bar{p}_j^f(x,0,\omega)$, and $\bar{p}_j^b(x,0,\omega)$, respectively. The elements of $\mathbf{R}_h^l(\omega)$ that correspond to the y-DOFs at the upstream face of the dam are zero, as are the elements of $\mathbf{Q}_h^l(\omega)$ that correspond to the x-DOFs at the reservoir bottom.

We now present a few ideas to reduce the computational effort in evaluating the hydrodynamic terms $\bar{\mathbf{R}}_0^l(\omega)$, and $\bar{\mathbf{R}}_j^f(\omega)$, which are obtained from the frequency response functions $\bar{p}_0^l(0,y,\omega)$ and $\bar{p}_j^f(0,y,\omega)$ by the principle of virtual displacements. The frequency response functions $\bar{p}_0^x(0,y,\omega)$ and $\bar{p}_j^f(0,y,\omega)$ defined in Eqs. (5.5.11a) and (5.5.11c) are summations of the contributions of an infinite number of natural vibration modes of the impounded water. The computational effort in evaluating the hydrodynamic terms is directly proportional to the number of vibration modes included in these summations, so only the significant vibration modes should be included. A criterion has been developed to determine the required number of modes (Fenves and Chopra 1984a).

The eigenvalues $\mu_n(\omega)$ and eigenfunctions $Y_n(y, \omega)$ of the impounded water, Eqs. (2.3.14) and (2.3.15), used in evaluating the frequency response functions $\bar{p}_0^x(0,y,\omega)$ and $\bar{p}_j^f(0,y,\omega)$ are frequency-dependent if the reservoir bottom is absorptive. The eigenfunctions $Y_n(y, \omega)$ enter the hydrodynamic terms only through the integrals $I_{0n}(\omega)$ and $I_{jn}(\omega)$ defined in Eq. (5.5.12). For a given excitation frequency, Eq. (2.3.14) is solved for the eigenvalues $\mu_n(\omega)$, using the Newton–Raphson algorithm; the integrals $I_{0n}(\omega)$ and $I_{jn}(\omega)$ are then evaluated in their discretized form. Fortunately, it is not necessary to compute $\mu_n(\omega)$, $I_{0n}(\omega)$, and $I_{jn}(\omega)$ for every excitation frequency ω because accurate results for dam response can be obtained when these functions are determined by linear interpolation between their exact values computed at widely separated values of ω (Fenves and Chopra 1984a).

5.6 FREQUENCY-DOMAIN EQUATIONS: DAM–WATER–FOUNDATION SYSTEM

The hydrodynamic forces at the upstream face of the dam and at the reservoir bottom have been expressed in terms of the acceleration $\ddot{\bar{Z}}_j^l(\omega)$ of the generalized coordinates. Substitution of Eq. (5.5.13b) and $\ddot{\bar{Z}}_j^l(\omega) = -\omega^2 \bar{Z}_j^l(\omega)$ into Eqs. (5.4.12) and (5.5.13b) leads to

$$\tilde{\mathbf{S}}(\omega)\,\bar{\mathbf{Z}}^l(\omega) = \tilde{\mathbf{L}}^l(\omega) \tag{5.6.1}$$

where the elements of the matrix $\tilde{\mathbf{S}}$ and the vector $\tilde{\mathbf{L}}$, after rearrangement of terms, are given by

$$\begin{aligned}\tilde{S}_{nj}(\omega) = {} & \left[-\omega^2 + (1+i\eta_s)\,\lambda_n^2\right]\delta_{nj} + \boldsymbol{\psi}_n^T\left[\tilde{\boldsymbol{\mathcal{S}}}_f(\omega) - (1+i\eta_s)\tilde{\boldsymbol{\mathcal{S}}}_f(0)\right]\boldsymbol{\psi}_j \\ & + \omega^2\boldsymbol{\psi}_{fn}^T\left[\bar{\mathbf{r}}_j^f(\omega) + \bar{\mathbf{R}}_j^b(\omega)\right] - \omega^2\boldsymbol{\psi}_{bn}^T\tilde{\boldsymbol{\mathcal{S}}}_{rq}(\omega)\tilde{\boldsymbol{\mathcal{S}}}_{qq}^{-1}(\omega)\left[\bar{\mathbf{Q}}_j^f(\omega) + \bar{\mathbf{Q}}_j^b(\omega)\right]\end{aligned} \tag{5.6.2a}$$

$$\tilde{L}_n^l(\omega) = -\boldsymbol{\psi}_n^T\mathbf{m}_c\boldsymbol{\iota}_c^l + \boldsymbol{\psi}_{fn}^T\,\bar{\mathbf{R}}_0^l(\omega) - \boldsymbol{\psi}_{bn}^T\tilde{\boldsymbol{\mathcal{S}}}_{rq}(\omega)\tilde{\boldsymbol{\mathcal{S}}}_{qq}^{-1}(\omega)\,\bar{\mathbf{Q}}_0^l(\omega) \tag{5.6.2b}$$

Equations (5.6.1) and (5.6.2b) contain effects of dam–water–interaction, dam–foundation interaction, and dam–water–foundation rock interaction in various forms: (i) dam–foundation interaction effects appear in eigenvalues λ_n and eigenvectors $\boldsymbol{\psi}_n$ of the associated dam–foundation system and through the foundation stiffness matrix $\boldsymbol{\mathcal{S}}_f(\omega)$ [Eqs. (5.3.4) and (5.4.8)]; (ii) additional hydrodynamic forces $\mathbf{R}_0^l(\omega)$ on rigid dam due to the free-field ground motions; (iii) dam–water interaction effects appear through the hydrodynamic forces $\mathbf{R}_j^f(\omega)$ on the upstream face of the dam and $\mathbf{Q}_j^f(\omega)$ on the reservoir bottom due to deformational motions of the dam; (iv) water–foundation interaction effects appear through the hydrodynamic forces $\mathbf{R}_j^b(\omega)$ on the face of the dam and $\mathbf{Q}_j^b(\omega)$ on the reservoir bottom due to deformational motions of the reservoir bottom; and (v) hydrodynamic forces $\mathbf{Q}_0^l(\omega)$, $\mathbf{Q}_j^f(\omega)$, and $\mathbf{Q}_j^b(\omega)$ at the reservoir bottom, which influence the deformations at the dam–foundation interface. Observe that one group of terms arises from the hydrodynamic forces $\mathbf{Q}_0^l$, $\mathbf{Q}_j^f$, and $\mathbf{Q}_j^b$ at the reservoir bottom due to the various excitations mentioned earlier. The other such term involves the hydrodynamic forces $\mathbf{R}_j^b$ at the upstream face of the dam due to deformational motions of the reservoir bottom.

Water–foundation interaction, which gives rise to these terms may be neglected or, preferably, modeled in a simple way. Dropping these terms introduces errors but they are small, as demonstrated in Appendix 5.1. These errors can be essentially eliminated by a simple model that can conveniently be included in the substructure method. For this purpose, water–foundation interaction is modeled approximately by the wave-reflection coefficient α (first introduced in Chapter 2) determined from Eq. (2.3.5) using properties of rock underlying the reservoir. Such a model is able to predict the earthquake response of dams quite accurately (see Appendix 5.2). Thus, we conclude that the dam–water–foundation system to be analyzed by the substructure method should always include the wave-reflection coefficient α computed from the properties of rock. With such a model, there is no need to consider separately the effects of sediments at the reservoir bottom; see Section 11.10.5.

Dropping the terms arising from water–foundation interaction in Eq. (5.6.2) leads to the final form for the elements of $\tilde{\mathbf{S}}$ and $\tilde{\mathbf{L}}$:

$$\tilde{S}_{nj}(\omega) = \left[-\omega^2 + \left(1 + \eta_s\right)\lambda_n^2\right]\delta_{nj} + \boldsymbol{\psi}_n^T\left[\tilde{\boldsymbol{\mathcal{S}}}_f(\omega) - \left(1 + \eta_s\right)\tilde{\boldsymbol{\mathcal{S}}}_f(0)\right]\boldsymbol{\psi}_j + \omega^2\boldsymbol{\psi}_{fn}^T\bar{\mathbf{r}}_j^f(\omega) \tag{5.6.3a}$$

$$\tilde{L}_n^l = -\boldsymbol{\psi}_n^T\mathbf{m}_c\boldsymbol{\iota}_c^l + \boldsymbol{\psi}_{fn}^T\bar{\mathbf{R}}_0^l(\omega) \tag{5.6.3b}$$

The frequency-dependent hydrodynamic terms may be interpreted as follows: the real-valued part of the last term on the right side of Eq. (5.6.3a) represents an added mass matrix and its imaginary-valued part is a damping matrix; the last term on the right side of Eq. (5.6.3b) is an added force vector.

Equations (5.6.1) and (5.6.3b) represent J complex-valued equations in the frequency response functions $\bar{Z}_j^l(\omega)$, $j = 1, 2, \ldots J$, for the generalized coordinates that correspond to the Ritz vectors included in the analysis. The matrix $\tilde{\mathbf{S}}(\omega)$ and vector $\tilde{\mathbf{L}}(\omega)$ are computed according to Eq. (5.6.3) for each excitation frequency ω, and Eq. (5.6.1) is solved to obtain $\bar{Z}_j^l(\omega)$. Repeated solution for the excitation frequencies covering the range over which the earthquake ground motion and structural response have significant components leads to the complete frequency response functions for the generalized coordinates.

5.7 RESPONSE HISTORY ANALYSIS

The response history of the dam to arbitrary ground motion can be computed once the complex-valued frequency response functions $\bar{Z}_j^l(\omega)$ – $l = x, y$ and $j = 1, 2, \cdots J$ – for the generalized coordinates have been obtained by solving Eqs. (5.6.1) and (5.6.3) for all excitation frequencies in the range of interest. The time functions for generalized coordinates are given by the Fourier integral as the superposition of responses to individual harmonic components of the ground motion:

$$Z_j^l(t) = \frac{1}{2\pi}\int_{-\infty}^{\infty}\bar{Z}_j^l(\omega)\,A_g^l(\omega)\,e^{i\omega t}\,d\omega \tag{5.7.1}$$

where $A_g^l(\omega)$ is the Fourier transform of the l-component of the specified free-field ground acceleration $a_g^l(t)$:

$$A_g^l(\omega) = \int_0^d a_g^l(t)\, e^{i\omega t}\, dt \tag{5.7.2}$$

in which d is the duration of the ground motion. The Fourier integrals in Eqs. (5.7.1) and (5.7.2) are computed in their discrete forms using the Fast Fourier Transform (FFT) algorithm. The displacement response to the horizontal and vertical components of ground motion, simultaneously, is obtained by transforming the generalized coordinates to the nodal displacements according to Eq. (5.4.5):

$$\mathbf{r}_c(t) = \sum_{j=1}^{J} \left[Z_j^x(t) + Z_j^y(t) \right] \boldsymbol{\psi}_j \tag{5.7.3}$$

The stresses in the dam at any instant of time can be determined from the nodal displacements. The vector $\boldsymbol{\sigma}_p(t)$ of the stress components in finite element p is related to the nodal displacement vector $\mathbf{r}_p(t)$ for that element by

$$\boldsymbol{\sigma}_p(t) = \mathbf{T}_p \mathbf{r}_p(t) \tag{5.7.4}$$

where $\mathbf{T}_p$ is the stress-displacement transformation matrix for the element.

5.8 EAGD-84 COMPUTER PROGRAM

The substructure method for earthquake analysis of gravity dams developed in this chapter has been implemented in the computer program EAGD-84 (Fenves and Chopra 1984c). In this report the development of an appropriate idealization of the system is discussed, the required input data to the computer program are described, the output is explained, and the response results from a sample analysis are presented.

This computer program enables complete analysis of the dynamic response of the dam to the simultaneous action of the horizontal and vertical components of ground motion. The dynamic stresses are combined with the initial, static stresses in the dam due to the weight of the dam and hydrostatic pressures. However, the user may perform a separate static analysis that includes thermal, creep, construction sequence, and other effects, and input the resulting initial stresses in the computer program. The output from the computer program includes the complete response history of (i) the horizontal and vertical displacements at all nodal points, and (ii) the three components of the two-dimensional stress state in all the finite elements. From these results, the user can plot the distribution of stresses in the dam at selected time instants, and the distribution of envelope values of maximum principal stresses in the dam; see Chapters 6 and 7. Such results aid in identifying areas of the dam that may crack during an earthquake.

Two enhancements of the EAGD-84 program implemented recently should facilitate use of the program and expand the range of its applicability (Løkke 2013). MATLAB modules were developed to facilitate development of the finite element model (FEM) for the dam and to prepare data to be input into the program. Secondly, compliance data necessary to construct the dynamic stiffness matrix for the foundation were expanded. The original version of the program included such data for five values of the constant hysteretic damping factor $\eta_f = 0.01, 0.10, 0.25$, and 0.50, which in retrospect turned out to be too coarse. Compliance data have now been added for a closely spaced set of η_f values.

Developed as a computer program for research purposes, EAGD-84 lacks the convenient user interfaces characteristic of commercial finite-element computer codes. However, it has been used for many design and evaluation projects worldwide. It may be accessed from NISEE Library: http://nisee.berkeley.edu/elibrary/getpkg?id=EAGD84.

APPENDIX 5: WATER–FOUNDATION INTERACTION

A5.1 Effects of Water–Foundation Interaction

Documented in this section are the errors arising from dropping hydrodynamic forces $\mathbf{Q}_0^l$, $\mathbf{Q}_j^f$, and $\mathbf{Q}_j^b$ at the reservoir bottom and $\mathbf{R}_j^b$ at the upstream face of the dam. This approximation in the substructure method implies that the effects of water–foundation interaction are neglected.

Determined by the direct FEM (Chapter 11), the frequency response functions for the idealized dam described in Section 2.5.2† with full reservoir but no sediments at the reservoir bottom are presented in Figure A5.1.1 for two cases: water–foundation interaction is included in one

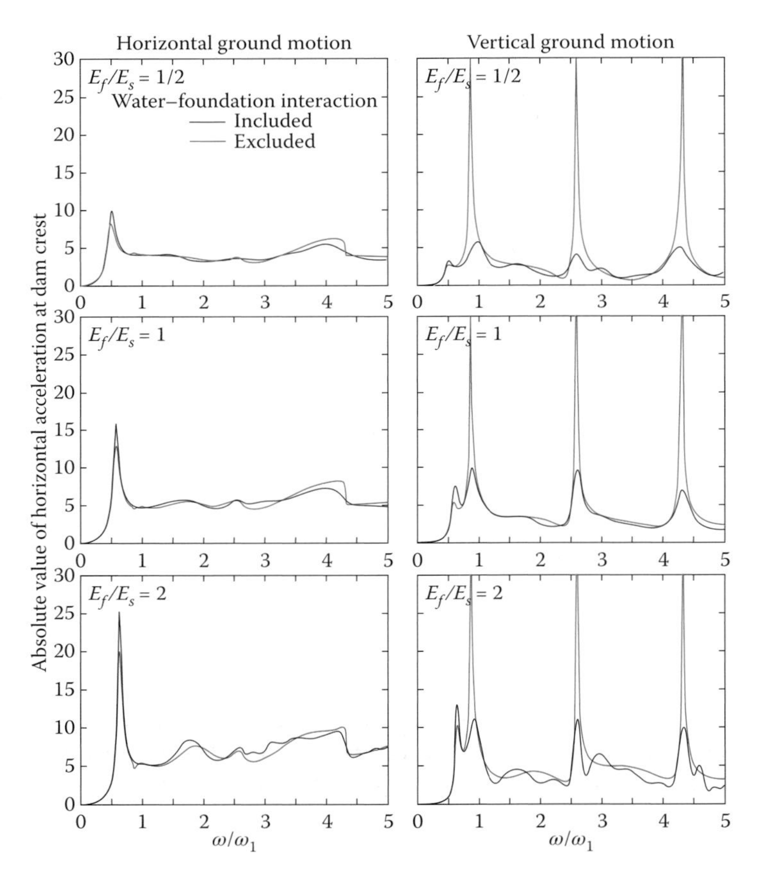

Figure A5.1.1 Effects of water–foundation interaction on frequency response function for dams with full reservoir but no sediments, due to horizontal and vertical ground motions. Results are presented for three values of E_f/E_s = ½, 1, and 2.

† System properties are as follows; Concrete: E_s = 22.4 GPa, density = 2483 kg/m^3, Poisson's ratio = 0.2, viscous damping = 2%. Rock: E_s/E_s = ½, 1 and 2; = 22.4 GPa, density = 2643 kg/m^3, Poisson's ratio = 0.33, viscous damping = 2%. Height of dam = 120 m. Reservoir depth = 120 m.

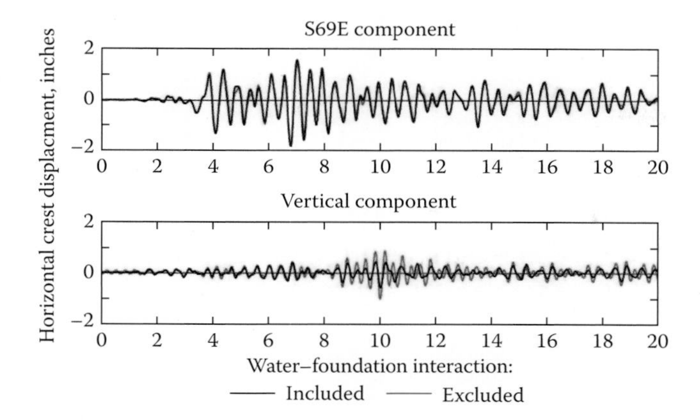

Figure A5.1.2 Effects of water–foundation interaction on earthquake response of a dam with full reservoir but no sediments, $E_f/E_s = 1$. Responses presented are for two excitations: S69E component and vertical component of Taft (1952) ground motion.

case but excluded in the other. Results are presented for both components of ground motion and three values of E_f/E_s.

These results demonstrate that water–foundation interaction has negligible influence on the response to horizontal ground motion at all excitation frequencies, except those very close to the resonant frequencies of the system, where the influence is noticeable. However this influence is not significant as confirmed by the two sets of earthquake response histories presented in Figure A5.1.2a, which are essentially identical.

The preceding observation based on frequency response functions is generally valid also for response to vertical ground motion but for one important exception. At the resonant frequencies ω_n^r of the reservoir alone, the response is greatly overestimated if water–foundation interaction is ignored. Because this discrepancy occurs within a narrow range of frequencies around ω_n^r, the overestimate is not as large in the earthquake response of the dam (Figure A5.1.2b) to this particular excitation; this observation is expected to be valid for most ground motions. These results support the approximation of neglecting water–foundation interaction in the substructure method.

A5.2 Modeling Water–Foundation Interaction

Although the errors introduced by neglecting water–foundation interaction are not large, they can be essentially eliminated by a simple model that can conveniently be included in the substructure method. For this purpose, water–foundation interaction is modeled approximately by the wave-reflection coefficient α introduced in Chapter 2 to model a wave absorptive reservoir bottom:

$$\alpha = \frac{1 - \xi C}{1 + \xi C} \tag{A5.2.1}$$

where

$$\xi = \frac{\rho}{\rho_r C_r} \tag{A5.2.2}$$

C is the speed of pressure waves in water = 4720 fps or 1480 mps; ρ is the density of water; $C_r = \sqrt{E_r/\rho_r}$ is the compression wave velocity in rock; E_r is the Young's modulus; and ρ_r is the density. Observe that α is now determined from the properties of rock, not the sediments.

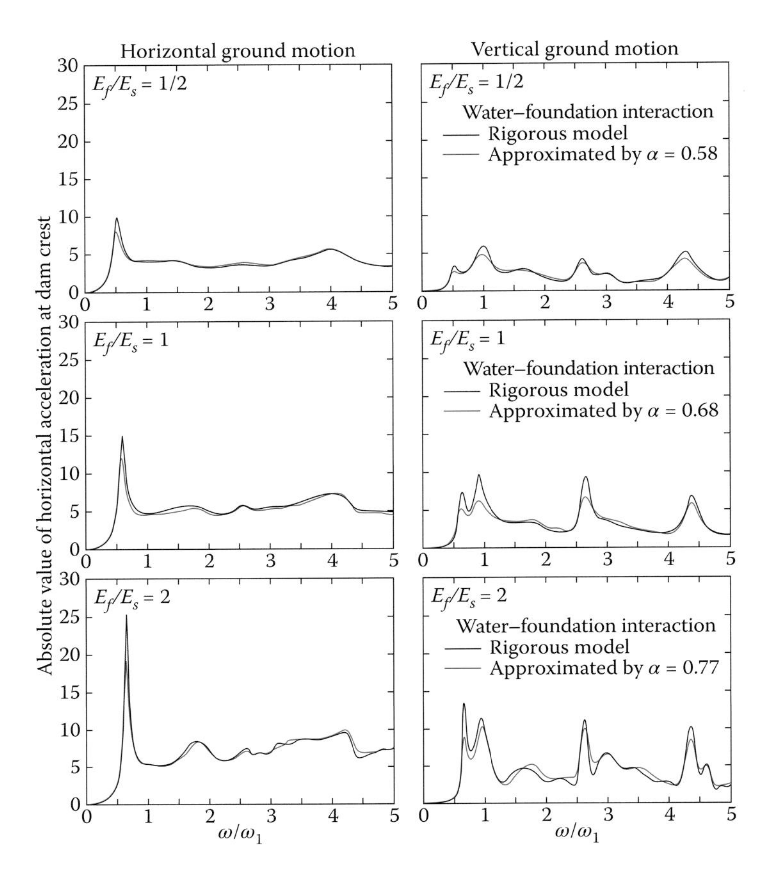

Figure A5.2.1 Frequency response functions for dam with full reservoir but no sediments due to horizontal and vertical ground motions. Results are presented for two models for water–foundation interaction: (i) rigorous and (ii) approximate α-model.

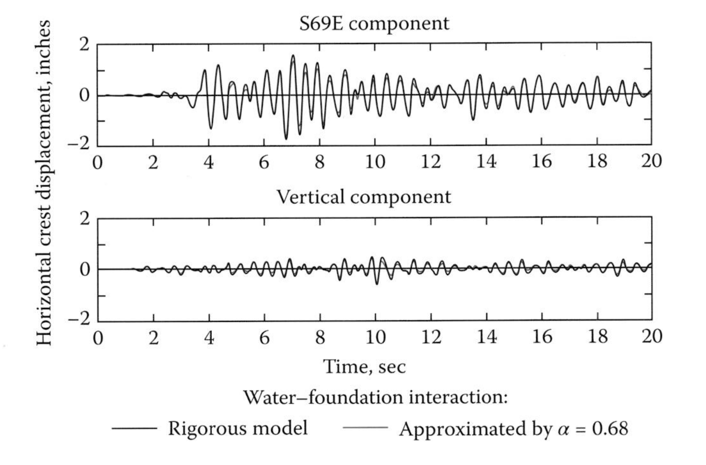

Figure A5.2.2 Earthquake response of dam with full reservoir but no sediments, $E_f/E_s = 1$ to S69E and vertical components of Taft (1952) ground motion. Results are presented for two models for water–foundation interaction: (i) rigorous and (ii) approximate α-model.

Determined by the direct FEM (Chapter 11), the frequency response functions for a gravity dam with full reservoir but no sediments at the reservoir bottom are presented in Figure A5.2.1 for two models for water–foundation interaction: (i) rigorous and (ii) approximate α-model with α values as noted in the figure: $\alpha = 0.58$, 0.68, and 0.77 for $E_f/E_s = ½$, 1, and 2 respectively. These results demonstrate that, near the resonant frequencies, the response is underestimated by the α-model; at all other excitation frequencies, the α-model is highly accurate. However, the α-model – with the value of α based on rock properties – is able to predict the response to earthquake ground motion quite accurately. This is apparent in Figure A5.2.2.

6

Dam–Water–Foundation Interaction Effects in Earthquake Response

PREVIEW

The effects of dam–water–foundation interaction on frequency response functions at lower excitation frequencies – a range including the fundamental natural frequencies of the dam alone and of impounded water – were identified in Chapters 2 and 3. The analysis procedure presented in Chapter 5 includes the contributions of all significant vibration modes, thus making it possible to identify the effects of dam–water–foundation interaction over the complete range of excitation frequencies (Fenves and Chopra 1985a), not included herein, and in response to earthquake ground motion, which is the subject of this chapter.

The responses of Pine Flat Dam to Taft ground motion are presented under different assumptions for the foundation – rigid or flexible – and for the reservoir – empty or full – and a range of properties of the wave-absorptive reservoir bottom. Based on these results, we identify the effects of dam–water interaction, reservoir-bottom absorption, and dam–foundation interaction on the earthquake response of concrete gravity dams.

6.1 SYSTEM, GROUND MOTION, CASES ANALYZED, AND SPECTRAL ORDINATES

6.1.1 Pine Flat Dam

Located on King's River in central California, Pine Flat Dam, a 400-ft-high concrete gravity dam, is constructed of 36 monoliths and has a total crest length of 1850 ft (Figure 6.1.1a). The system analyzed is the tallest non-overflow monolith of this dam (Figure 6.1.1b), with the dam and foundation assumed to be in a state of plane stress; a basis for this assumption was presented in Section 5.1.1. The two-dimensional finite element idealization for this monolith, shown in Figure 6.1.2, consists of 136 quadrilateral elements with 162 nodal points. The finite

Earthquake Engineering for Concrete Dams: Analysis, Design, and Evaluation, First Edition. Anil K. Chopra.

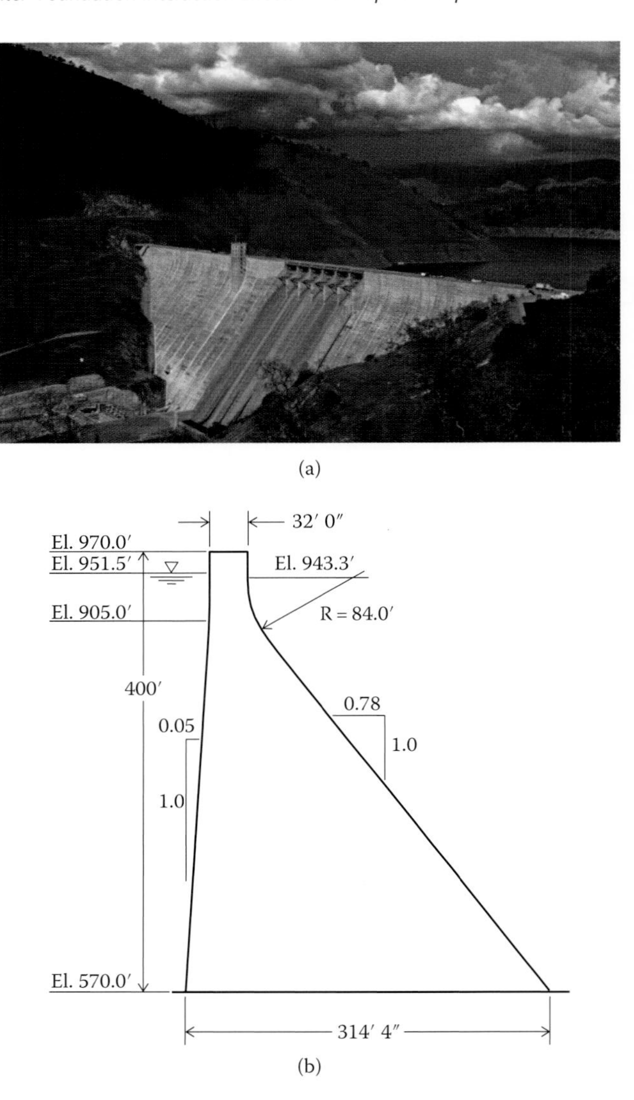

Figure 6.1.1 (a) Pine Flat Dam, near Fresno, California; and (b) tallest, non-overflow monolith of Pine Flat Dam.

element idealization has 306 degrees of freedom (DOFs) if the foundation is assumed to be rigid, and 324 DOF if dam–foundation interaction is considered. The mass concrete in the dam is assumed to be a homogeneous, isotropic, linear elastic solid with the following properties based, in part, on forced vibration tests of the dam (Rea et al. 1975); Young's modulus of elasticity E_s = 3.25 million psi, unit weight = 155 lb/ft^3, and Poisson's ratio = 0.2. Energy dissipation in the dam is represented by a constant hysteretic damping factor, η_s = 0.04, which corresponds to a viscous damping ratio of 2% in all natural vibration modes of the dam (without impounded water) on rigid foundation.

The material properties assumed for the foundation rock are as follows: Young's modulus of elasticity E_f = 3.25 million psi, unit weight = 165 lb/ft^3, Poisson's ratio = 1⁄3, and a constant

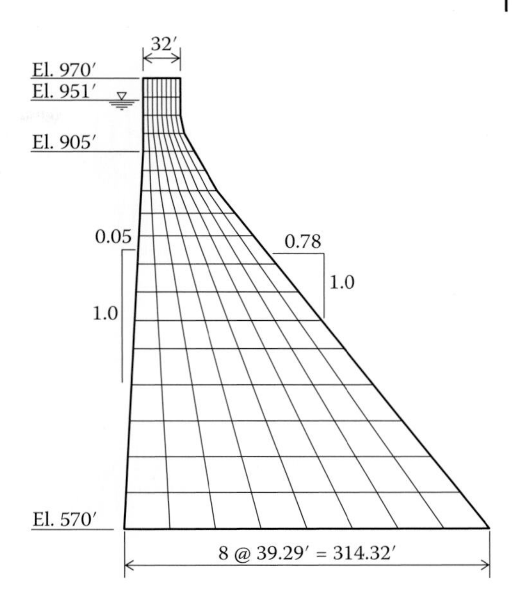

Figure 6.1.2 Finite element idealization of tallest, non-overflow monolith of Pine Flat Dam.

hysteretic damping factor of 0.04. The impounded water is assumed to have the following properties; velocity of pressure waves $C = 4720$ ft/sec, and unit weight = 62.4 lb/ft^3. With the water surface at El. 951 (Figure 6.1.2), the depth of water is 381 ft. Three different values of the wave reflection coefficient characterizing reservoir bottom absorption are considered; in order of increasing absorptiveness, they are $\alpha = 1.0$ (non-absorptive reservoir bottom), 0.50, and zero.

6.1.2 Ground Motion

The ground motion recorded at Taft Lincoln School Tunnel during the Kern County, California, earthquake of July 21, 1952, is selected as the free-field ground acceleration for analysis of Pine Flat Dam. The ground motion acting in the horizontal direction, transverse to the axis of the dam, and in the vertical direction is defined as the S69E and vertical components of the recorded ground motion, respectively. These two components and their peak values of acceleration are shown in Figure 6.1.3.

6.1.3 Cases Analyzed and Response Results

To identify the effects of dam–water interaction, reservoir-bottom absorption, and dam–foundation interaction, the selected dam–water–foundation system was analyzed – using EAGD-84 computer program – for eight sets of assumptions and conditions listed in Table 6.1.1. For each of the eight cases the response of the dam was computed for two excitations: horizontal component only, and vertical component only (Figure 6.1.3); initial stresses prior to the earthquake were excluded.

The fundamental resonant period and effective damping ratio, determined by the half-power bandwidth method – both obtained from the frequency response function for horizontal ground motion – are listed in Table 6.1.1, along with the corresponding pseudo-acceleration $A(\widetilde{T}_1, \widetilde{\zeta}_1)$ obtained from the response spectrum for S69E component of Taft ground motion.

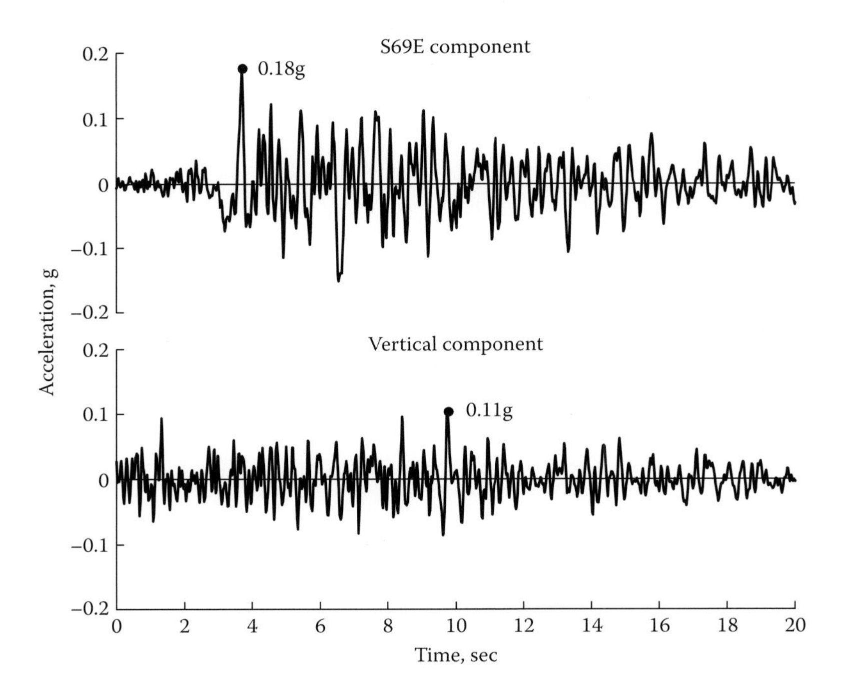

Figure 6.1.3 S69E and vertical components of ground motion recorded at Taft Lincoln School Tunnel, during the Kern County, California, earthquake July 21, 1952.

Table 6.1.1 Cases of Pine Flat Dam analyzed, fundamental mode properties, and pseudo-acceleration ordinates.

Case	Foundation	Reservoir	α	Resonant period $\tilde{T}_1$, sec	Damping ratio, $\tilde{\zeta}_1$, %	$A\left(\tilde{T}_1, \tilde{\zeta}_1\right) \div g$
1	Rigid	Empty	—	0.318	2.0	0.67
2	Rigid	Full	1.0	0.397	1.6	0.52
3	Rigid	Full	0.5	0.390	4.8	0.35
4	Rigid	Full	0	0.378	4.6	0.35
5	Flexible	Empty	—	0.390	8.7	0.30
6	Flexible	Full	1.0	0.493	8.7	0.32
7	Flexible	Full	0.5	0.488	11.3	0.30
8	Flexible	Full	0	0.471	12.8	0.30

The results of the computer analyses consist of the response history of horizontal and vertical displacements at all the nodal points, and the three components of plane stress in all the finite elements. We will be especially interested in the peak value of a response quantity, or for brevity, *peak response* defined as the maximum of the absolute value of the response quantify: $r_0 \equiv \max_t |r(t)|$. The peak values of horizontal displacement at the crest of the dam and of the maximum principal stresses at three critical locations in the dam monolith are summarized in

Table 6.2.1 for the dam supported on rigid foundation (Cases 1–4), and in Table 6.3.1 for the dam supported on flexible foundation (Cases 5–8); for the latter cases dam–foundation interaction was included in the analysis. The history of horizontal displacement at the dam crest is shown in Figures 6.2.1 and 6.3.1, and the distribution of envelope values of the maximum principal stresses in the dam monolith (positive is maximum tensile stress, negative is minimum compressive stress) in Figures 6.2.2 and 6.3.2. By envelope values we mean the peak values of stress at every location; obviously these do not represent the state of stress at any one time instant because, in general, the peak stresses at different locations occur at different times.

6.2 DAM–WATER INTERACTION

6.2.1 Hydrodynamic Effects

As mentioned in Chapters 2 and 5, interaction between the dam and water impounded in the reservoir introduces frequency-dependent hydrodynamic terms in the equations of motion that affect the dynamic response of the dam. The hydrodynamic terms can be interpreted as an added force (different for horizontal and vertical ground motion), an added mass, and an added damping (Sections 2.4 and 5.6). The added hydrodynamic mass for a full reservoir and non-absorptive reservoir bottom lengthens the fundamental resonant period of the dam from 0.318 to 0.397 sec, and decreases the effective damping ratio at the fundamental resonant period from 2.0% to 1.6% (Table 6.1.1), for reasons explained in Sections 2.5.3 and 2.6.1. The interaction effects on the response of a dam to an earthquake ground motion are controlled, in part, by the change in the response spectrum ordinate corresponding to the change in the fundamental resonant period and damping; these spectral ordinates are shown in Table 6.1.1. The added hydrodynamic force has less effect on the response to horizontal ground motion because it is relatively small compared to the effective earthquake force associated with the mass of the dam. It has been shown (Chakrabarti and Chopra 1974) that dam–water interaction has little effect on the higher resonant frequencies, but reduces the amplitude of the higher resonant peaks because of the added hydrodynamic damping at these higher frequencies.

The earthquake response of the dam shown in Figures 6.2.1 and 6.2.2 includes results for empty reservoir and full reservoir with non-absorptive reservoir bottom. In part, because of the lengthened fundamental resonant period due to dam–water interaction, the peak crest displacement increases from 1.53 to 2.10 in. (Figure 6.2.1), the peak value of maximum principal stress increases from 344 to 397 psi at the upstream face, from 399 to 481 psi at the downstream face, and from 354 to 527 psi at the heel (Table 6.2.1 and Figure 6.2.2a,b). The area enclosed by a particular stress contour increases, indicating that tensile stresses exceed the value corresponding to that contour over a larger portion of the monolith because of dam–water interaction effects (Figure 6.2.2). The general distribution pattern of maximum principal stresses over the dam monolith does not change substantially.

In contrast to horizontal ground motion, the added hydrodynamic force associated with vertical ground motion, which acts on the dam in the lateral direction, is large over the entire range of excitation frequencies compared to the effective earthquake force associated with the mass of the dam (Section 2.5.3). In particular, if the reservoir bottom is non-absorptive, the added force due to vertical ground motion is infinite at the natural vibration frequencies of the impounded water, leading to unbounded peaks in the frequency response function for the dam (see Section 2.5.3; Chakrabarti and Chopra 1974). The response of the dam with full reservoir and

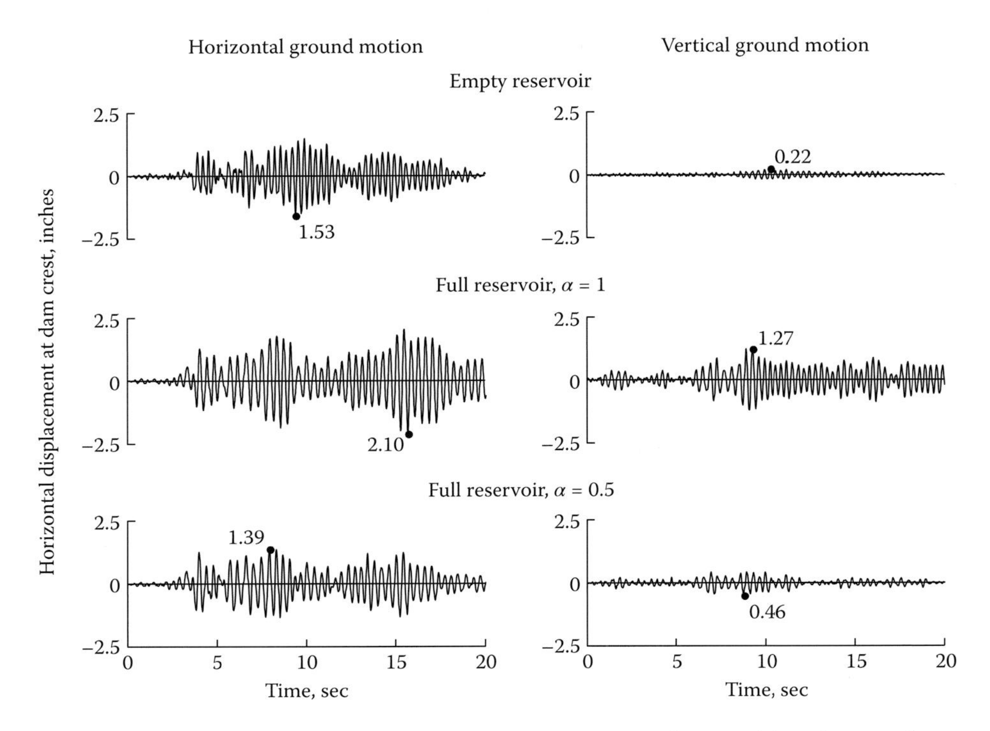

Figure 6.2.1 Displacement response of Pine Flat Dam supported on rigid foundation to horizontal and vertical components of Taft ground motion.

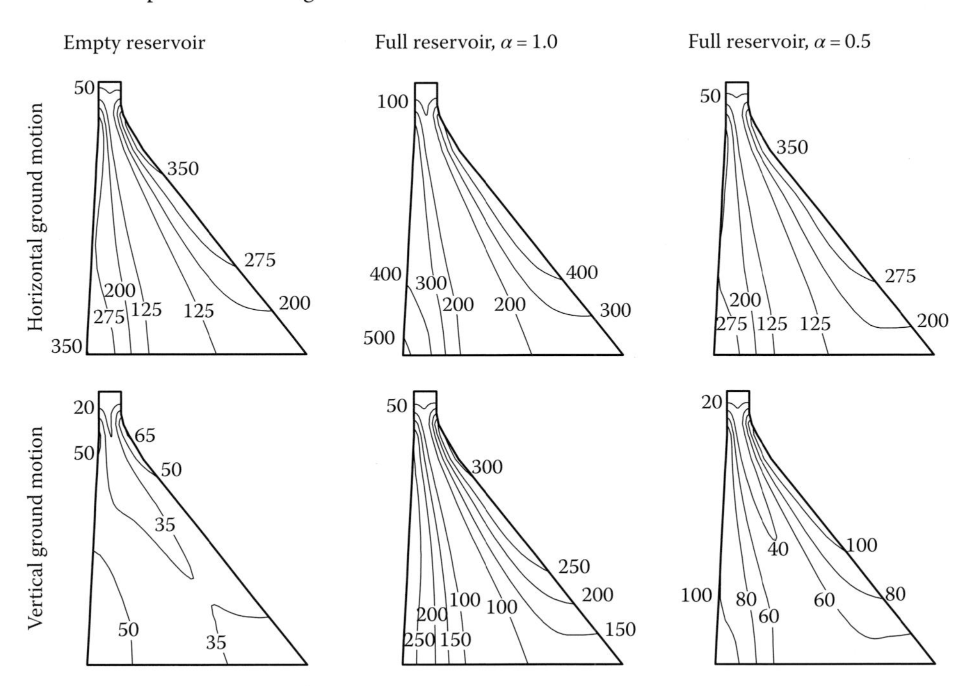

Figure 6.2.2 Envelope values of maximum principal stresses (in psi) in Pine Flat Dam supported on rigid foundation due to horizontal and vertical components of Taft ground motion.

Table 6.2.1 Peak responses of Pine Flat Dam supported on rigid foundation to Taft ground motion.

Case	Reservoir	α	Horizontal crest displ., in.	Maximum principal stress, psi		
				Upstream face	Downstream face	Heel
Horizontal ground motion, only						
1	Empty	—	1.53	344	399	354
2	Full	1.0	2.10	397	481	527
3	Full	0.5	1.39	298	354	346
4	Full	0	1.38	290	308	330
Vertical ground motion, only						
1	Empty	—	0.22	54	68	64
2	Full	1.0	1.27	284	321	272
3	Full	0.5	0.46	100	119	116
4	Full	0	0.26	66	79	66

non-absorptive reservoir bottom to the vertical component of Taft ground motion, clearly shows the large increase in response due to impounded water, with the peak value of crest displacement increasing from 0.22 to 1.27 in. (Figure 6.2.1) accompanied by a similarly large increase in the maximum principal stresses in the dam (Figure 6.2.2 and Table 6.2.1).

The aforementioned results, and others not presented here, demonstrate that generally the earthquake response of dams is increased significantly because of the impounded water, with the magnitude of the hydrodynamic effects depending on the component of ground motion. In particular, the response to vertical ground motion increases greatly, thus becoming a larger percentage of the response to horizontal ground motion.

6.2.2 Reservoir Bottom Absorption Effects

As first noted in Section 2.5, reservoir bottom absorption affects all the hydrodynamic terms in the equations of motion, and, therefore, the response of the dam to earthquake ground motion. As the reservoir bottom becomes more absorptive, i.e. as the wave reflection coefficient α decreases, the added damping at the fundamental resonant period increases because of increasing absorption of hydrodynamic pressure waves at the reservoir bottom, resulting in increased effective damping ratio (Table 6.1.1). Reservoir bottom absorption is known primarily to affect the fundamental resonant response of the dam, but has little effect on the response at higher excitation frequencies (Fenves and Chopra 1985a).

The earthquake response of the dam with full reservoir and absorptive reservoir bottom to the horizontal component of Taft ground motion is shown in Figure 6.2.1. These results demonstrate that the principal effect of reservoir bottom absorption is to reduce the peaks in the displacement response without a significant change in the frequency content of the response. Because of added damping due to reservoir bottom absorption, the peak displacement of the dam decreases from 2.10 to 1.39 in. for $\alpha = 0.5$; the maximum principal stress at the downstream face decreases from 481 to 354 psi, and the maximum principal stress at the upstream face decreases similarly (Figure 6.2.2 and Table 6.2.1). The area enclosed by a particular stress contour decreases, indicating that tensile stresses exceed the value corresponding to that contour over a smaller portion of the monolith because of reservoir bottom absorption; however, the general pattern of maximum principal stresses is not substantially altered.

Reservoir bottom absorption reduces the added hydrodynamic force and the dam response to vertical ground motion at all excitation frequencies, and, in particular, eliminates the unbounded peaks in the added hydrodynamic force (Section 2.3.3) and in the dam response at excitation frequencies equal to the natural vibration frequencies of the impounded water (see Section 2.5.3; Fenves and Chopra 1985a). Consequently, reservoir bottom absorption drastically reduces the peak displacement from 1.27 to 0.46 in. for $\alpha = 0.5$ (Figure 6.2.1), and similarly reduces the peak stresses in the dam monolith (Figure 6.2.2 and Table 6.2.1). For example, the maximum principal stress at the downstream face is reduced from 321 to 119 psi.

It is therefore concluded that reservoir bottom absorption moderates the increase in response of the dam due to its interaction with the impounded water, increases the effective damping ratio, and reduces the response to vertical ground motion as a percentage of the response to horizontal ground motion. This implies that a non-absorptive reservoir bottom leads to unrealistically large response for dams with impounded water, particularly due to vertical ground motion.

6.2.3 Implications of Ignoring Water Compressibility

As seen in Chapter 5, if compressibility of water is considered, analysis techniques based on the substructure method are formulated in the frequency domain; thus specialized computer software, such as EAGD-84, is required for such analyses. On the other hand, if water compressibility is ignored, hydrodynamic effects may be modeled approximately by an added mass of water moving with the dam (Section 2.3.4); consequently, standard finite-element analysis software is applicable, and that is attractive. But is it reasonable to ignore compressibility of water?

As mentioned in Section 2.5.4, the key parameter that determines the significance of water compressibility is $\Omega_r = \omega_1^r/\omega_1$, where ω_1 and ω_1^r are the fundamental natural vibration frequencies of the dam and impounded water, respectively. Therein, the frequency response functions for the dam considering only its fundamental mode of vibration demonstrated that water compressibility effects are significant in the response of concrete gravity dams to harmonic ground motion if $\Omega_r < 2$, but they become insignificant for $\Omega_r > 2$. This conclusion is reconfirmed in this section where we present the response to earthquake ground motion of two hypothetical dams; both with a cross section the same as Pine Flat Dam, one with a realistic value of elastic modulus, E_s = 4 million psi, and the other with an unrealistically small value of E_s = 0.65 million psi. For each dam, results are presented for two cases: (i) including compressibility of water (with $\alpha = 0.9$ in the latter case); and (ii) neglecting compressibility of water.

Water compressibility plays an important role in the response of dams with realistic value of elastic modulus. For the Pine Flat Dam cross section with E_s = 4.0 million psi, neglecting water compressibility overestimates the crest displacement due to horizontal ground motion by 47% (2.34 in. versus 1.59 in.; see Figure 6.2.3), and stresses by 31–57%, depending on the location (Figure 6.2.4). In contrast, the response to vertical ground motion is underestimated by a factor of 5 if compressibility of water is neglected; such underestimation is likely to occur in most cases.

In some cases, response to horizontal ground motion may be underestimated if water compressibility is neglected. This is demonstrated in Figures 6.2.5 and 6.2.6 wherein the earthquake response of a 150-ft-high dam with E_s = 4.0 million psi is presented. Neglecting water compressibility underestimates the crest displacement by 20% (0.139 in. versus 0.111 in.) and stresses by 13–21%, depending on the location.

As expected from the discussion in Section 2.5, the effects of water compressibility become much smaller in the earthquake response of dams with very low values of elastic modulus.

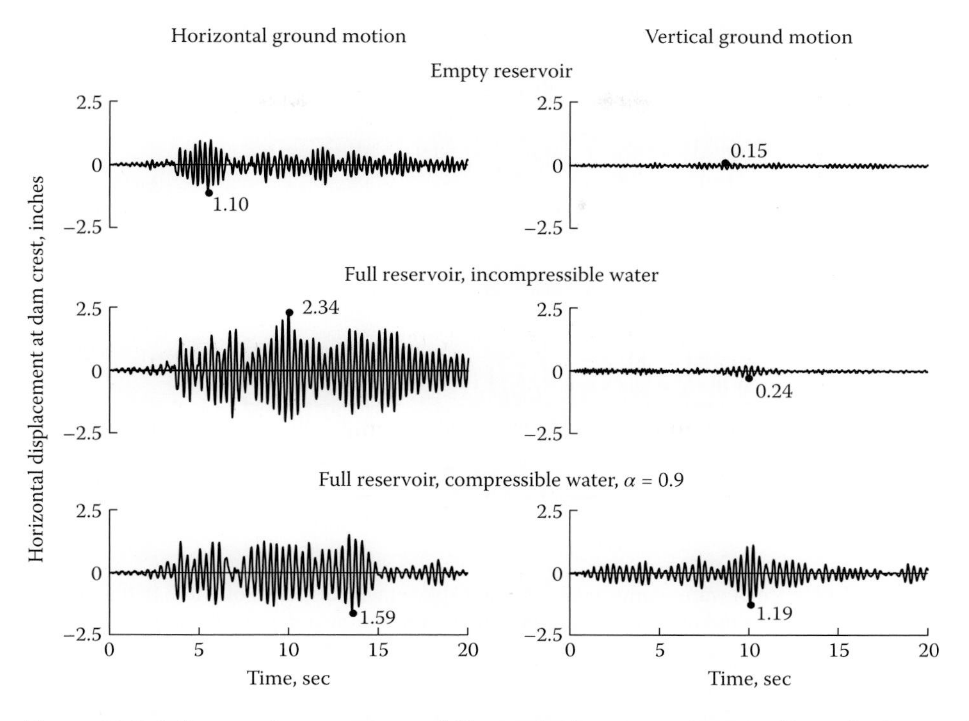

Figure 6.2.3 Influence of water compressibility on displacement response of Pine Flat Dam with $E_s = 4$ million psi supported on rigid foundation to horizontal and vertical components of Taft ground motion; empty reservoir case is included for reference.

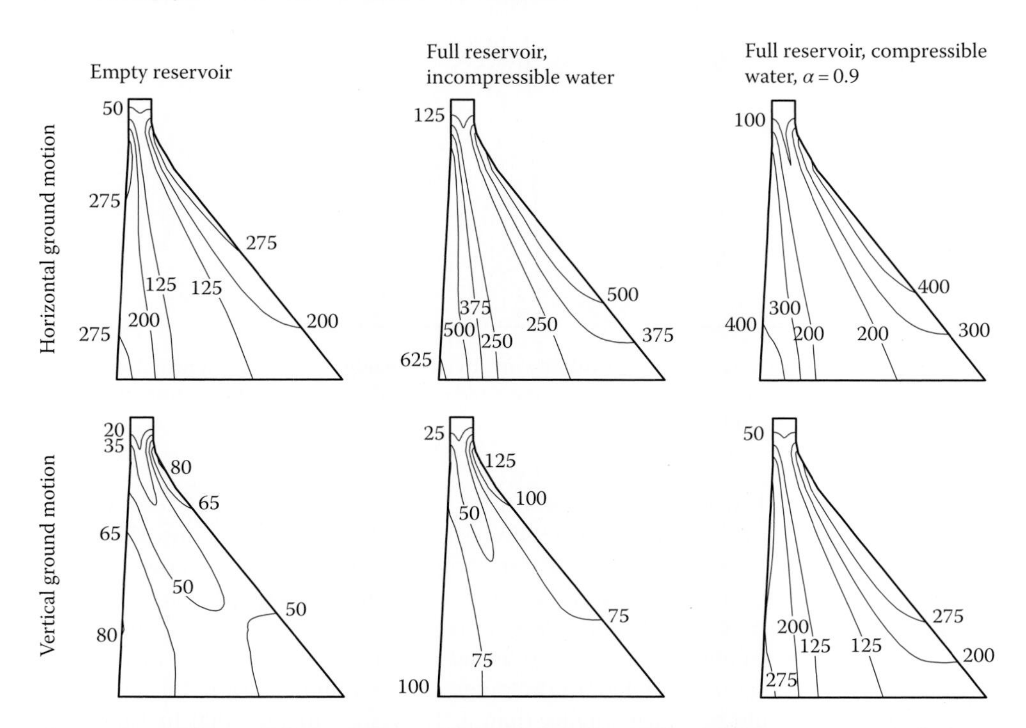

Figure 6.2.4 Influence of water compressibility on envelope values of maximum principal stresses, in psi, in Pine Flat Dam with $E_s = 4$ million psi supported on rigid foundation due to horizontal and vertical components of Taft ground motion; empty reservoir case is included for reference.

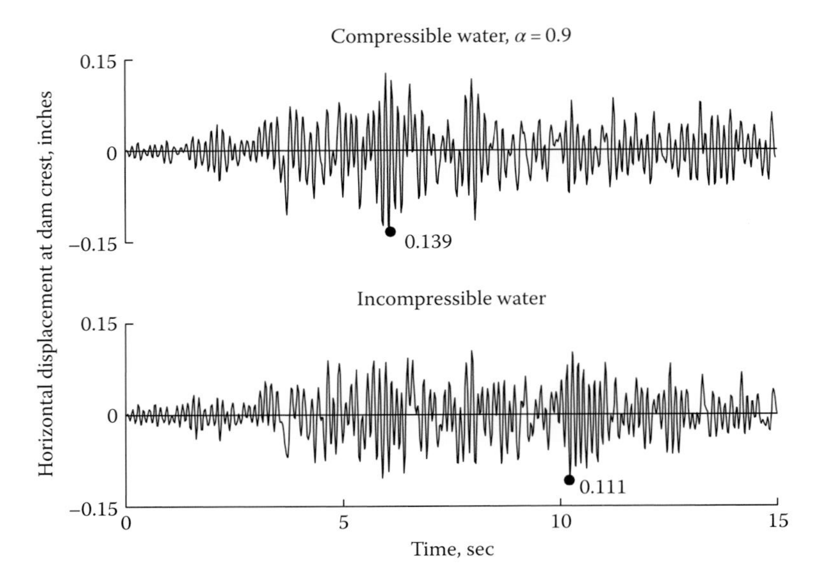

Figure 6.2.5 Influence of water compressibility on displacement response of a 150-ft-high dam with $E_s = 4$ million psi supported on rigid foundation due to horizontal and vertical components of Taft ground motion.

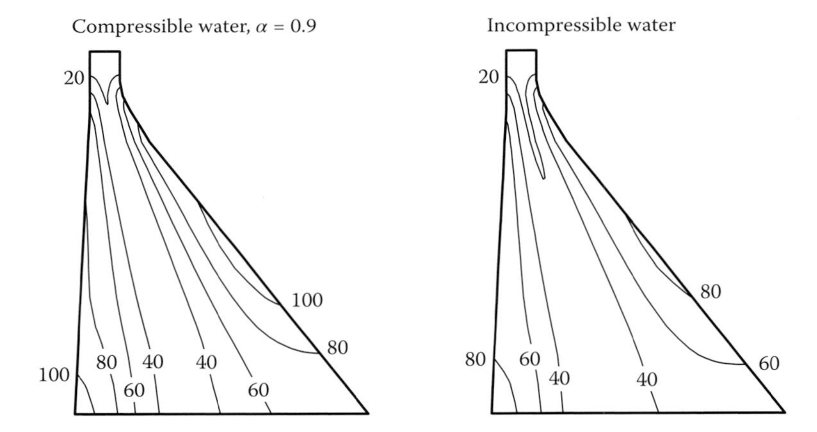

Figure 6.2.6 Influence of water compressibility on envelope values of maximum principal stresses, in psi, in a 150-ft-high dam with $E_s = 4$ million psi supported on rigid foundation due to horizontal and vertical components of Taft ground motion.

This is demonstrated in Figures 6.2.7 and 6.2.8 wherein the earthquake response of the Pine Flat Dam cross section with $E_s = 0.65$ million psi is presented; for this unrealistically small E_s, the corresponding value of $\Omega_r = 2.2$. Water compressibility effects are now very small in the response to upstream ground motion; however, even for this very low elastic modulus, water compressibility influences significantly the dam response to vertical ground motion.

Water compressibility is expected to be significant in the response of most gravity dams because E_s is generally much higher, in the range of 2–5 million psi, and for dams with close to full reservoir, Ω_r would be much smaller than 2. Thus, we conclude that hydrodynamic effects should not be modeled by an added mass of water vibrating with the dam – a model that ignores water incompressibility – in standard finite element analysis of dams.

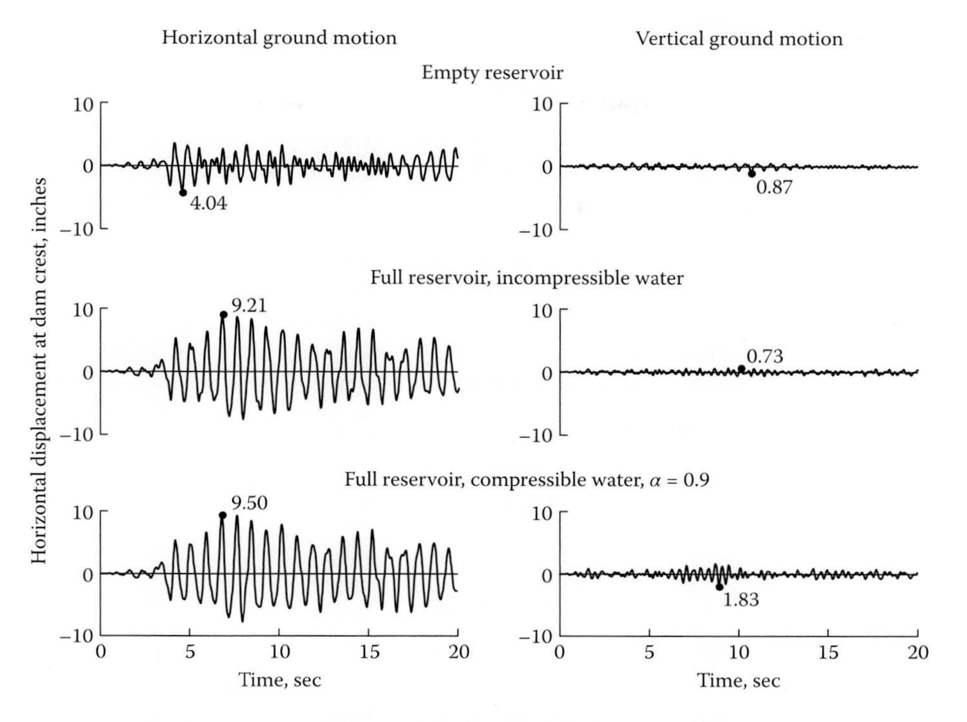

Figure 6.2.7 Influence of water compressibility on displacement response of Pine Flat Dam with $E_s = 0.65$ million psi supported on rigid foundation to horizontal and vertical components of Taft ground motion; empty reservoir case is included for reference.

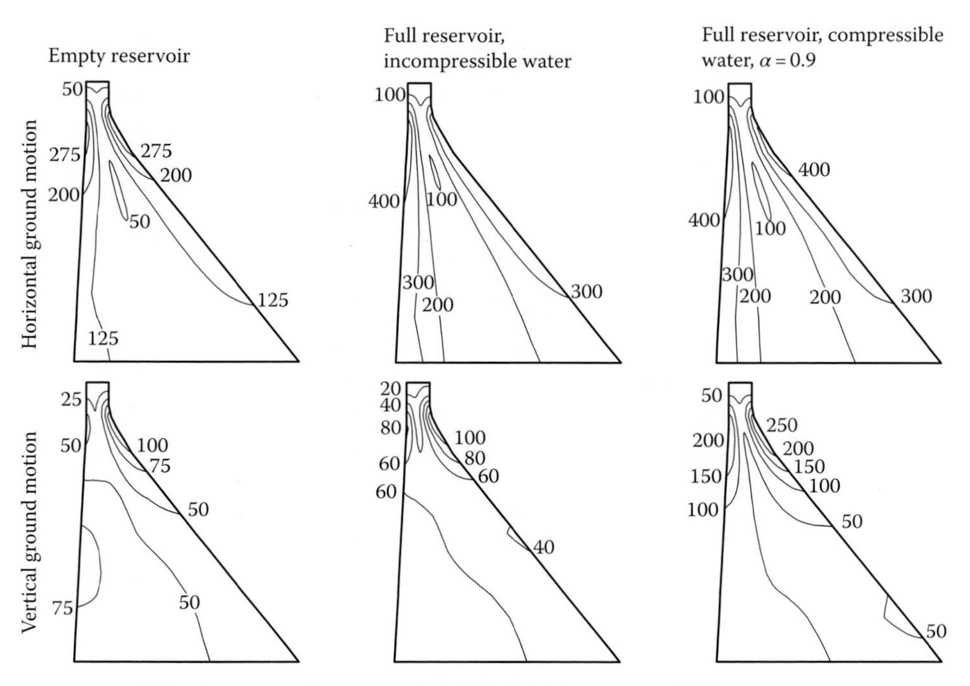

Figure 6.2.8 Influence of water compressibility on envelope values of maximum principal stresses, in psi, in Pine Flat Dam with $E_s = 0.65$ million psi supported on rigid foundation due to horizontal and vertical components of Taft ground motion; empty reservoir case is included for reference.

6.3 DAM–FOUNDATION INTERACTION

6.3.1 Dam–Foundation Interaction Effects

Dam–foundation interaction lengthens the fundamental resonant period of Pine Flat Dam from 0.318 to 0.390 sec because of foundation-rock flexibility and mass; and increases the effective damping from 2.0% to 8.7% at that period because of energy radiation and material damping in the foundation (Table 6.1.1). The effects of increased damping and lengthened fundamental resonant period can be seen in the response of the dam with an empty reservoir to horizontal ground motion. Dam–foundation interaction reduces the crest displacement from 1.53 in. (Figure 6.2.1) to 1.14 in. (Table 6.3.1, Figure 6.3.1) and also the maximum principal stresses throughout the dam monolith, as seen by comparing Figure 6.3.2 to Figure 6.2.2 and Case 5 in Table 6.3.1 to Case 1 in Table 6.2.1. The maximum principal stress reduces from 344 to 185 psi at the upstream face, from 399 to 199 psi at the downstream face, and from 354 to 228 psi at the heel.

Dam–foundation interaction reduces significantly the response of the dam to vertical ground motion whether the reservoir is empty or full, and the reservoir bottom is absorptive or non-absorptive, as seen by comparing (Figure 6.3.2 to Figure 6.2.2 and Case 5 in Table 6.3.1 to Case 1 in Table 6.2.1).

6.3.2 Implications of Ignoring Foundation Mass

As seen in Chapter 5, dam–foundation interaction introduces frequency-dependent (dynamic) stiffness matrix for the foundation, thus requiring analysis of the dam by a substructure method formulated in the frequency domain, implemented in specialized computer software, such as EAGD-84. The temptation to use commercial finite-element analysis software instead has motivated the practice of ignoring the mass of the foundation rock (Clough 1980). Can this seemingly unrealistic assumption be justified?

When rock is assumed to have no mass, radiation and material damping mechanisms characteristic of dam–foundation interaction (see Section 3.3) do not develop, resulting in

Table 6.3.1 Peak responses of Pine Flat Dam including dam–foundation interaction to Taft ground motion.

				Maximum principal stress, psi		
Case	Reservoir	α	Horizontal crest displ, in.	Upstream face	Downstream face	Heel
			Horizontal ground motion, only			
1	Empty	—	1.14	185	199	228
2	Full	1.0	1.89	277	288	417
3	Full	0.5	1.76	259	266	391
4	Full	0	1.74	256	270	385
			Vertical ground motion, only			
5	Empty	—	0.17	27	39	34
6	Full	1.0	1.04	219	222	155
7	Full	0.5	0.45	88	88	71
8	Full	0	0.25	47	64	44

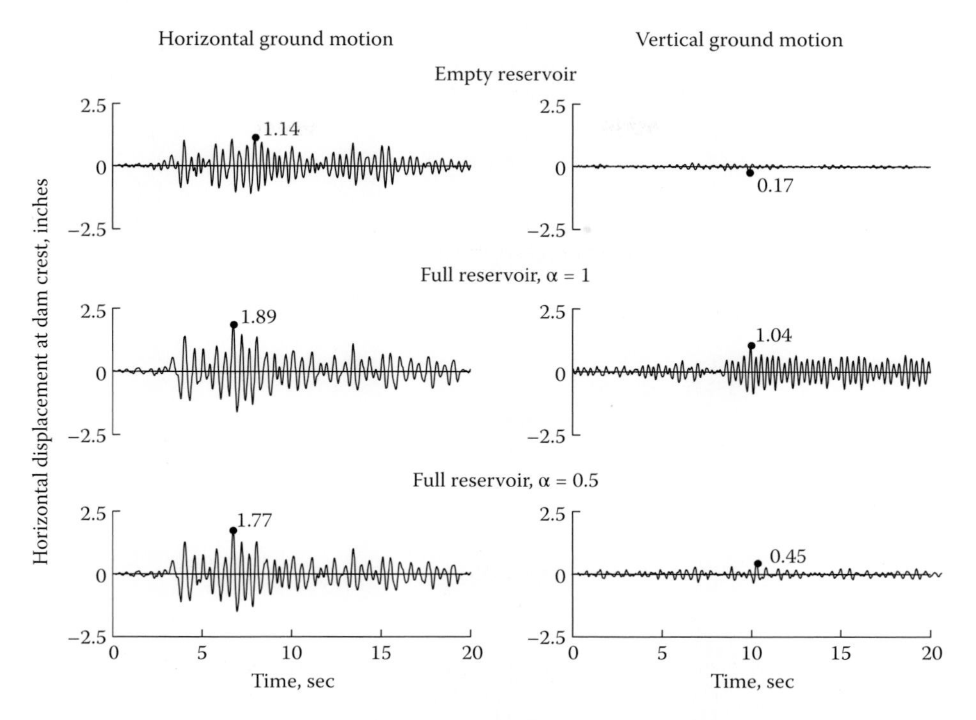

Figure 6.3.1 Displacement response of Pine Flat Dam including dam–foundation interaction to horizontal and vertical components of Taft ground motion.

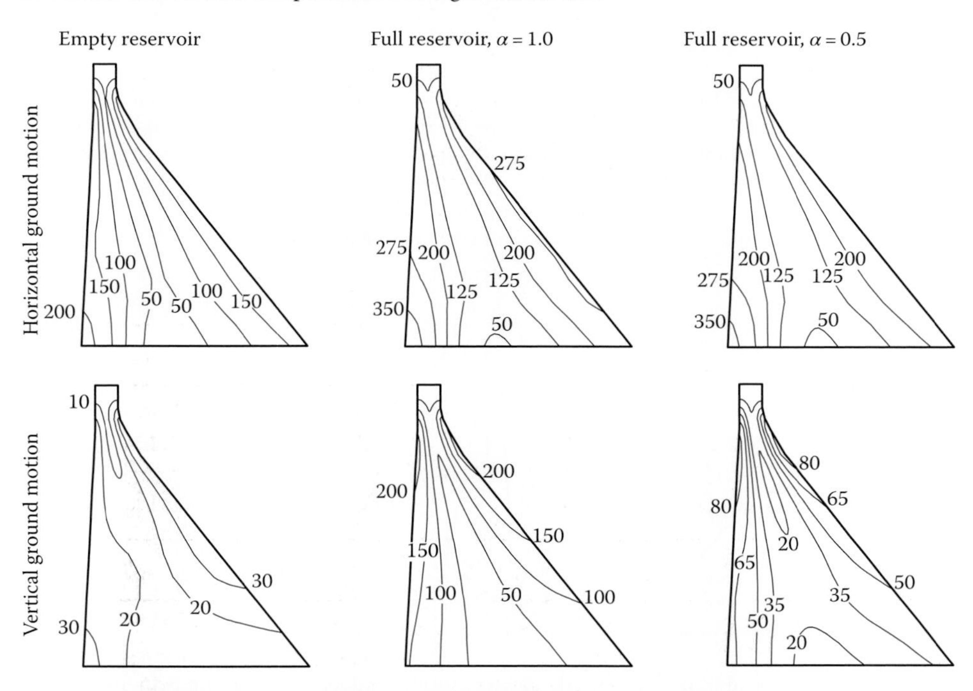

Figure 6.3.2 Envelope values of maximum principal stresses (in psi) in Pine Flat Dam including dam–foundation interaction due to horizontal and vertical components of Taft ground motion.

lower damping, and hence larger spectral ordinate. This is evident from the results presented for two cases: (i) including all effects of dam–foundation interaction (Case 5 in Table 6.1.1); and (ii) ignoring the mass of foundation rock, i.e. considering foundation flexibility only. The resonant period, $\widetilde{T}_1 = 0.415$ sec, damping ratio $\widetilde{\zeta}_1 = 6.3\%$, and the pseudo-acceleration ordinate, $A(\widetilde{T}_1, \widetilde{\zeta}_1) = 0.43$ g if rock is assumed massless, in contrast to $\widetilde{T}_1 = 0.390$ sec, $\widetilde{\zeta}_1 = 8.7\%$, and $A(\widetilde{T}_1, \widetilde{\zeta}_1) = 0.30$ g if dam–foundation interaction is included. These data suggest that the response of the dam will be overestimated if foundation mass is ignored. This expectation is confirmed by the results of response history analysis for the two cases: Figures 6.3.3 and 6.3.4

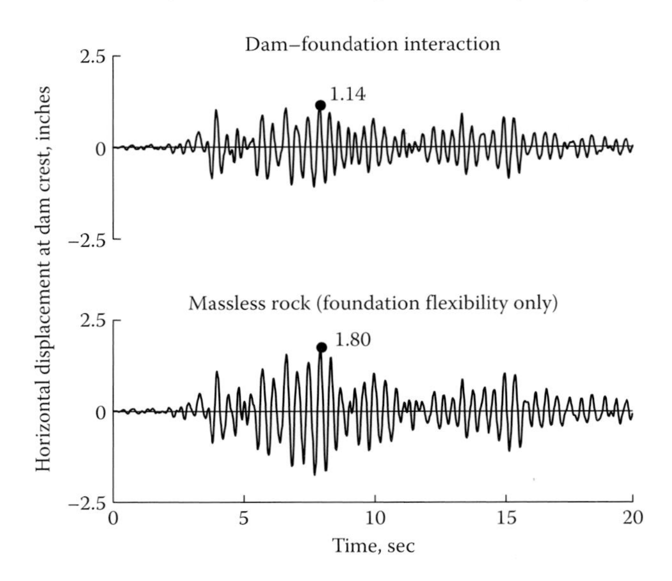

Figure 6.3.3 Influence of foundation modeling on displacement response of Pine Flat Dam to horizontal component of Taft ground motion. Results are presented for two cases: (a) including all effects of dam–foundation interaction; and (b) assuming rock to be massless, i.e. considering foundation flexibility only.

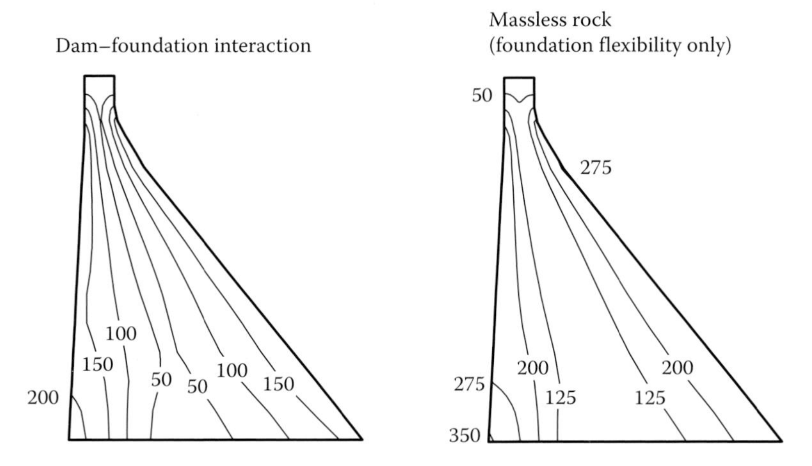

Figure 6.3.4 Influence of foundation modeling on envelope values of maximum principal stresses, in psi, in Pine Flat Dam due to horizontal ground motion. Results are presented for two cases: (a) including all effects of dam–foundation interaction; (b) and assuming rock to be massless, i.e. considering foundation flexibility only.

show the crest displacement history and envelope of maximum principal stresses, respectively. By assuming rock to be massless, the crest displacement is overestimated by 58% and the principal stresses by 23–82%, depending on the location; obviously, the degree of overestimation will vary with the dam and ground motion.

In many cases such excessive overestimation of stresses will lead to overly conservative – and hence unnecessarily expensive – designs for new dams, and to the erroneous conclusion that an existing dam is unsafe, thus requiring unnecessary retrofit that is invariably very expensive.

6.4 DAM–WATER–FOUNDATION INTERACTION EFFECTS

As noted in the preceding sections, the fundamental resonant period of the dam is lengthened because of dam–water interaction and also because of dam–foundation interaction. Simultaneous consideration of the two sources of interaction results in a fundamental resonant period of the dam that is longer than the period including either interaction individually. Consistent with Eq. (4.1.4c), the data in Table 6.1.1 indicates that dam–water interaction with an absorptive reservoir bottom lengthens the fundamental resonant period of the dam by almost the same percentage, whether the foundation rock is rigid or flexible.

The response of the dam–foundation system with full reservoir and non-absorptive reservoir bottom due to horizontal ground motion is shown in Figures 6.3.1 and 6.3.2, and Table 6.3.1. Because of dam–water interaction the peak crest displacement increases from 1.14 to 1.89 in.; and the peak value of maximum principal stress increases from 185 to 277 psi at the upstream face, from 199 to 288 psi at the downstream face, and from 228 to 417 psi at the heel. The area enclosed by a particular stress contour increases because of the dam–water interaction, with no substantial change in the general pattern of the contours.

As stated in Section 6.2, where the foundation was assumed to be rigid, if the reservoir bottom is non-absorptive, the added hydrodynamic force for vertical ground motion is infinite at the natural vibration frequencies of the impounded water, leading to unbounded peaks in the frequency response function for the dam; these observations remain valid for dam–foundation interacting systems (Fenves and Chopra 1985a). These unbounded peaks cause large amplification in the response of the dam to vertical ground motion, with the peak crest displacement increasing from 0.17 to 1.04 in. (see Figure 6.3.1), accompanied by similarly large increases in the maximum principal stresses (Figure 6.3.2 and Table 6.3.1).

The effects of reservoir bottom absorption on the response of the dam including dam–foundation interaction to horizontal ground motion can be gleaned from Figures 6.3.1 and 6.3.2, and Table 6.3.1. The peak crest displacement of 1.89 in. for non-absorptive reservoir bottom is reduced to 1.77 in. for an absorptive bottom with $\alpha = 0.5$; the time-variation pattern is essentially unaffected, confirming that reservoir bottom absorption increases the effective damping, but the resonant period is essentially unaffected (Table 6.1.1). The reduction in maximum principal stresses in the dam monolith due to reservoir bottom absorption is much less, approximately 6% when dam–foundation interaction is included (Figure 6.3.2 and Table 6.3.1) compared to approximately 25% if the foundation is rigid (Figure 6.2.2 and Table 6.2.1). This observation, which is consistent with the earlier statements concerning the fundamental resonant response, results from the fact that the added damping due to reservoir bottom absorption is less effective because it is in addition to significant damping due to dam–foundation interaction arising from radiation and material damping in the foundation. In general, the effects of reservoir bottom absorption on the response of dams to horizontal ground motion are less important in the presence of dam–foundation interaction than if the foundation is rigid.

Reservoir bottom absorption drastically reduces the response to vertical ground motion irrespective of the foundation condition. With dam–foundation interaction included, for $\alpha = 0.5$ the crest displacement is reduced by 57%, and the maximum principal stresses in the monolith by 55–60% (Figure 6.3.2 and Table 6.3.1); the corresponding reductions for the dam on rigid foundation are only slightly larger. These results, just like those for dams on rigid foundation, demonstrate that reservoir bottom absorption has a greater effect on the response of dams to vertical ground motion than on their response to horizontal ground motion.

In summary, the earthquake response of Pine Flat dam is increased by dam–water interaction and decreased by reservoir bottom absorption with the magnitude of these effects depending on the condition of foundation, rigid or flexible, and on the component of ground motion, horizontal or vertical. In particular, both dam–water interaction and reservoir bottom absorption have a profound effect on the dam response to vertical ground motion irrespective of the foundation condition, but relatively less, although still very significant, effect on the dam response to horizontal ground motion with the magnitude of the effect decreasing further if dam-foundation interaction is included. Stated differently, the response of the dam with an empty reservoir due to vertical ground motion expressed as a percentage of the response to horizontal ground motion is small; the percentage greatly increases because of dam–water interaction with a non-absorptive reservoir bottom, and from this increased value it decreases significantly because of reservoir bottom absorption.

7

Comparison of Computed and Recorded Earthquake Responses of Dams

PREVIEW

The ultimate test of the linear analysis procedure that includes dam–water–foundation interaction, presented in Chapter 6, is to compare the computed responses with motions of dams recorded during earthquakes. After discussing the difficulties of finding good examples, we present one such comparison for Tsuruda Dam in Japan subjected to an earthquake and subsequent aftershocks that caused no damage; hence, this is a case that is amenable to linear analysis. We demonstrate that the analysis procedure can very closely match the recorded motions of the dam. Thereafter, we revive an old (1970s) investigation of linear analysis of Koyna Dam that sustained significant damage from an earthquake. The earthquake performance of Koyna Dam and the subsequent dynamic analysis of the dam are of historical interest because they had an enormous influence on the development of earthquake engineering for concrete gravity dams.

7.1 COMPARISON OF COMPUTED AND RECORDED MOTIONS

7.1.1 Choice of Example

Required for such a comparison is a gravity dam that (i) is tall enough, say, over 100 m, resulting in significant amplification of motion from the base to the crest; (ii) is located in a wide valley, and thus is amenable to two-dimensional analysis of a monolith; (iii) remains essentially elastic during an earthquake that causes significant motions at the site, thus permitting linear analysis; and (iv) includes motions recorded at a minimum of two locations – base and crest – of the dam. Information required for analysis of the dam includes plan, elevation, and cross sections; unit weight and elastic modulus of concrete and rock; and overall damping in the dam–water–foundation

Earthquake Engineering for Concrete Dams: Analysis, Design, and Evaluation, First Edition. Anil K. Chopra.

system. We have not been able to find a single example satisfying all these requirements, so we have chosen one that partially serves our purpose.

7.1.2 Tsuruda Dam and Earthquake Records

Completed in 1966, Tsuruda Dam is straight in plan except for a kink near the right abutment, 450 m long at the crest and 117.5 m high above the deepest part of the foundation; see Figure 7.1.1. Cross sections of the tallest non-overflow and overflow monoliths are shown in Figure 7.1.2.

The site was instrumented with three accelerographs; one in the free field near the abutments (top of the canyon), and two located on the dam (one in the gallery of the dam close to the foundation level and one at the crest of the dam). Unfortunately, the precise location of these accelerographs is unknown. Lacking a free-field record near the base of the dam, the only avenue remaining is to treat the motion recorded in the lower gallery as the free-field motion.

This instrument array recorded motions during an earthquake of magnitude 6.6 centered at a distance of 13 km from the dam, and five aftershocks of magnitude in the range 4.3–6.4. The

Figure 7.1.1 Tsuruda Dam, Japan. Source: Credit; Sendaigawa River Office, Kyushu Bureau, Ministry of Land, Infrastructure, Transportation, and Tourism.

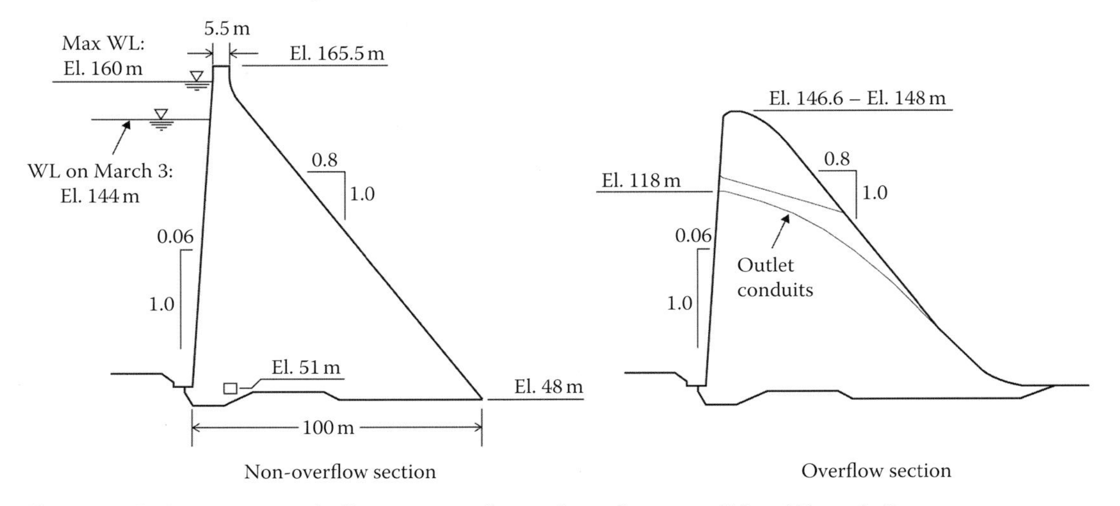

Figure 7.1.2 Cross section of tallest non-overflow and overflow monoliths of Tsuruda Dam.

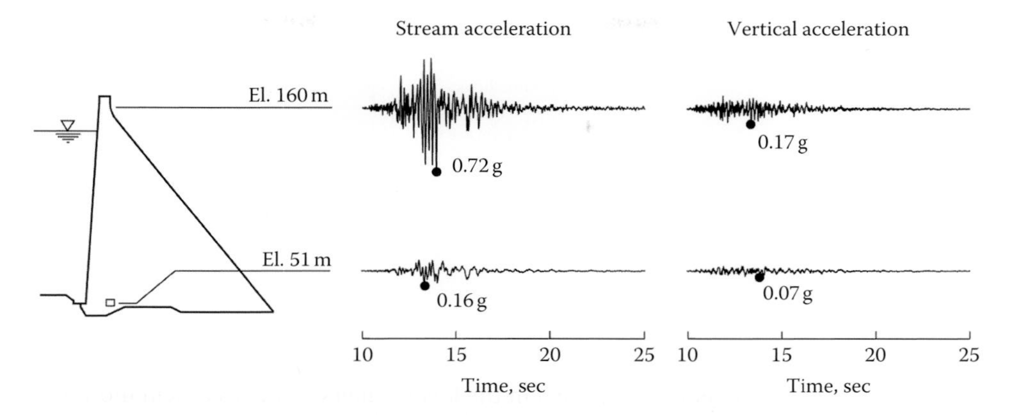

Figure 7.1.3 Accelerations recorded during main earthquake event, March 26, 1997, at the elevations shown.

horizontal and vertical components of acceleration recorded during the main event at the base and crest of the dam are shown in Figure 7.1.3, where the peak values are noted.

7.1.3 System Analyzed

Figure 7.1.4 presents a schematic of the dam–water–foundation system and a finite element model of the tallest non-overflow monolith of the dam. As in Chapter 6, the foundation is treated as a viscoelastic half-space and the fluid domain as a uniform channel, unbounded in the upstream direction.

We could not find any information on the elastic moduli of concrete or rock at this dam. Thus, these properties were determined indirectly by matching the resonant frequencies of the model with those determined from the transfer function between the motions recorded at the crest of the dam and near the base of the dam. Determined from such records for the main event and five aftershocks, the transfer functions are shown in Figure 7.1.5. The resonant frequencies are not unique because the water level was not the same over the six events. The average values of the first three resonant frequencies are 4.2, 10.0, and 14.5 Hz.

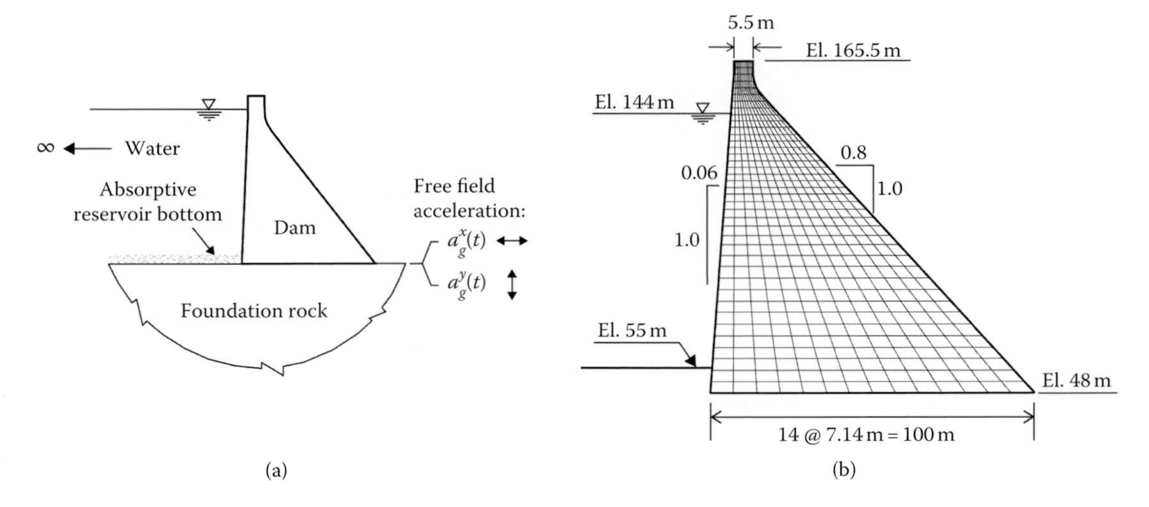

Figure 7.1.4 (a) Schematic of EAGD model for dam–water–foundation system; and (b) FE model for the dam.

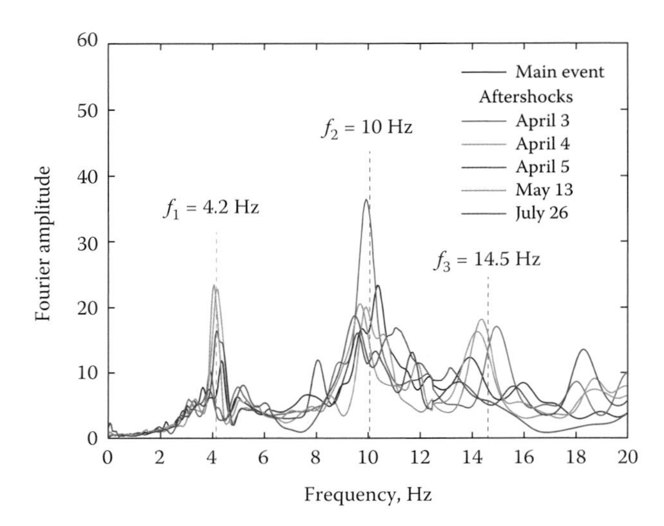

Figure 7.1.5 Transfer functions.

Typical values were chosen for some of the material properties; concrete: Poisson's ratio = 0.20 and unit mass = 2400 kg/m^3; rock: Poisson's ratio = 0.33 and unit mass = 2700 kg/m^3. With these values fixed, the moduli of elasticity for concrete and rock in the 2D model of Figure 7.1.4 were selected to achieve the closest match with the identified frequencies. This investigation resulted in elastic moduli of 47 GPa (6.75 million psi) for concrete – an unusually high value compared to typical values of elastic modulus of concrete – and 70 GPa (10 million psi) for rock. This apparent discrepancy arises because the resonant frequencies of the 2D model have been matched to the measured values for the actual dam, a 3D structure. The ratio of the fundamental frequencies of 2D and 3D models of the dam is 1.25, implying that the values of elastic moduli for a 3D model would be reduced by a factor $(1.25)^2 = 1.57$ to match the measured frequencies. This adjustment would result in an elastic modulus of 30 GPa (4.33 million psi), which is typical of mass concrete.

We could not find any reports on forced vibration tests on the dam that would have provided reliable values for measured damping in the overall system. Lacking such results, we explored the possibility of estimating damping from the transfer functions (Figure 7.1.5) but were not able to obtain reliable results. The results were too sensitive to how the recorded motions were filtered to smooth the transfer function. One choice of filters led to 3–4% damping and another resulted in 8–10% damping.

Given the lack of data, constant hysteretic damping factors for the dam alone and foundation rock separately were selected within the range of values based on calibration of numerical models for other dams against measured values of damping (Proulx et al. 2001, 2004; Alves and Hall 2006b; Chopra and Wang 2008). We assumed $\eta_s = 0.02$ (i.e. viscous damping ratio of 1% in all vibration modes of the dam alone) and $\eta_f = 0.02$ (i.e. viscous damping ratio of 1% in the foundation rock). The wave reflection coefficient was selected as 0.50, which seems appropriate for a 50-year-old dam with considerable build-up of sediments at the bottom of the reservoir.

7.1.4 Comparison of Computed and Recorded Responses

Determined by the EAGD-84 computer program, the response of the system of Figure 7.1.4 with the aforementioned properties to the main earthquake event is presented in Figure 7.1.6.

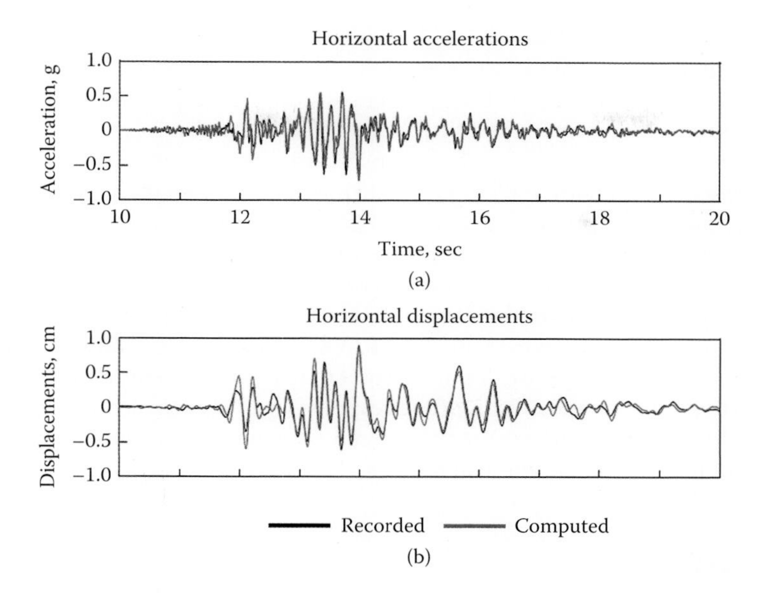

Figure 7.1.6 Comparison of computed and recorded motion in the horizontal (stream) direction at El. 160 m during the main earthquake event: (a) accelerations; and (b) displacements.

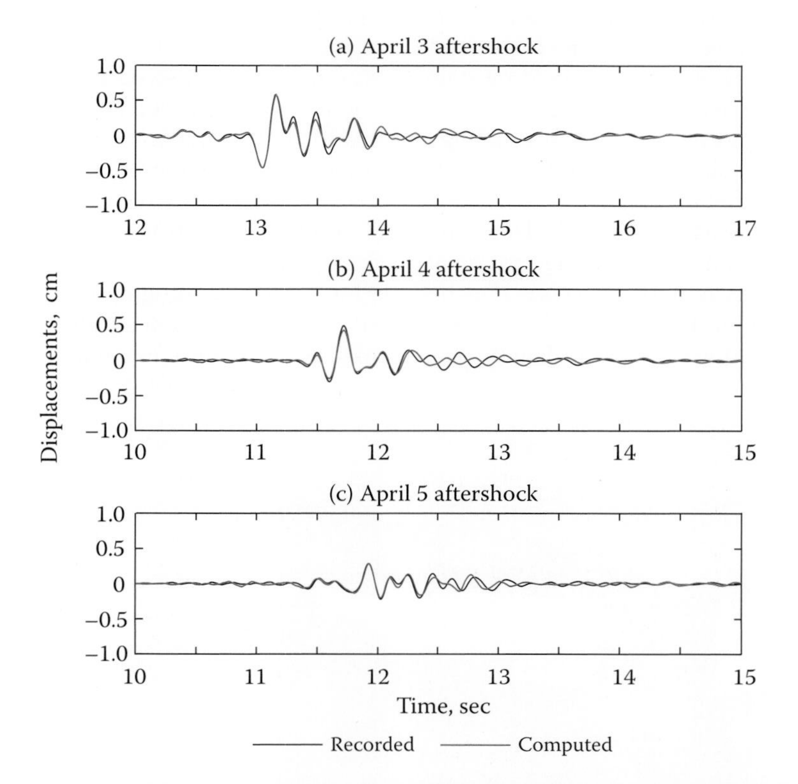

Figure 7.1.7 Comparison of computed and recorded motion in the horizontal (stream) direction at El. 160 m during three aftershocks that occurred on (a) April 3; (b) April 4; and (c) April 5.

The computed displacements and accelerations at the crest of the dam compare very well with the recorded accelerations and associated displacements. The time variation of the computed displacements and accelerations is close to the recorded response over the entire duration of shaking; the peak accelerations differ by 2% and the peak displacements by 5%.

Similar comparison between the computed responses and motions recorded during aftershock events 1–3 are presented in Figure 7.1.7. It is reassuring that the agreement between computed and recorded responses is similarly excellent for all four earthquake events.

However, the almost perfect agreement achieved here cannot be expected in general. Lack of information on the material properties of concrete and rock forced us to determine these properties to match the resonant frequencies of the dam, obtained by system identification applied to the recorded motions. Obviously, such calibration of the EAGD-84 model contributed to the high degree of accuracy of the computed response. It is important to instrument many dams, develop a database of their properties, and to validate computed response against recorded motions. Such investigations would help document the accuracy that can be expected of these analyses and identify improvements required in the numerical models.

7.2 KOYNA DAM CASE HISTORY

7.2.1 Koyna Dam and Earthquake Damage

Constructed during the years 1954 to 1963, Koyna Dam is a straight gravity structure (Figure 7.2.1), about 2800 ft (853 m) long and 338 ft (103 m) high above the deepest foundation. Located in western India, the dam was designed according to the standard practice of the time; earthquake effects were represented by static lateral forces with a seismic coefficient of 0.05, uniform over the height. The design criteria were; no tension in the section, compressive stresses to be within allowable values, and sliding and overturning of the monoliths to be prevented. The resulting cross sections for the non-overflow and overflow monoliths are shown in Figure 7.2.2.

A magnitude 6.5 earthquake on 11 December 1967, centered 8 mi from the dam, caused intense shaking at the dam site. A strong-motion accelerograph located in one of the very short monoliths produced an important record that may be interpreted as roughly representative of the ground motion; in the original uncorrected record the peak values of acceleration were 0.63 g along the longitudinal axis of the dam (cross-stream direction), 0.46 g in the transverse (or

Figure 7.2.1 Koyna Dam.

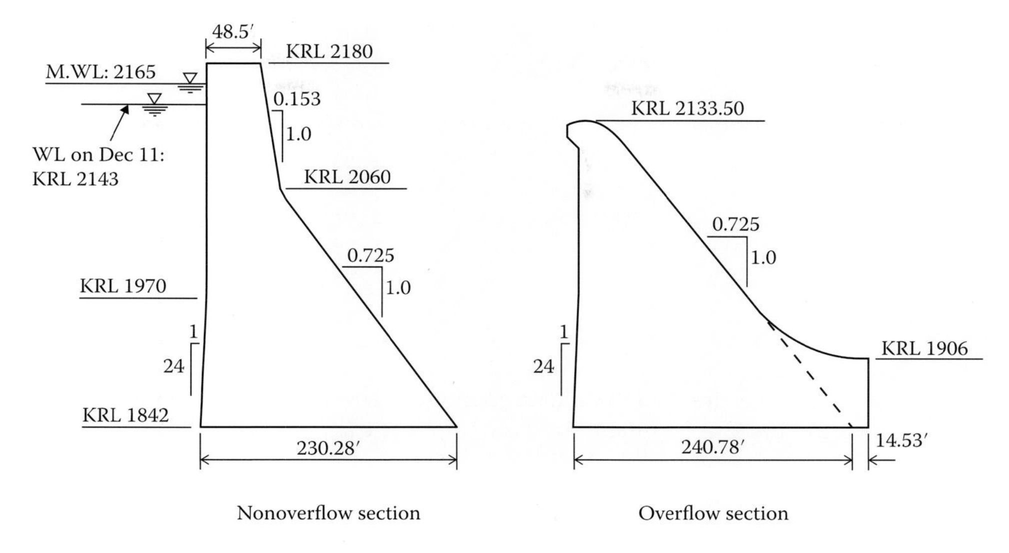

Figure 7.2.2 Koyna Dam: cross sections.

stream) direction, and 0.36 g in the vertical direction. In the processed and base-line corrected version of the record, the peak ground acceleration (PGA) values were reduced to 0.49, 0.38, and 0.24 g, respectively [NOAA database]. Unfortunately, the motion at the crest of the dam was not recorded for lack of an accelerograph.

The dam sustained significant damage, the most important being horizontal cracking on either the upstream or downstream face, or on both faces, of a number of monoliths. The principal cracking was in the taller non-overflow monoliths on both sides of the overflow (spillway) section around K.R.L. 2016, the level where the slope of the downstream face changes abruptly (Figures 1.1.2 and 7.2.2). The overflow monoliths were not damaged. Relative movement between adjacent monoliths was evident from spalling of concrete along the vertical joints, and from increased seepage through the joints after the earthquake.

The damage was repaired soon after the earthquake in two ways: first, the major cracks were repaired by injecting epoxy resin; and second, the taller non-overflow monoliths were pre-stressed in the vertical direction from the crest down to an elevation well below major cracks. Subsequently, buttresses were added on the downstream face of the non-overflow monoliths (Figure 7.2.3).

The structural damage sustained by Koyna Dam was of profound significance to the development of earthquake engineering for concrete dams. A modern dam, designed according to analysis procedures and design criteria that represented "standard" practice worldwide at the time, had been damaged by ground shaking that was intense, but by no means extreme. It was clear that the design forces had little resemblance to how the dam responded during the earthquake. For example, the criterion of no tension had been "satisfied" in designing the dam, implying that no cracking was anticipated, but significant cracking occurred. The experience at Koyna Dam was a watershed event in the sense that it eventually revolutionized dam engineering and prompted the development of dynamic analysis procedures for concrete gravity dams, some of which were presented in Chapters 2–6.

7.2.2 Computed Response of Koyna Dam

Using the substructure method presented in Chapter 5, the response of the tallest non-overflow monolith including dam–water–foundation interaction to the ground motion recorded during

Figure 7.2.3 Koyna Dam after the addition of buttresses.

the 1967 earthquake was computed. The finite element model for the monolith is presented in Figure 7.2.4 and the transverse and vertical components of the ground motion in Figure 7.2.5; material properties and other data are included in Appendix 7.

The time-variation of computed displacements at the base and crest of the dam, relative to the free-field ground displacements, are presented in Figure 7.2.6, and the distribution of maximum principal stresses over the monolith at selected instants of time are shown in Figure 7.2.7, wherein positive values denote tensile stresses; initial displacements and stresses prior to the earthquake are included. These results indicate larger tensile stresses in the upper part of the dam, especially at the elevation where the slope of the downstream face changes abruptly. These stresses exceed 600 psi on the upstream face and 1000 psi on the downstream face; values that are two to three times the estimated tensile strength (350 psi) of the concrete used in upper parts of the dam. Clearly these results suggest significant cracking at locations consistent with the damage during the earthquake.

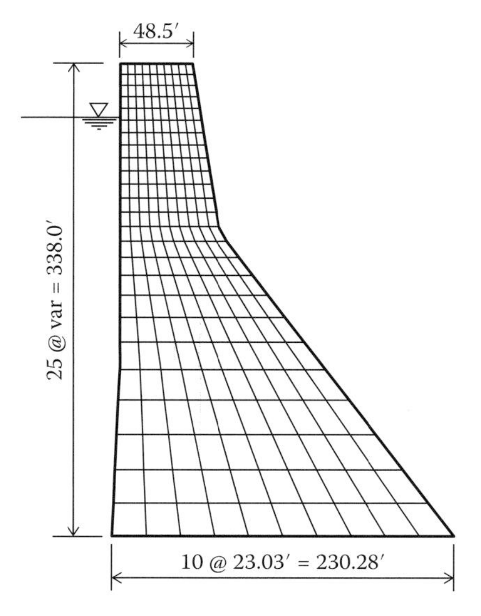

Figure 7.2.4 Finite-element model of non-overflow monolith.

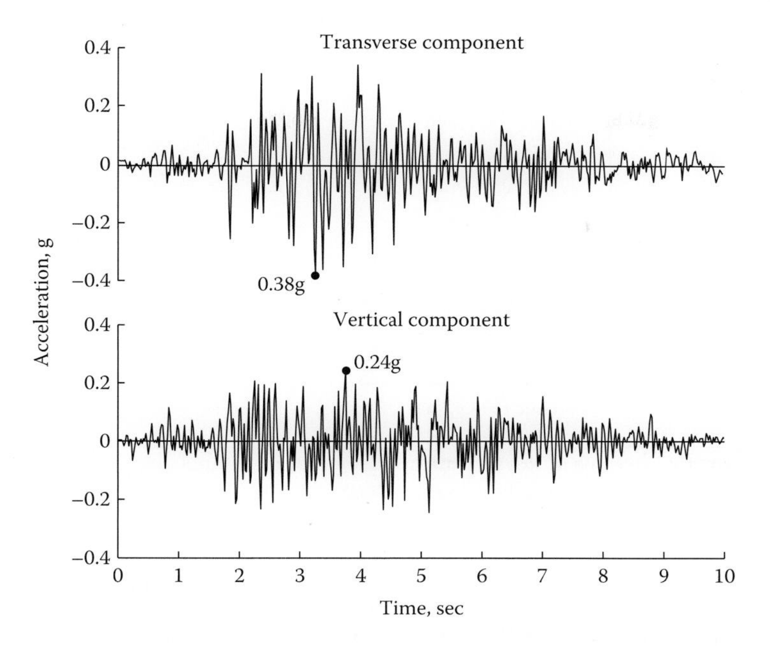

Figure 7.2.5 Transverse and vertical components of ground motion recorded at block 1-A of Koyna Dam on December 11, 1967.

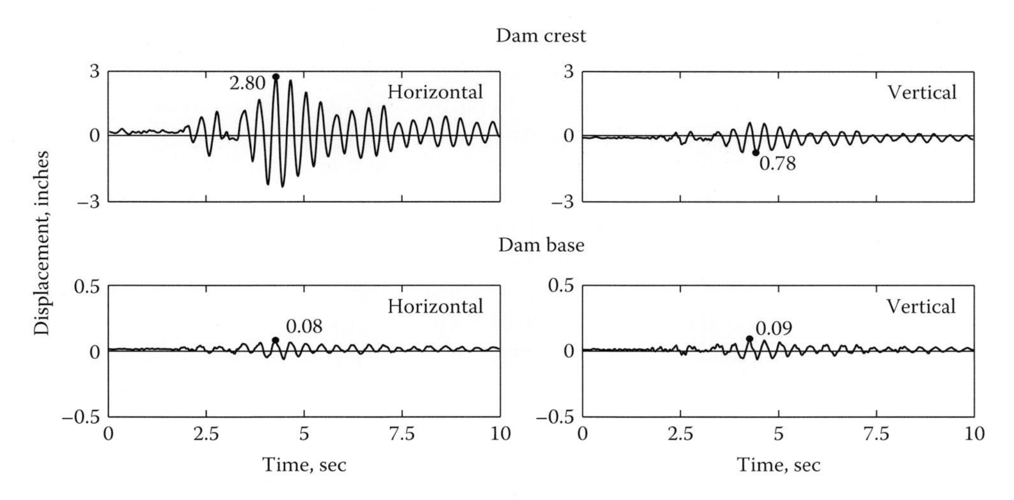

Figure 7.2.6 Displacement response of Koyna Dam to transverse and vertical components of ground motion recorded during the December 11, 1967 earthquake; initial static displacements are included.

These response results demonstrate the fallacy in the practice of decreasing the concrete strength with increase in elevation within some dams, for example, Koyna Dam in India (Chopra and Chakrabarti 1973) and the 717-ft-high Dworshak Dam in Idaho, USA. This practice seems to be motivated by the observation that traditional design analyses (Section 1.3.1), which ignored the dynamics of the system, predicting largest stresses near the base of the dam and decreasing stresses at higher elevations. However, as indicated by dynamic analysis (Figure 7.2.7) and by the location of earthquake-induced cracks in Koyna Dam, higher-strength concrete should be

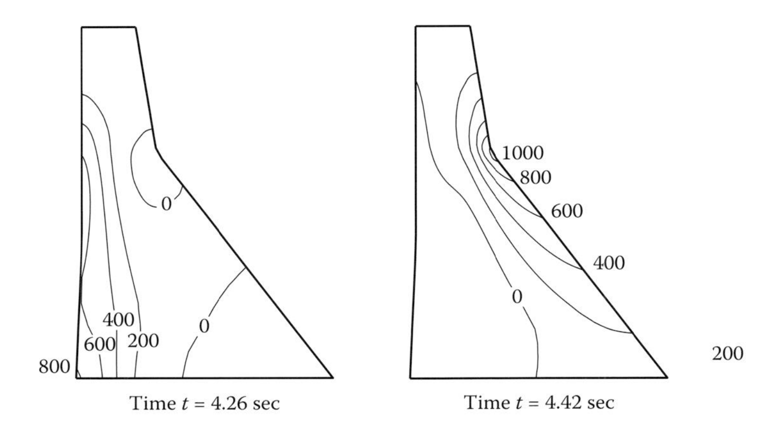

Figure 7.2.7 Maximum principal stresses in Koyna Dam at selected time instants due to transverse and vertical components of ground motion recorded during the December 11, 1967 earthquake; initial static stresses are included.

provided in the upper part of the dam near the upstream and downstream faces – if the designer chooses to vary the concrete strength over the dam.

7.2.3 Response of Typical Gravity Dam Sections

Because the cross section of the non-overflow monoliths of Koyna Dam is not typical of gravity dams (Figure 7.2.8), some dam engineers speculated that dams with conventional cross sections would not have been damaged by the Koyna earthquake. To explore this issue, the response of Pine Flat Dam to the ground motion recorded at Koyna Dam is presented in Figure 7.2.9; also shown for comparison are stresses in Koyna Dam. The maximum tensile stresses in Pine Flat Dam are 532 and 746 psi at the upstream face and downstream faces, respectively, compared to 750 and 1050 psi in Koyna Dam. Larger stresses develop in Koyna Dam because the upper neck portion is taller and wider, the slope of the downstream face is steeper, and the change in the downstream face slope is abrupt relative to typical cross-sectional shapes. Thus, these results support the speculation that the Koyna cross section is more vulnerable to earthquake damage than a typical gravity dam section. However, the results presented also indicate that the Koyna

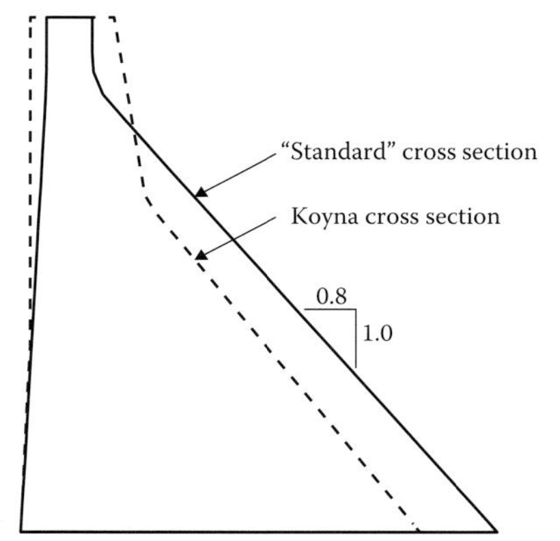

Figure 7.2.8 Comparison of Koyna Dam section with "standard" cross section.

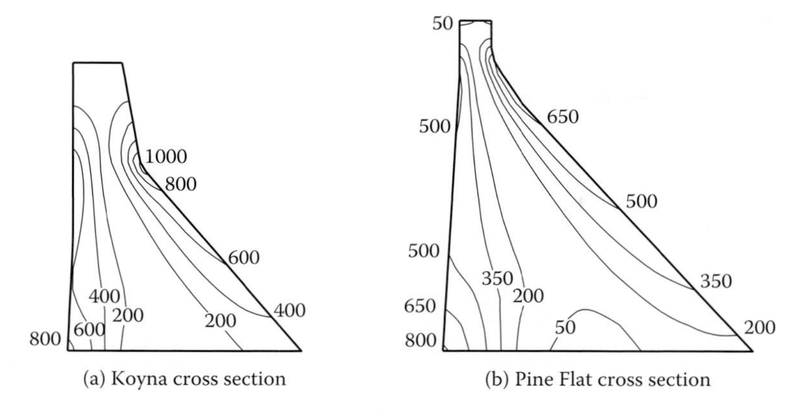

Figure 7.2.9 Envelope values of maximum principal stresses in (a) Koyna Dam; and (b) Pine Flat Dam due to transverse and vertical components of ground motion recorded during the December 11, 1967, earthquake; initial static stresses are included.

ground motion would have caused large tensile stresses even in typical gravity dam sections. The gradual transition in the slope of the downstream face, a shorter and narrower neck section, and a flatter downstream face lead to some reduction in tensile stresses. However, the stresses are still large enough to suggest significant – albeit less – cracking.

7.2.4 Response of Dams with Modified Profiles

Most existing gravity dams employ a similar cross-sectional shape that satisfies traditional design criteria with earthquake effects represented by unrealistically small static forces (Section 1.3). During an earthquake, the motion of the upper region of a concrete gravity dam of typical profile is amplified, resulting in large tensile stresses in the neck region as shown in Chapter 6. To demonstrate that these stresses can be significantly reduced by shaping the dam cross section, two modified profiles are examined. The first is obtained by removing a portion of concrete identified in Figure 7.2.10a near the crest of Pine Flat Dam. We will refer to this as the "structural section" because it satisfies traditional design criteria; however, concrete is added to provide some thickness at the crest to resist impact of any floating objects, to afford a roadway, and to provide a free board above the maximum water level. The second modified profile, shown in Figure 7.2.10b, has straight upstream and downstream faces with slopes of 0.05–1.0 (horizontal to vertical) and

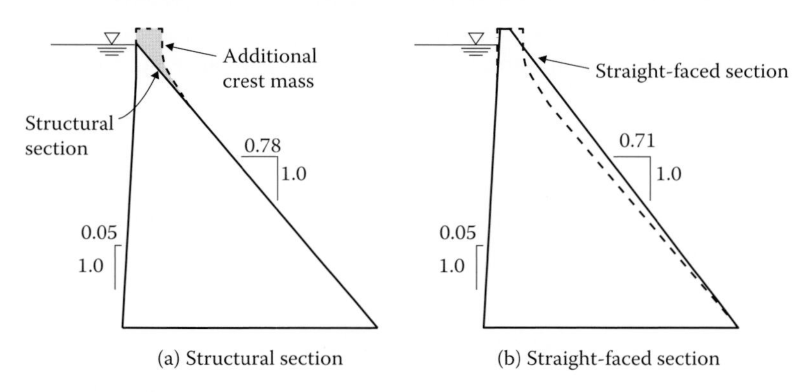

Figure 7.2.10 Modified cross sections.

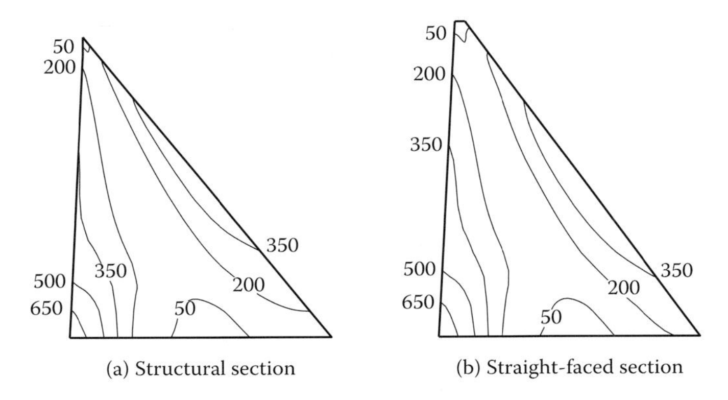

Figure 7.2.11 Envelope values of maximum principal stresses in modified cross sections due to transverse and vertical components of ground motion recorded during the December 11, 1967 earthquake; initial static stresses are included.

0.71–1.0, respectively. The crest width is 12 ft, and the base width and height are the same as Pine Flat Dam. Material properties for the two cases are listed in Appendix 7.

The stresses in the dam are increased considerably by the additional crest mass. The stress reductions in the structural section (Figure 7.2.11a) compared to Pine Flat cross section (Figure 7.2.9b) are 34% and 39% on the upstream and downstream faces, respectively. The stresses in the structural section are much smaller for two reasons: first, removal of the crest mass modifies the vibration modes, reducing the displacement gradients in the upper part of the dam (Chopra and Chakrabarti 1973); second, the inertia forces associated with the additional crest mass undergoing amplified acceleration are eliminated. It is of interest to note that monolith 18 of Koyna Dam that suffered the worst damage during the earthquake supported an elevator tower that extended 50 ft above the crest of the dam (Chopra and Chakrabarti 1973).

These results suggest that lightweight structural systems should be developed to support a roadway and meet other functional requirements. The benefits of a lightweight crest have been demonstrated (Hall et al. 1991); stress reductions achieved are similar to those achieved for the structural section.

The stress reductions in the straight-faced section are 30–34% compared to the Pine Flat cross section (Figures 7.2.11b and 7.2.9b). The stresses are smaller because the neck region where the width of the section is reduced is eliminated. Cross sections with a downstream face that is almost straight have been employed for some modern dams, especially roller-compared concrete (RCC) dams; one example is Olivenhain Dam, a 318-ft-high RCC dam near San Diego, California, USA (Figure 7.2.12).

Determination of an optimum cross-sectional shape for a concrete gravity dam must consider costs as well as earthquake performance. The volume of the straight-faced dam exceeds that of Pine Flat Dam by 6.9%, while the lightweight crest may be more expensive to build. However, a small cost increase may be a prudent investment in light of the much better performance of the modified dams expected during earthquakes.

Figure 7.2.12 Olivenhain Dam, a 318-ft-high RCC dam near San Diego, California, completed in 2003.

APPENDIX 7: SYSTEM PROPERTIES

	Koyna	Pine Flat	Pine Flat structural	Pine Flat straight-faced
		Cross section		
E_s (million psi)	4.5	4.5	4.5	4.5
E_f/E_s	2	2	2	2
H_s (ft)	338	400	381	400
H (ft)	301	381	381	381
α	0.75	0.75	0.75	0.75
		Unit weight		
Concrete	165	165	165	165
Rock (pcf)	165	165	165	165
		Poisson's ratio		
Concrete	0.2	0.2	0.2	0.2
Rock	0.33	0.33	0.33	0.33
		Hysteretic damping factor		
Concrete	0.4	0.4	0.4	0.4
Rock (pcf)	0.4	0.4	0.4	0.4

Part II

ARCH DAMS

8

Response History Analysis of Arch Dams Including Dam–Water–Foundation Interaction

PREVIEW

A substructure method to determine the earthquake response of arch dams as a function of time is presented in this chapter; it is a generalization of the analysis procedure for gravity dams developed in Chapter 5. Included are all significant effects of dam–water–foundation interaction. Representing the dam, the impounded water, and the supporting foundation as three substructures permits selecting the idealization most appropriate for each substructure: finite-element model for the dam; finite-element model for a small part of the fluid domain adjacent to the dam connected to sub-channels of infinite length; and a boundary element model for the foundation idealized as a uniform canyon cut in viscoelastic half-space. The substructure method is formulated in the frequency domain to determine the complex-valued frequency response functions, followed by Fourier synthesis of the responses to individual harmonic components to determine the response of the dam to free-field ground motion specified at the dam–foundation interface. The analysis procedure is developed first for spatially-uniform ground motion and later extended to include spatial variations. The chapter ends with a brief mention of EACD-3D-96 and EACD-3D-2008, computer programs that implement the analysis procedure.

8.1 SYSTEM AND GROUND MOTION

The three-dimensional system considered consists of a concrete arch dam supported by flexible foundation in a canyon and impounding a reservoir of water with sediments deposited at the bottom and sides of the reservoir (Figure 8.1.1). The canyon is assumed to be infinitely long, with an arbitrary but uniform cross section cut in a homogeneous viscoelastic half-space. The cross section of the canyon is defined by the dam–foundation interface (Figure 8.1.2). The system is

Earthquake Engineering for Concrete Dams: Analysis, Design, and Evaluation, First Edition. Anil K. Chopra.

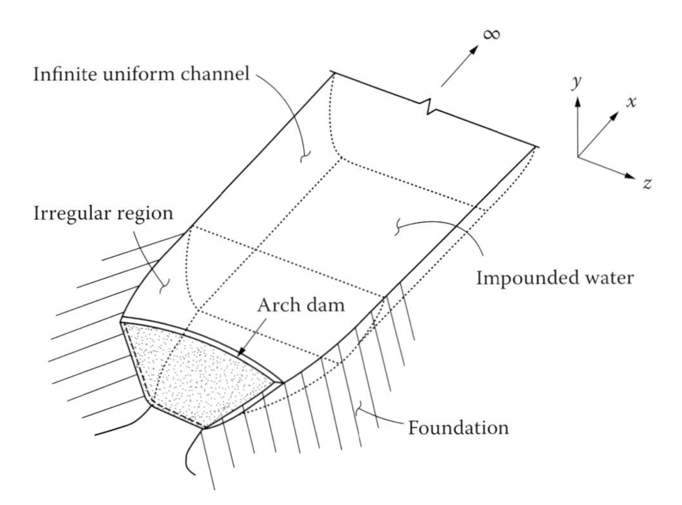

Figure 8.1.1 Arch dam–water–foundation system.

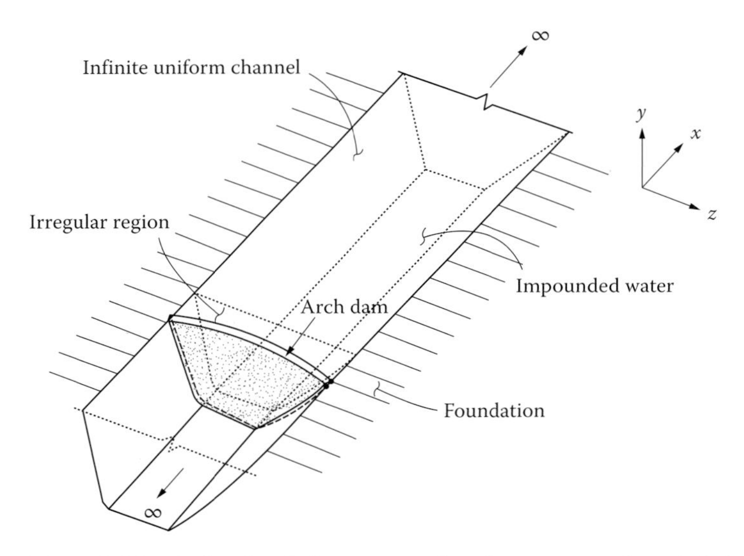

Figure 8.1.2 Idealized arch dam–water–foundation system in an infinitely-long canyon with arbitrary but uniform cross section.

analyzed under the assumption of linear behavior for the concrete dam, impounded water, and foundation rock. Thus the possibility of concrete cracking, sliding, or opening of contraction joints, opening of lift joints, or any other source of nonlinearity is not considered.

The dam–water–foundation system is idealized as shown in Figures 8.1.3a–c. The arch dam is idealized as an assemblage of finite elements (FE) (Figure 8.1.3a). The dam–foundation interface is discretized by boundary elements with their nodal points matching the finite element idealization of the dam (Figure 8.1.3b). The assumption of a uniform canyon may introduce minor incompatibility between the dam abutment and the canyon (Figure 8.1.3d), which requires special treatment as described in Tan and Chopra (1995a,b), where the errors associated with this approximation are shown to be small.

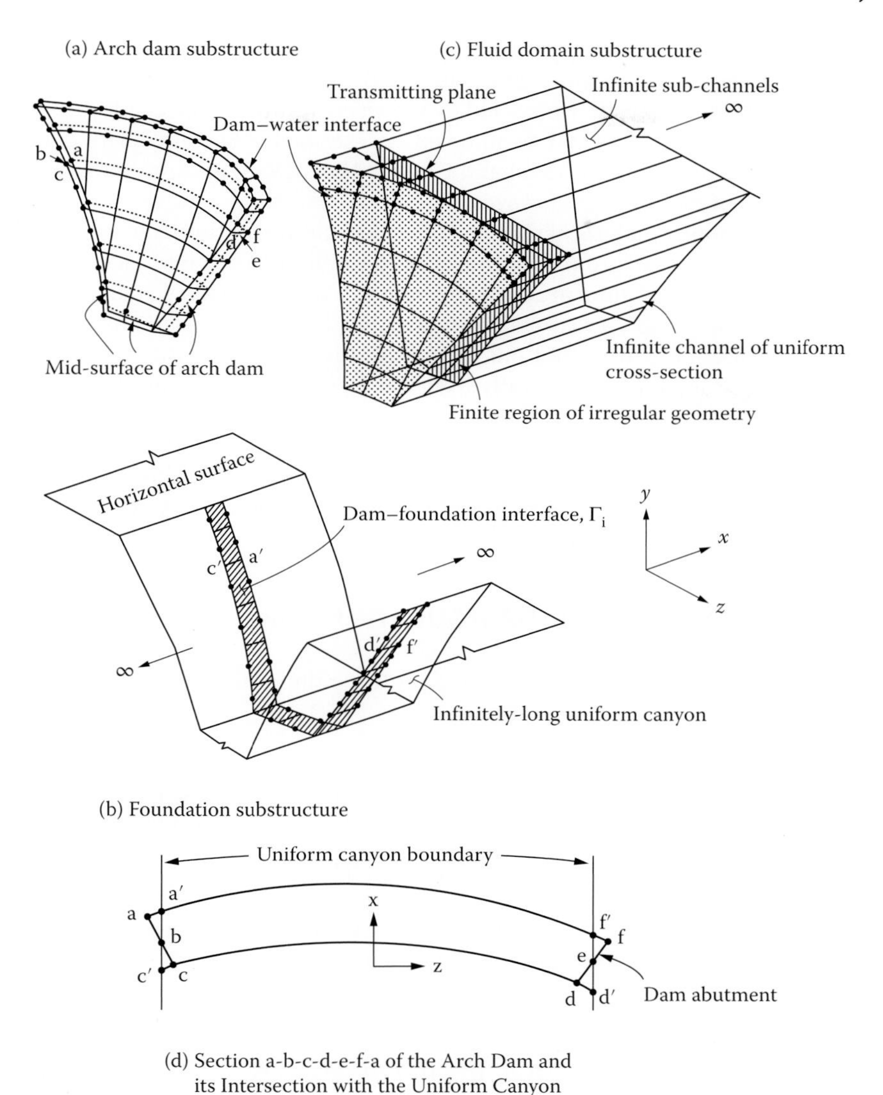

Figure 8.1.3 (a and c) Finite-element models of dam and fluid domain; (b) boundary element model of foundation; and (d) incompatibility between the dam abutment and canyon.

The reservoir impounded by a dam is of complicated shape, as dictated by the natural topography of the canyon, and often extends several miles in the upstream direction. To efficiently recognize the long extent of the reservoir in the upstream direction, the fluid domain is idealized as a finite region of irregular geometry connected to an infinitely long channel with uniform cross section; we will refer to the latter as "uniform channel." The finite region is idealized as an assemblage of three-dimensional finite elements (Figure 8.1.3c) compatible with the finite element mesh for the dam at its upstream face. The channel is discretized into sub-channels compatible with the finite element mesh of the finite region over their common cross section, the transmitting plane (Figure 8.1.3c). Modeling of the fluid domain is described in detail elsewhere (Tan and Chopra 1995a).

The bottom and sides of the reservoir may consist of highly variable layers of exposed bedrock, alluvium silt, and other sedimentary material. Because these materials deposited at the reservoir boundary are not adequately modeled by the viscoelastic half-space idealization of the foundation domain, they are approximately modeled as described in Section 2.3 for gravity dams and extended elsewhere to arch dams (Hall and Chopra 1982, 1983) as a reservoir bottom that partially absorbs incident hydrodynamic waves.

In earthquake response analysis of dams by the substructure method, the earthquake excitation is specified as the free-field ground motion at the dam–foundation interface. This free-field ground motion was taken to be uniform across the base in two-dimensional analyses of concrete gravity dams (Chapter 5), possibly a reasonable assumption although it has not been investigated comprehensively for lack of recorded motions at multiple locations along the base. However, the ground motion is known to vary significantly over the dam–foundation interface at arch dam sites (Chapter 10). Initially, these spatial variations in ground motion are not included in developing the analysis procedure (Sections 8.2–8.7). The three components of ground acceleration are defined by: $a_g^x(t)$ in the upstream direction, $a_g^z(t)$ in the cross-stream direction, and $a_g^y(t)$ in the vertical direction. Near the end of this chapter the analysis procedure is extended to include spatial variations in ground motion.

8.2 FREQUENCY-DOMAIN EQUATIONS: DAM SUBSTRUCTURE

The equations of motion for the dam idealized as a three-dimensional finite element system (Figure 8.2.1a) are

$$\mathbf{m}_c\ddot{\mathbf{r}}_c + \mathbf{c}_c\dot{\mathbf{r}}_c + \mathbf{k}_c\mathbf{r}_c = -\mathbf{m}_c\boldsymbol{\iota}_c^x a_g^x(t) - \mathbf{m}_c\boldsymbol{\iota}_c^y a_g^y(t) - \mathbf{m}_c\boldsymbol{\iota}_c^z a_g^z(t) + \mathbf{R}_c(t) \tag{8.2.1}$$

in which $\mathbf{m}_c$, $\mathbf{c}_c$, and $\mathbf{k}_c$ are the mass, damping, and stiffness matrices for the finite element system; $\mathbf{r}_c$ is the vector of nodal displacements relative to the free-field ground displacement:

$$\mathbf{r}_c^T = \left\langle r_1^x \;\; r_1^y \;\; r_1^z \;\; r_2^x \;\; r_2^y \;\; r_2^z \ldots r_n^x \;\; r_n^y \;\; r_n^z \ldots r_{N+N_b}^x \;\; r_{N+N_b}^y \;\; r_{N+N_b}^z \right\rangle \tag{8.2.2}$$

where r_n^x, r_n^y, and r_n^z are the x, y, and z components of the displacement of node n; N is the number of nodes in the FE model for the dam; N_b is the number of nodes on the dam–foundation interface; and

$$\begin{aligned} \boldsymbol{\iota}_c^x &= \langle 1 \;\; 0 \;\; 0 \;\; 1 \;\; 0 \;\; 0 \ldots 1 \;\; 0 \;\; 0 \ldots 1 \;\; 0 \;\; 0 \rangle^T \\ \boldsymbol{\iota}_c^y &= \langle 0 \;\; 1 \;\; 0 \;\; 0 \;\; 1 \;\; 0 \ldots 0 \;\; 1 \;\; 0 \ldots 0 \;\; 1 \;\; 0 \rangle^T \\ \boldsymbol{\iota}_c^z &= \langle 0 \;\; 0 \;\; 1 \;\; 0 \;\; 0 \;\; 1 \ldots 0 \;\; 0 \;\; 1 \ldots 0 \;\; 0 \;\; 1 \rangle^T \end{aligned} \tag{8.2.3}$$

The force vector $\mathbf{R}_c(t)$ includes hydrodynamic forces $\mathbf{R}_h(t)$ at the upstream face of the dam due to dam–water interaction, and forces $\mathbf{R}_b(t)$ at the dam–foundation interface due to interaction between the dam and the foundation.

For harmonic ground acceleration $a_g^l(t) = e^{i\omega t}$ in the $l = x$ (stream), y (vertical), or z (cross-stream) direction, the displacements and forces can be expressed in terms of their complex-valued frequency response functions: $\mathbf{r}_c(t) = \bar{\mathbf{r}}_c^l(\omega)e^{i\omega t}$, $\mathbf{R}_c(t) = \bar{\mathbf{R}}_c^l(\omega)e^{i\omega t}$, $\mathbf{R}_h(t) = \bar{\mathbf{R}}_h^l(\omega)e^{i\omega t}$, and $\mathbf{R}_b(t) = \bar{\mathbf{R}}_b^l(\omega)e^{i\omega t}$. The governing Eq. (8.2.1) then becomes

$$\left[-\omega^2\mathbf{m}_c + i\omega\mathbf{c}_c + \mathbf{k}_c\right]\bar{\mathbf{r}}_c^l(\omega) = -\mathbf{m}_c\boldsymbol{\iota}_c^l + \bar{\mathbf{R}}_c^l(\omega); \quad l = x, \; y, \; \text{or} \; z \tag{8.2.4}$$

If the energy dissipation in the dam is modeled by constant hysteretic damping, Eq. (8.2.4) becomes

$$\left[-\omega^2\mathbf{m}_c + (1 + i\eta_s)\mathbf{k}_c\right]\bar{\mathbf{r}}_c^l(\omega) = -\mathbf{m}_c\boldsymbol{\iota}_c^l + \bar{\mathbf{R}}_c^l(\omega); \quad l = x, \; y, \; \text{or} \; z \tag{8.2.5}$$

where η_s is the constant hysteretic damping factor for the dam.

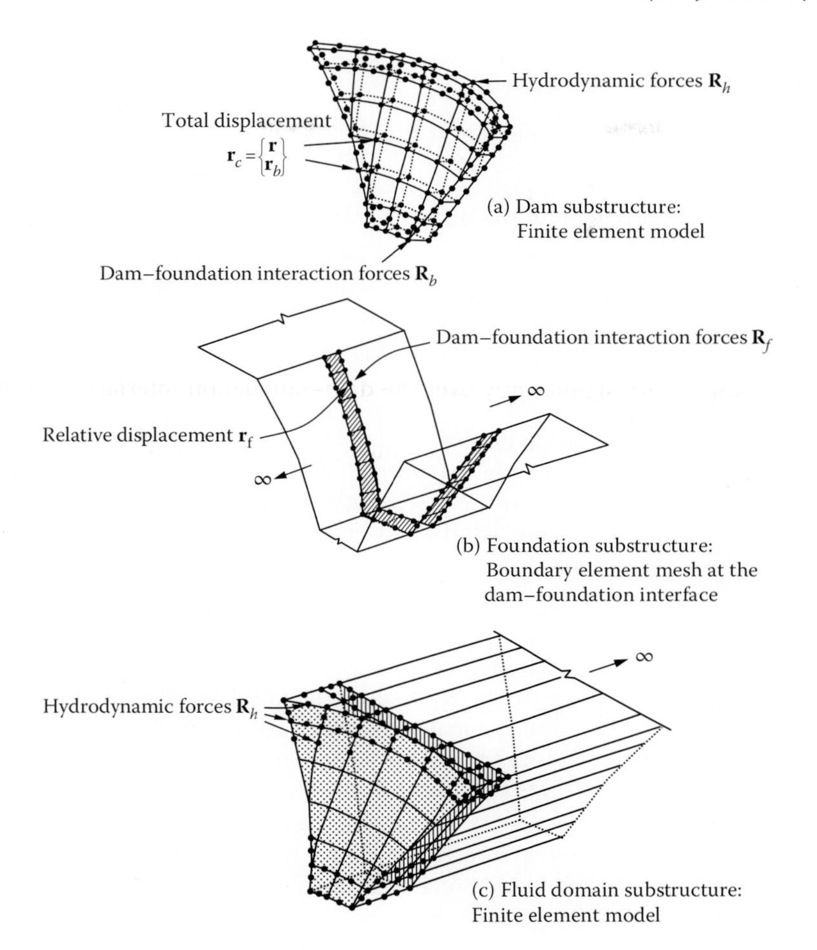

Figure 8.2.1 Substructure representation of the dam–water–foundation system.

Partitioning $\mathbf{r}_c$ into $\mathbf{r}$ for nodal points not located on the dam–foundation interface, and $\mathbf{r}_b$ for nodal points located on this interface, Eq. (8.2.5) can be rewritten as

$$\left(-\omega^2 \begin{bmatrix} \mathbf{m} & \mathbf{0} \\ \mathbf{0} & \mathbf{m}_b \end{bmatrix} + (1 + i\eta_s) \begin{bmatrix} \mathbf{k} & \mathbf{k}_b \\ \mathbf{k}_b^T & \mathbf{k}_{bb} \end{bmatrix}\right) \begin{Bmatrix} \bar{\mathbf{r}}^l(\omega) \\ \bar{\mathbf{r}}_b^l(\omega) \end{Bmatrix} = -\begin{Bmatrix} \mathbf{m}\iota^l \\ \mathbf{m}_b \iota_b^l \end{Bmatrix} + \begin{Bmatrix} \bar{\mathbf{R}}_h^l(\omega) \\ \bar{\mathbf{R}}_b^l(\omega) \end{Bmatrix} \tag{8.2.6}$$

Dam–foundation interaction forces $\mathbf{R}_b$ will be expressed in terms of the displacements at the dam–foundation interface by analysis of the foundation substructure (Section 8.3), and the hydrodynamic forces $\mathbf{R}_h$ will be expressed in terms of the acceleration of the upstream face of the dam by analysis of the fluid domain substructure (Section 8.5).

8.3 FREQUENCY-DOMAIN EQUATIONS: FOUNDATION SUBSTRUCTURE

The interaction forces acting at the foundation surface are denoted by $\mathbf{R}_f$, and the corresponding displacements relative to the free-field ground displacement by $\mathbf{r}_f$ (Figure 8.2.1b). For unit harmonic ground acceleration, these forces and displacements can be expressed in terms of their complex-valued frequency response functions; thus, $\mathbf{R}_f(t) = \bar{\mathbf{R}}_f(\omega)e^{i\omega t}$ and $\mathbf{r}_f(t) = \bar{\mathbf{r}}_f(\omega)e^{i\omega t}$. The forces and displacements are related as follows:

$$\mathcal{S}_f(\omega)\bar{\mathbf{r}}_f(\omega) = \bar{\mathbf{R}}_f(\omega) \tag{8.3.1}$$

where $\mathcal{S}_f(\omega)$ is the complex-valued, frequency-dependent (dynamic) stiffness matrix for the foundation defined at the nodal points on the dam–foundation interface, Γ_i (see Section 5.3). The size of the square matrix, $\mathcal{S}_f(\omega)$, is equal to the number of degrees of freedom (DOFs) in the finite element idealization of the dam at its interface with the foundation. The jth column of this matrix multiplied by $e^{i\omega t}$ is the set of complex-valued forces required at the interface DOFs to maintain a unit harmonic displacement, $e^{i\omega t}$, in the jth DOF with zero displacements in all other DOFs (see Figure 5.3.1).

Evaluation of these forces requires solution of a series of mixed boundary value problems (BVPs) with displacements prescribed at the interface, Γ_i, and tractions prescribed as zero outside Γ_i – on the canyon wall and the half-space surface. Instead of directly solving this mixed BVP, it is more convenient to solve a stress BVP in which non-zero tractions are specified at the interface, Γ_i, and the resulting displacements at Γ_i are determined. Assembled from these displacements, the dynamic flexibility matrix is inverted to determine the dynamic stiffness matrix $\mathcal{S}_f(\omega)$.

A direct boundary element procedure has been developed to determine $\mathcal{S}_f(\omega)$ for a canyon cut in a viscoelastic half-space (Zhang and Chopra 1991b). The assumption of uniform cross section of the canyon permits analytical integration along the canyon axis of the three-dimensional boundary integral equation. Thus, the original three-dimensional problem is reduced to an infinite series of two-dimensional problems, each of which corresponds to a particular wave number and involves Fourier transforms of full-space Green's functions. Appropriate superposition of the solutions of these two-dimensional BVPs leads to a dynamic flexibility matrix that is inverted to determine $\mathcal{S}_f(\omega)$. It has been demonstrated that this procedure is more accurate and efficient than the general three-dimensional boundary element method.

As in Chapter 5, water-foundation interaction effects are neglected because their influence on dam response is known to be small; see Appendix 5 and Fok and Chopra (1986a).

8.4 DAM–FOUNDATION SYSTEM

8.4.1 Frequency-Domain Equations

Equilibrium of the interaction forces between the dam and the foundation substructures at the dam–foundation interface requires that

$$\bar{\mathbf{R}}_b^l(\omega) + \bar{\mathbf{R}}_f^l(\omega) = 0 \tag{8.4.1}$$

and compatibility of interaction displacements at the interface requires that

$$\bar{\mathbf{r}}_b^l(\omega) = \bar{\mathbf{r}}_f^l(\omega) \tag{8.4.2}$$

Upon use of Eqs. (8.4.1) and (8.4.2), Eq. (8.3.1) becomes

$$\bar{\mathbf{R}}_b^l(\omega) = -\mathcal{S}_f(\omega)\bar{\mathbf{r}}_b^l(\omega) \tag{8.4.3}$$

which upon substituting into Eq. (8.2.6), gives

$$\left\{-\omega^2\begin{bmatrix}\mathbf{m} & 0\\ 0 & \mathbf{m}_b\end{bmatrix} + (1+i\eta_s)\begin{bmatrix}\mathbf{k} & \mathbf{k}_b\\ \mathbf{k}_b^T & \mathbf{k}_{bb}\end{bmatrix} + \begin{bmatrix}\mathbf{0} & \mathbf{0}\\ \mathbf{0} & \mathcal{S}_f(\omega)\end{bmatrix}\right\}\begin{Bmatrix}\bar{\mathbf{r}}^l(\omega)\\ \bar{\mathbf{r}}_b^l(\omega)\end{Bmatrix} = -\begin{Bmatrix}\mathbf{m}\boldsymbol{\iota}^l\\ \mathbf{m}_b\boldsymbol{\iota}_b^l\end{Bmatrix} + \begin{Bmatrix}\bar{\mathbf{R}}_h^l(\omega)\\ 0\end{Bmatrix} \tag{8.4.4}$$

The vector $\bar{\mathbf{R}}_h(\omega)$ of frequency response functions for hydrodynamic forces, which contains non-zero terms corresponding only to the DOFs at the nodal points on the upstream face of the dam, will be expressed in terms of the corresponding accelerations by analysis of the fluid domain substructure (Section 8.5).

8.4.2 Reduction of Degrees of Freedom

Equation (8.4.4), which represents a set of 3 $(N + N_b)$ frequency-dependent, complex-valued equations, is to be solved for many values of the excitation frequency. To reduce the computational effort, the displacements $\mathbf{r}_c$ relative to the free-field ground motion are expressed as a linear combination of J Ritz vectors, selected as the eigenvectors of an associated dam foundation system that remains to be defined:

$$\mathbf{r}_c(t) = \sum_{j=1}^{J} Z_j(t)\boldsymbol{\psi}_j \tag{8.4.5}$$

in which $Z_j(t)$ is the generalized coordinate that corresponds to the jth Ritz vector $\boldsymbol{\psi}_j$. For response to the l-component of ground acceleration, Eq. (8.4.5) can be expressed in terms of the complex-valued frequency response functions for the generalized coordinates:

$$\bar{\mathbf{r}}_c^l(\omega) = \sum_{j=1}^{J} \bar{Z}_j^l(\omega)\boldsymbol{\psi}_j \tag{8.4.6}$$

The eigenvalues λ_j and corresponding Ritz vectors $\boldsymbol{\psi}_j$ are solutions of the eigenvalue problem for an associated dam–foundation system, defined by replacing $\mathcal{S}_f(\omega)$ by its static value $\mathcal{S}_f(0)$:

$$\left[\mathbf{k}_c + \tilde{\mathcal{S}}_f(0)\right] \boldsymbol{\psi}_j = \lambda_j^2 \mathbf{m}_c \boldsymbol{\psi}_j \tag{8.4.7}$$

where

$$\tilde{\mathcal{S}}_f(\omega) = \begin{bmatrix} 0 & 0 \\ 0 & \mathcal{S}_f(\omega) \end{bmatrix} \tag{8.4.8}$$

The Ritz vectors are normalized, such that $\boldsymbol{\psi}_j^T \mathbf{m}_c \boldsymbol{\psi}_j = 1$.

Introducing the transformation Eq. (8.4.6) into Eq. (8.4.4), pre-multiplying by $\boldsymbol{\psi}_n^T$, and utilizing the orthogonality properties of the eigenvectors with respect to the "stiffness" and mass matrices of Eq. (8.4) leads to

$$\mathbf{S}(\omega)\,\bar{\mathbf{Z}}^l(\omega) = \mathbf{L}^l(\omega) \tag{8.4.9}$$

where the elements of the matrix $\mathbf{S}(\omega)$ and the vector $\mathbf{L}^l(\omega)$ are

$$S_{nj}(\omega) = \left[-\omega^2 + \left(1 + i\eta_s\right)\lambda_n^2\right]\delta_{nj} + \boldsymbol{\psi}_n^T \left[\tilde{\mathcal{S}}_f(\omega) - \left(1 + i\eta_s\right)\tilde{\mathcal{S}}_f(0)\right]\boldsymbol{\psi}_j \tag{8.4.10a}$$

$$L_n^l(\omega) = -\boldsymbol{\psi}_n^T \mathbf{m}_c \boldsymbol{\iota}_c^l + \left\{\boldsymbol{\psi}_{fn}^T\right\} \bar{\mathbf{R}}_h^l(\omega) \tag{8.4.10b}$$

for $n, j = 1, 2, 3, \ldots, J$; $\bar{\mathbf{Z}}^l(\omega)$ is the vector of frequency response functions $\bar{Z}_j^l(\omega)$ for the generalized coordinates $Z_j(t)$; δ_{nj} is the Kronecker delta function; $\boldsymbol{\psi}_{fn}$ is the subvector of $\boldsymbol{\psi}_n$ that contains only those elements corresponding to the nodal points at the upstream face of the dam.

Equations (8.4.9) and (8.4.10) represent J complex-valued, frequency-dependent equations in the unknown generalized coordinates. These equations need to be solved for several hundred to a few thousand values of the excitation frequency to determine the frequency response functions. Fortunately, accurate solutions can be obtained by including only a small number of Ritz vectors, typically less than 20, thus greatly reducing the computational effort (Tan and Chopra 1995a).

8.5 FREQUENCY-DOMAIN EQUATIONS: FLUID DOMAIN SUBSTRUCTURE

The unknown forces $\mathbf{R}_h(t)$, whose frequency response functions $\bar{\mathbf{R}}_h^l(\omega)$ appear in Eq. (8.4.10a), can be expressed in terms of the accelerations at the upstream face of the dam and the reservoir boundary by analyzing the fluid domain. The motion of water is governed by the three-dimensional wave equation:

$$\frac{\partial^2 p}{\partial x^2} + \frac{\partial^2 p}{\partial y^2} + \frac{\partial^2 p}{\partial z^2} = \frac{1}{C^2}\frac{\partial^2 p}{\partial t^2} \tag{8.5.1}$$

where $p(x, y, z, t)$ is the hydrodynamic pressure (in excess of hydrostatic pressure), and C is the velocity of pressure waves in water. For harmonic ground acceleration $a_g^l(t) = e^{i\omega t}$, the hydrodynamic pressure can be expressed as $p(x, y, z, t) = \bar{p}^l(x, y, z, \omega)e^{i\omega t}$, where $\bar{p}^l(x, y, z, \omega)$ is the complex-valued frequency response function for hydrodynamic pressure that is governed by the Helmholtz equation:

$$\frac{\partial^2 \bar{p}}{\partial x^2} + \frac{\partial^2 \bar{p}}{\partial y^2} + \frac{\partial^2 \bar{p}}{\partial z^2} + \frac{\omega^2}{C^2}\,\bar{p} = 0 \tag{8.5.2}$$

The linear form of the governing equation and the boundary conditions permits the hydrodynamic pressure to be expressed as

$$\bar{p}^l(x, y, z, \omega) = \bar{p}_0^l(x, y, z, \omega) + \sum_{j=1}^{J} \bar{\ddot{Z}}_j^l(\omega)\bar{p}_j^f(x, y, z, \omega) \tag{8.5.3}$$

In Eq. (8.5.3), $\bar{p}_0(x, y, z, \omega)$ is the frequency response for the hydrodynamic pressure due to the lth component of ground acceleration of a rigid dam and reservoir boundary (Figure 8.5.1a). This is the solution of Eq. (8.5.2) subjected to the radiation condition at $x = \infty$ and the following boundary conditions:

$$\begin{gathered}
\frac{\partial}{\partial n}\,\bar{p}_0^l(s, r, \omega) = -\rho\varepsilon^l(s, r) \\
\left[\frac{\partial}{\partial n} - i\omega\xi\right]\bar{p}_0^l(s, r, \omega) = -\rho\varepsilon^l(s', r') \\
\bar{p}_0^l(x, H, z, \omega) = 0
\end{gathered} \tag{8.5.4}$$

in which H is the y-coordinate of the free surface of water measured from the bottom of the reservoir; ρ is the mass density of water; ξ is the damping coefficient first introduced in Section 2.3.1; s and r are the localized spatial coordinates on the upstream face of the dam; s' and r' are the localized spatial coordinates on the reservoir boundary (Figure 8.5.2); $\varepsilon^l(s, r)$ is a function defined along accelerating boundaries ($s, r = s, r$ for the upstream face of the dam or $s, r = s', r'$ for the reservoir boundary) as the length of the l-component ($l = x$, y, or z) of a unit vector along the direction of the inward normal n (Figure 8.5.2).

In Eq. (8.5.3), $\bar{p}_j^f(x, y, z, \omega)$ is the frequency response function for the hydrodynamic pressure due to harmonic acceleration of the upstream face of the dam corresponding to the jth Ritz

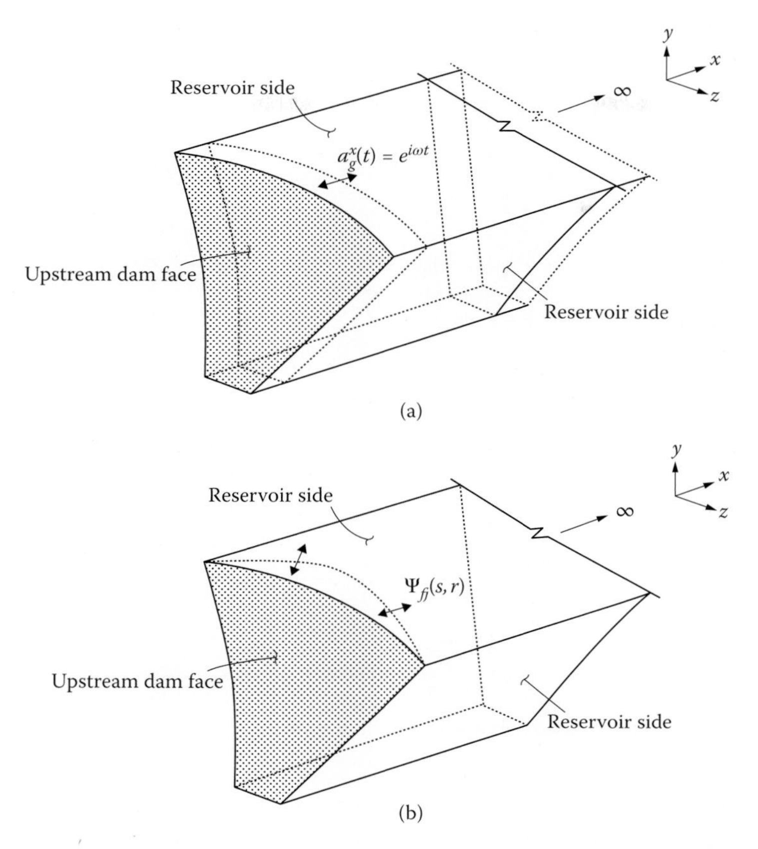

Figure 8.5.1 Reservoir boundary accelerations causing hydrodynamic pressures on the upstream face of the dam defined by frequency response functions: (a) $p_0(s, r, \omega)$ and (b) $p_j(s, r, \omega)$; part (a) is shown for the x-component of ground motion.

vector with no motion at the reservoir boundary (Figure 8.5.1b). This is the solution of Eq. (8.5.2) subjected to the radiation condition at $x = \infty$ and the following boundary conditions:

$$\frac{\partial}{\partial n}\,\overline{p}_j^f(s, r, \omega) = -\rho\psi_{fj}(s, r)$$

$$\left[\frac{\partial}{\partial n} - i\omega\xi\right]\overline{p}_j^f\left(s', r', \omega\right) = 0 \tag{8.5.5}$$

$$\overline{p}_j^f(x, H, z, \omega) = 0$$

where $\psi_{fj}(s, r)$ is the function representing the normal component of the jth Ritz vector at the dam–water interface.

Procedures for solving these BVPs and evaluating $\overline{p}_0(x, y, z, \omega)$ and $\overline{p}_j^f(x, y, z, \omega)$ for the fluid domain, idealized as shown in Figure 8.1.1, are available (Hall and Chopra 1983; Fok and Chopra 1986b).

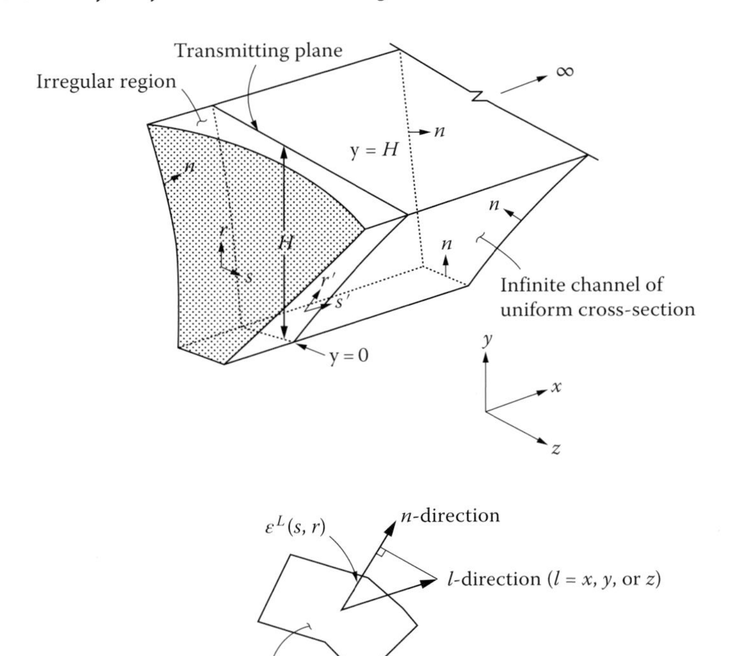

$\varepsilon^L(s, r)$

n-direction

l-direction ($l = x, y,$ or z)

Reservoir boundary defined by $s, r = s, r$ or s', r' coordinates

Figure 8.5.2 Definition of various terms associated with the fluid domain.

The frequency response functions for hydrodynamic forces $\mathbf{R}_h^l(t)$ associated with the hydrodynamic pressure $\bar{p}^l(x, y, z, \omega)$ are derived from Eq. (8.5.3):

$$\bar{\mathbf{R}}_h^l(\omega) = \bar{\mathbf{R}}_0^l(\omega) + \sum_{j=1}^{J} \bar{\ddot{Z}}_j^l(\omega)\ \bar{\mathbf{R}}_j^f(\omega) \tag{8.5.6}$$

where $\bar{\mathbf{R}}_0^l(\omega)$ and $\bar{\mathbf{R}}_j^f(\omega)$ are the nodal forces statically equivalent to the corresponding pressure functions $-\bar{p}_0^l(x, y, z, \omega)$ and $-\bar{p}_j^f(x, y, z, \omega)$, respectively, evaluated at the upstream face of the dam.[†]

8.6 FREQUENCY-DOMAIN EQUATIONS: DAM–WATER–FOUNDATION SYSTEM

The hydrodynamic forces at the upstream face of the dam have been expressed in terms of the acceleration $\bar{\ddot{Z}}_j^l(\omega)$ of the generalized coordinates. The substitution of Eq. (8.5.6) and $\bar{\ddot{Z}}_j^l(\omega) = -\omega^2 \bar{Z}_j(\omega)$, into Eqs. (8.4.9) and (8.4.10) leads to

$$\mathbf{S}(\omega)\, \bar{\mathbf{Z}}^l(\omega) = \mathbf{L}^l(\omega) \tag{8.6.1}$$

[†] The algebraic sign attached to the pressure functions differs from that in the text following Equation (5.5.13) because the coordinate systems were defined differently in the two cases. This inconsistency has not been eliminated herein to facilitate reference to the original publications that formed the basis for Chapters 5 and 8.

where the elements in the matrix $\mathbf{S}(\omega)$ and the vector $\mathbf{L}^l(\omega)$ – after rearrangement of terms – are given by

$$S_{nj}(\omega) = \left[-\omega^2 + \left(1 + i\eta_s\right) \lambda_n^2\right] \delta_{nj} + \boldsymbol{\psi}_n^T \left[\tilde{\mathcal{S}}_f(\omega) - \left(1 + i\eta_s\right) \tilde{\mathcal{S}}_f(0)\right] \boldsymbol{\psi}_j + \omega^2 \boldsymbol{\psi}_{fn}^T \bar{\mathbf{R}}_j^f(\omega) \quad (8.6.2a)$$

$$L_n^l(\omega) = -\boldsymbol{\psi}_n^T \mathbf{m}_c \boldsymbol{\iota}_c^l + \boldsymbol{\psi}_{fn}^T \bar{\mathbf{R}}_0^l(\omega) \quad (8.6.2b)$$

The frequency-dependent hydrodynamic terms may be interpreted as follows: the real-valued part of the last term on the right side of Eq. (8.6.2a) represents an added mass matrix, and its imaginary-valued part is a damping matrix; the last term on the right side of Eq. (8.6.2b) is an added force vector.

Equations (8.6.1) and (8.6.2) contain the effects of dam–water interaction and of dam–foundation interaction in various forms: (i) dam–foundation interaction effects appear in eigenvalues λ_n and eigenvectors $\boldsymbol{\psi}_n$ of the associated dam–foundation system through the foundation stiffness matrix $\tilde{\mathcal{S}}_f(\omega)$; (ii) additional hydrodynamic forces $\mathbf{R}_0^l(\omega)$ on a rigid dam due to the free-field ground motions; (iii) dam–water interaction effects appear through the hydrodynamic forces $\bar{\mathbf{R}}_j^f(\omega)$ on the upstream face of the dam; and (iv) water–foundation rock interaction effects have been neglected because they are known to be small; see Appendix 5 and Fok and Chopra (1986a).

8.7 RESPONSE HISTORY ANALYSIS

The response history of the dam to arbitrary ground motion can be computed once the complex-valued frequency response functions $\bar{Z}_j^l(\omega)$ – $l = x, y, z$ and $j = 1, 2, \ldots J$ – for the generalized coordinates have been obtained by solving Eqs. (8.6.1) and (8.6.2) for all excitation frequencies in the range of interest. The time-functions for generalized coordinates are given by the Fourier integral as the superposition of responses to individual harmonic components of the ground motion:

$$Z_j^l(t) = \frac{1}{2\pi} \int_{-\infty}^{\infty} \bar{Z}_j^l(\omega)\; A_g^l(\omega)\, e^{i\omega t} d\omega \quad (8.7.1)$$

where $A_g^l(\omega)$ is the Fourier transform of the l-component of the specified free-field ground acceleration $a_g^l(t)$:

$$A_g^l(\omega) = \int_0^d a_g^l(t)\, e^{i\omega t} dt \quad (8.7.2)$$

in which d is the duration of the ground motion. The Fourier integrals in Eqs. (8.7.1) and (8.7.2) are computed in their discrete form using the Fast Fourier Transform (FFT) algorithm. The displacement response to the stream, vertical, and cross-stream components of ground motion, simultaneously, is obtained by transforming the generalized coordinates to the nodal displacements according to Eq. (8.4.5):

$$\mathbf{r}_c(t) = \sum_{j=1}^{J} \left[Z_j^x(t) + Z_j^y(t) + Z_j^z(t)\right] \boldsymbol{\psi}_j \quad (8.7.3)$$

The stresses in the dam at any instant of time can be determined from the nodal displacements. The vector $\boldsymbol{\sigma}_p(t)$ of stress components in finite element p is related to the nodal displacement vector $\mathbf{r}_p(t)$ for that element by

$$\boldsymbol{\sigma}_p(t) = \mathbf{T}_p \mathbf{r}_p(t) \quad (8.7.4)$$

where $\mathbf{T}_p$ is the stress-displacement transformation matrix for the element.

A step-by-step summary of the analytical procedure developed in Sections 8.2 through 8.7 is available in Tan and Chopra (1995a).

8.8 EXTENSION TO SPATIALLY VARYING GROUND MOTION

Ground motions recorded at arch dams vary along the dam–foundation interface. Such data include records obtained at three dams: (i) Pacoima Dam (California, USA) during the magnitude 4.3 earthquake on January 13, 2001, and the magnitude 6.9 Northridge earthquake on January 17, 1994 (Hall 1996; Alves and Hall 2006a); (ii) Mauvoisin Dam (Switzerland) during the magnitude 4.6 Valpelline earthquake on March 31, 1996 (Proulx et al. 2004); and Fei-Tsui Dam (Taiwan) during the September 21, 1999, magnitude 7.3 Chi-Chi earthquake (Loh and Wu 2000).

Rarely are these spatial variations in ground motion considered in earthquake analysis of arch dams, and when they are included, dam–water–foundation interaction is usually oversimplified. Water compressibility, foundation mass, and foundation damping (material and radiation) are all ignored, and the semi-unbounded extent of the foundation and fluid domains is not recognized (e.g. Alves and Hall 2006b; Mojtahedi and Fenves 2000). All these factors are known to be important in determining the earthquake response of arch dams to spatially-uniform ground motion, as will be demonstrated in Chapters 9 and 10; therefore, they should also be included in analyzing dam response to spatially-varying excitation.

The substructure method presented in the preceding sections of this chapter has been generalized to determine the response of the dam to spatially-varying ground motion (Wang and Chopra 2010). Without developing the extended procedure in this section, we summarize the principal changes in the final equations. The equations of motion of the dam must now be formulated in terms of the total displacements $\mathbf{r}_c^t$; its two subvectors are $\mathbf{r}_b^t$ for the nodes on the dam–foundation interface and $\mathbf{r}^t$ for all other nodes. The total displacements are separated into two parts:

$$\left\{ \begin{array}{c} \mathbf{r}^t(t) \\ \mathbf{r}_b^t(t) \end{array} \right\} = \left\{ \begin{array}{c} \mathbf{r}^s(t) \\ \mathbf{r}_b^f(t) \end{array} \right\} + \left\{ \begin{array}{c} \mathbf{r}(t) \\ \mathbf{r}_b(t) \end{array} \right\} \tag{8.8.1}$$

where $\mathbf{r}^s$, known as quasi-static displacement, are the structural displacements due to the static application of the earthquake-induced free-field displacements $\mathbf{r}_b^f$ at the dam–foundation interface. Thus the two are related through

$$\left[\mathbf{k} \;\; \mathbf{k}_b \right] \left\{ \begin{array}{c} \mathbf{r}^s \\ \mathbf{r}_b^f \end{array} \right\} = \mathbf{0} \tag{8.8.2}$$

and $\mathbf{r}^s$ can be expressed as

$$\mathbf{r}^s = -\mathbf{k}^{-1}\mathbf{k}_b\mathbf{r}_b^f \tag{8.8.3}$$

The frequency domain analysis procedure is now formulated in terms of the Fourier transforms† of the response quantities instead of their complex frequency response functions when the excitation was spatially uniform (Sections 8.2–8.7). For spatially-varying excitations, Eqs. (8.6.1) and (8.6.2) have been generalized to

$$\mathbf{S}(\omega)\hat{\mathbf{Z}}(\omega) = \mathbf{L}(\omega) \tag{8.8.4}$$

where the elements in the matrix $\mathbf{S}(\omega)$ and the vector $\mathbf{L}(\omega)$ are given by

$$S_{nj}(\omega) = \left[-\omega^2 + \left(1 + i\eta_s\right) \lambda_n^2 \right] \delta_{nj} + \boldsymbol{\psi}_n^T \left[\tilde{\boldsymbol{\mathcal{S}}}_f(\omega) - \left(1 + i\eta_s\right) \; \tilde{\boldsymbol{\mathcal{S}}}_f(0) \right] \boldsymbol{\psi}_j + \omega^2 \boldsymbol{\psi}_{fn}^T \hat{\mathbf{R}}_j^f(\omega) \tag{8.8.5a}$$

$$L_n(\omega) = \boldsymbol{\psi}_{fn}^T \hat{\mathbf{R}}_0(\omega) + \omega^2 \boldsymbol{\psi}_n^T \mathbf{m}_c \hat{\mathbf{r}}_c^s(\omega) - \left(1 + i\eta_s\right) \boldsymbol{\psi}_{bn}^T \left[\mathbf{k}_b^T \mathbf{k}^{-1} \mathbf{k}_b + \mathbf{k}_{bb} \right] \hat{\mathbf{r}}_b^f(\omega) \tag{8.8.5b}$$

† The Fourier transform of $x(t)$. is defined as $\hat{x}(\omega) = \int_{-\infty}^{\infty} x(t) e^{-i\omega t} dt$, where $i = \sqrt{-1}$.

for $n, j = 1, 2, 3, \ldots, J$; $\hat{\mathbf{Z}}(\omega)$ is the vector of Fourier transforms $\hat{Z}_j(\omega)$ of the generalized coordinates $Z_j(t)$, δ_{nj} is the Kronecker delta function; $\boldsymbol{\psi}_{bn}$ is the subvector of $\boldsymbol{\psi}_n$ corresponding to $3N_b$ DOFs at the dam–foundation interface; $\boldsymbol{\psi}_{fn}$ is the subvector of $\boldsymbol{\psi}_n$ corresponding to nodal points at the upstream face of the dam.

By comparing Eqs. (8.6.1) and (8.6.2) with Eqs. (8.8.4) and (8.8.5), observe that the coefficient matrix $\mathbf{S}(\omega)$ on the left side remains unchanged, but spatially varying ground motion leads to major changes in the terms contained in $\mathbf{L}_n(\omega)$. The principal changes are in the terms associated directly with the ground motion and the resulting hydrodynamic pressures. The vector $\hat{\mathbf{R}}_0(\omega)$ represents the hydrodynamic forces equivalent to the hydrodynamic pressures $-\hat{p}_0(x, y, z, \omega)$ due to the quasi-static motion $\hat{\mathbf{r}}^s(\omega)$ of the upstream face of the dam and the free-field ground motion $\hat{\mathbf{r}}^f_b(\omega)$ of the reservoir boundary, which may vary spatially around the canyon, but are assumed to be spatially uniform along the upstream direction (Figure 8.8.1). This is the solution of Eq. (8.5.2) with $\bar{p}$ replaced by $\hat{p}$ subjected to the radiation condition at $x = \infty$ and the following boundary conditions:

$$\frac{\partial}{\partial n}\,\hat{p}_0(s, r, \omega) = -\rho\,\hat{\ddot{r}}^s_n(s, r, \omega)$$

$$\left[\frac{\partial}{\partial n} - i\omega q\right]\hat{p}_0(s, r, \omega) = -\rho\,\hat{\ddot{r}}^f_{bn}\,(s', r', \omega)$$

$$\bar{p}_0(x, H, z, \omega) = 0 \tag{8.8.6}$$

where $\hat{\ddot{r}}^s_n(s, r, \omega)$ is the function representing the component of $\hat{\ddot{\mathbf{r}}}^s(\omega)$ in the direction of the inward normal n to the upstream face; and $\hat{\ddot{r}}^f_{bn}\,(s', r', \omega)$ is the component of the free-field ground motion $\hat{\ddot{\mathbf{r}}}^f_b(\omega)$ in the direction of the inward normal n (Figure 8.8.2). The vector $\hat{\mathbf{R}}^f_j(\omega)$ contains the nodal forces statically equivalent to the hydrodynamic pressures $-\hat{p}^f_j(x, y, z, \omega)$ due to the harmonic acceleration of the dam in the jth Ritz vector without any reservoir boundary motion (Figure 8.5.1b). This is the solution of Eq. (8.5.2) with $\bar{p}$ replaced by $\hat{p}$ subjected to the radiation condition at $x = \infty$ and the boundary conditions of Eq. (8.5.5).

Once the Fourier transforms, $\hat{Z}_j(\omega)$, $j = 1, 2, \ldots, J$, of the generalized coordinates are obtained by solving Eqs. (8.8.4) and (8.8.5) for all excitation frequencies in the range of interest, the response of the dam to arbitrary ground motion can be computed. The generalized coordinate-time functions are given by the Fourier integral

$$Z_j(t) = \frac{1}{2\pi}\int_{-\infty}^{\infty} \hat{Z}_j(\omega)\, e^{i\omega t} d\omega \tag{8.8.7}$$

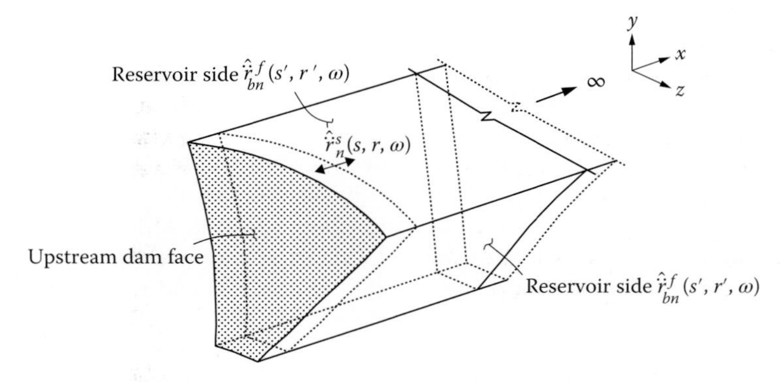

Figure 8.8.1 Reservoir boundary accelerations causing hydrodynamic pressures on the upstream face of the dam defined by Fourier transform $\hat{p}_0(s, r, \omega)$; for convenience, the figure is shown for spatially-uniform ground motion along the x-direction.

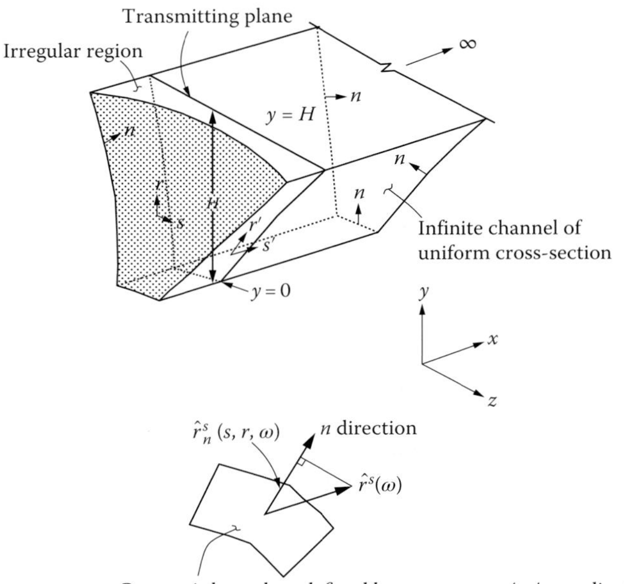

Figure 8.8.2 Definition of various terms associated with the fluid domain.

and the displacements $\mathbf{r}_c(t)$ by Eq. (8.4.5). The total (quasi-static plus dynamic) displacements are then given by Eq. (8.8.1) and the element stresses by Eq. (8.7.4) wherein $\mathbf{r}_p(t)$ now denotes the total displacements.

Before closing this section, we note that if the ground motion is spatially uniform: (i) the boundary conditions of Eq. (8.8.6) reduce to Eq. (8.5.4); and (ii) the free-field displacements $\mathbf{r}_b^f(t)$ at the dam–foundation interface and the quasi-static displacements $\mathbf{r}^s(t)$ reduce to (Chopra 2012, Section 9.7)

$$\left\{ \begin{array}{c} \mathbf{r}^s(t) \\ \mathbf{r}_b^f(t) \end{array} \right\} = \boldsymbol{\iota}_c^l u_g^l(t) \tag{8.8.8}$$

where the influence vector $\boldsymbol{\iota}_c^l$ was defined in Eq. (8.2.3) with the superscript $l = x, y,$ or z denoting the component of ground motion; and (iii) Eq. (8.8.5b) reduces to Eq. (8.6.2b).

The above-described procedure for analysis of the earthquake response of arch dams due to spatially varying excitation is summarized as a sequence of steps in Chopra and Wang (2008).

8.9 EACD-3D-2008 COMPUTER PROGRAM

The analytical procedure for the three-dimensional analysis of earthquake response of concrete dams presented in Sections 8.2–8.7 (Tan and Chopra 1995a,b) was implemented in the computer program EACD-3D-96 developed in 1996 (Tan and Chopra 1996a). Although restricted to spatially-uniform ground motion specified at the dam–foundation interface, this procedure recognized the semi-unbounded size of the foundation and fluid domains and included the effects of dam–water interaction with water compressibility and reservoir boundary absorption considered; and of dam–foundation interaction with inertia and damping (material and radiation) of the foundation considered. The EACD-3D-96 software may be accessed at the following URL: http://nisee.berkeley.edu/elibrary/Software/EACD3D96ZIP.

To overcome the aforementioned restriction, EACD-3D-96 was extended to implement the procedure developed in Wang and Chopra (2010), which was summarized in Section 8.8, for analysis of dam response to spatially-varying excitation, resulting in the computer program EACD-3D-2008. This software may be accessed from NISEE at the following URL: http://nisee.berkeley.edu/elibrary/Software/EACD-3D-2008ZIP.

The analytical procedure underlying EACD-3D-2008 assumes linear behavior for three substructures: concrete dam, fluid domain, and foundation domain; thus the possibility of concrete cracking, contraction joints of the dam opening during vibration, or water cavitation are not considered. The initial static analysis considers only the effects of the self-weight of the dam and hydrostatic pressures on the upstream face of the dam; thermal effects in the concrete or construction sequence of the dam are not included. However, the user may perform a separate static analysis that includes thermal, creep, construction sequence, and other effects, and import the resulting initial stresses in the computer program. The output from the computer program includes the complete response history of (i) the stream, cross-stream, and vertical components of displacements at all nodal points; and (ii) arch, cantilever, and principal stresses in all the finite elements. From these results, the user can plot the distribution of stresses in the dam at selected time instants and the distribution of envelope values of stresses in the dam; see Chapters 9–10. Such results aid in identifying areas of the dam that may crack during an earthquake.

9

Earthquake Analysis of Arch Dams: Factors to Be Included

PREVIEW

The principal effects of dam–water–foundation interaction, gleaned from frequency response functions for arch dams computed for a wide range of system parameters are identified first. Thereafter, the significance of these effects in earthquake response of arch dams is investigated with the objective of identifying the factors that must be included to obtain realistic estimates of response in designing new dams and evaluating existing dams. For this purpose, results of earthquake response analyses of four arch dams, conducted by the US Bureau of Reclamation, are presented. Based on these results, we identify the implications of neglecting the following factors: (i) mass of the foundation rock; and (ii) compressibility of water, assumptions that are not uncommon in analyses of dams implemented by professional engineers.

In the second part of this chapter, we explore the influence of spatial variation in ground motions around the canyon in earthquake response of dams. Analysis of Pacoima Dam for ground motions recorded during two earthquakes demonstrates that the spatial variation, typically ignored in dam engineering practice, can – in some cases – have profound influence on the earthquake-induced stresses in the dam.

9.1 DAM–WATER–FOUNDATION INTERACTION EFFECTS

Utilizing the analytical procedure presented in Chapter 8, comprehensive investigations of the effects of dam–foundation interaction, dam–water interaction, and reservoir boundary absorption on the response of Morrow Point Dam to harmonic ground motion have been reported (Fok and Chopra 1986c; Tan and Chopra 1995c). Frequency-response functions and earthquake response histories presented for a wide range of system parameters lead to the several observations and conclusions, summarized in this section

9.1.1 Dam–Water Interaction

Dam–water interaction lengthens the fundamental resonant period of the dam. Presented in Figure 9.1.1 is the ratio $\tilde{T}_r/T_1$ of the resonant period $\tilde{T}_r$ of the dam with reservoir filled to depth H to the period T_1 with empty reservoir plotted against normalized water depth H/H_s. Results are presented for the resonant period of the fundamental symmetrical and anti-symmetrical modes of vibration. Also presented for reference is the period ratio for a concrete gravity dam. These results are applicable to dams of any height with the specified geometry, and chosen values for Poisson's ratio, E_S, and H/H_s, and they are presented for a non-absorptive reservoir boundary ($\alpha = 1$). As for gravity dams (Chapter 2), dam–water interaction lengthens the vibration period also of arch dams, with the effect being negligible for H/H_s less than 0.5, but increases rapidly with water depth for H/H_s greater than 0.5. Dam–water interaction lengthens the vibration period of the symmetrical mode of arch dams more than that of gravity dams because the added hydrodynamic mass is a larger fraction of the mass of a slender arch dam than of a massive gravity dam. However, the period of the fundamental anti-symmetrical vibration mode of an arch dam is lengthened to a lesser degree than the symmetric vibration mode of the arch dam or a gravity dam.

Dam–water interaction has the effect of increasing the fundamental mode resonant response of arch dams to upstream or vertical ground motion, but decreasing the response to cross-stream ground motion. These effects are reduced in the presence of reservoir boundary absorption.

Since the classic paper by Westergaard (1933), many have argued that compressibility of water may be ignored, and hydrodynamic effects can be modeled by an added mass of water moving with the dam, thus simplifying the analysis greatly. To investigate the validity of this assertion, the effects of water compressibility on the earthquake response of arch dams were studied in depth (Fok and Chopra 1987). It was demonstrated that, as in the case of gravity dams, the key parameter that determines the significance of water compressibility effects in the response of arch dams is Ω_r, the ratio of the fundamental natural vibration frequencies of the impounded water

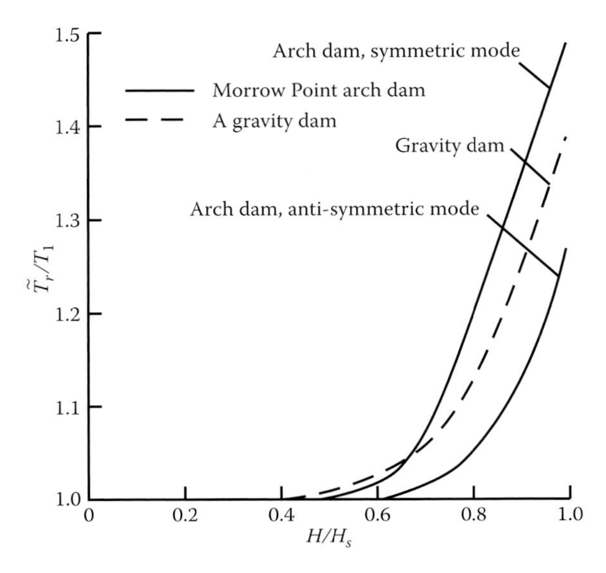

Figure 9.1.1 Variation of the fundamental period ratio, $\tilde{T}_r/T_1$, with water depth ratio H/H_s, for an arch dam and a gravity dam, both supported on a rigid foundation.

and dam alone. The effects of water compressibility become insignificant for larger values of Ω_r, i.e. smaller values of the Young's modulus E_s for mass concrete used in dams. For the range of $E_s = 2$–5 million psi, which is representative of mass concrete, the effects of water compressibility were demonstrated to be significant. Later in this chapter, the large errors introduced by ignoring these effects will be documented in the context of response of two dams to earthquake excitation.

9.1.2 Dam–Foundation Interaction

Dam–foundation interaction lengthens the fundamental resonant period of the dam primarily because of foundation flexibility, and widens the frequency bandwidth at the fundamental resonance because of material damping in the rock and radiation damping associated with wave propagation away from the dam in the semi-unbounded foundation domain (Tan and Chopra 1995c). As a result, the fundamental resonant response of the dam is reduced. These effects of dam–foundation interaction increase as E_f/E_s decreases, which for a fixed concrete modulus E_s implies decreasing foundation modulus E_f.

The ratio $\tilde{T}_f/T_1$ of the fundamental resonant period $\tilde{T}_f$ of the dam supported on flexible foundation to T_1 on rigid foundation is plotted in Figure 9.1.2 as a function of E_f/E_s; also included is the period ratio for a gravity dam. Presented in this form, these results are applicable to dams of any height with the specific geometry and chosen values for Poisson's ratio and density of concrete and rock. For a fixed E_f/E_s value, the period of the symmetric mode is lengthened more than that of the anti-symmetric mode. In particular, for E_f/E_s = ¼, the period of the symmetric mode is lengthened by 32% compared to 25% for the anti-symmetric mode. Dam–foundation interaction lengthens the vibration period of arch dams much less than that of gravity dams because they are less massive and the area of the dam–rock interface is smaller.

The effective damping ratio of the dam, estimated by the half-power bandwidth method applied to the frequency-response function near the fundamental resonance, is plotted against the E_f/E_s ratio in Figure 9.1.3. Dam–foundation interaction has the effect of increasing the effective damping ratio – compared to 5% for the dam on rigid foundation – as E_f/E_s decreases, suggesting that radiation damping associated with interaction increases as the foundation becomes more flexible. For a fixed E_f/E_s value, the effective damping ratio of the symmetric

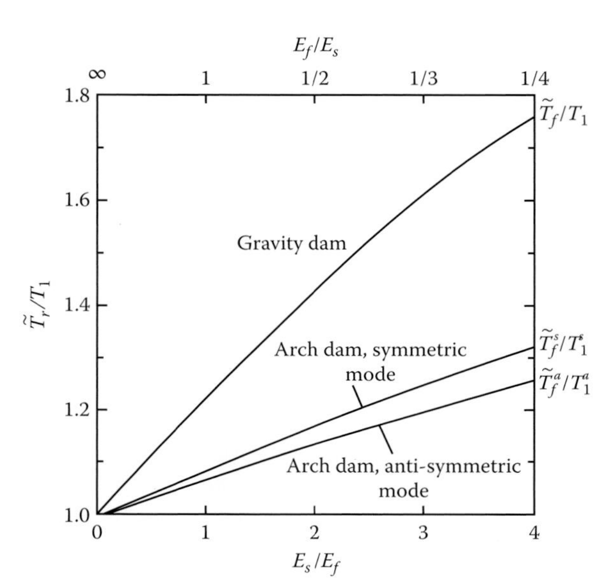

Figure 9.1.2 Variation of the fundamental period ratio, $\tilde{T}_f/T_1$, with the moduli ratio E_f/E_s, for an arch dam and a gravity dam, both with an empty reservoir.

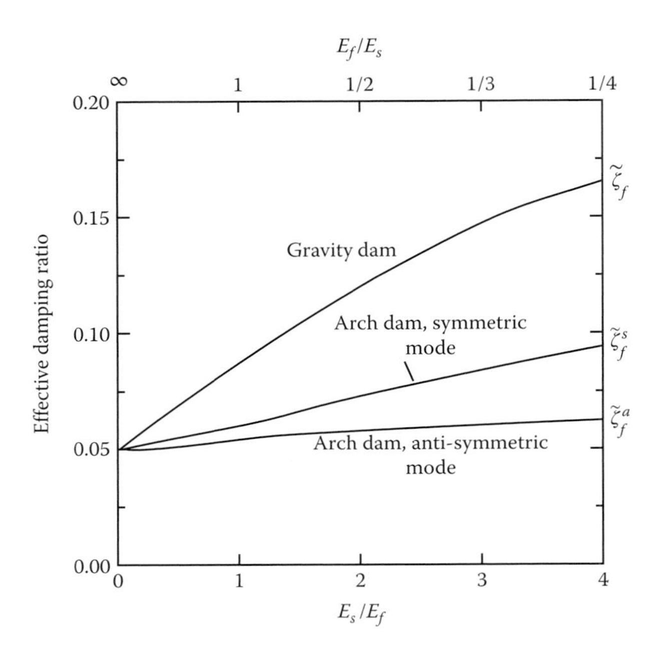

Figure 9.1.3 Variation of the effective damping ratio with the moduli ratio E_f/E_s, for an arch dam and a gravity dam, both with an empty reservoir.

mode is increased more than that of the anti-symmetric mode. In particular, for $E_f/E_s = 1/4$, the effective damping ratio increases from 5.0% to 9.5% in the symmetric vibration mode; and from 5.0% to 6.3% in the anti-symmetric vibration mode. This increase in effective damping of arch dams is much less than for gravity dams, thus confirming the earlier observation that dam–foundation interaction effects are relatively less significant for arch dams.

The preceding results indicate that dam–foundation interaction affects the response of the dam in its symmetric vibration modes, excited by upstream and vertical ground motions, more than in its anti-symmetric modes, excited by cross-stream ground motion. This observation is valid even for dams that are not perfectly symmetric so long as the notion of "symmetric" and "anti-symmetric" nodes is interpreted liberally.

Earthquake analyses of arch dams is greatly simplified if the foundation rock is assumed to have no mass because then only flexibility of the foundation needs to be considered, which can be computed by analyzing a bounded size model of the foundation as part of a standard finite element analysis. Can this seemingly unrealistic assumption be justified?

An earlier investigation (Tan and Chopra 1995a,c) compared the frequency-response functions for the dam computed for two cases: (i) all effects of dam–foundation interaction are included; and (ii) the mass of foundation rock is ignored, implying that flexibility of the foundation is included but foundation mass, material damping, and radiation damping are all ignored. This investigation demonstrated that foundation flexibility accounts for the shift in the fundamental resonant frequency, although not always accurately. However, the response amplitudes at the fundamental and higher resonant frequencies are overestimated when mass of the foundation rock is ignored, because then foundation material and radiation damping is implicity ignored. This overestimation of response is especially large for the smaller values of E_f/E_s.

These results suggest that the earthquake response of the dam will be overestimated if the foundation rock is assumed to be massless. We will demonstrate in Section 9.2.1 that this overestimation can be unacceptably large.

9.1.3 Dam–Water–Foundation Interaction

The simultaneous effects of dam–foundation interaction and of dam–water interaction on the response of the dam to upstream, vertical, and cross-stream ground motion have been investigated (Tan and Chopra 1995a,c), but those results and their detailed interpretation are not included here. This presentation is limited to three overall observations: (i) numerical results for arch dams satisfy Eq. (3.5.1) presented earlier for gravity dams, implying that reductions in ω_1 due to dam–water interaction and due to dam–foundation interaction may be determined independently and combined to determine $\tilde{\omega}_1$ for the dam–water–foundation system; (ii) the energy loss due to reservoir boundary absorption is more effective in reducing the response of the dam supported on stiffer foundations; and (iii) the damping – material and radiation – due to dam–foundation interaction is more effective in reducing the response of the dam as the reservoir boundary becomes less absorptive.

9.1.4 Earthquake Responses

The effects of dam–foundation interaction, dam–water interaction, and reservoir boundary absorption on the dynamics of arch dams, gleaned from frequency response functions have been confirmed by a detailed investigation of the response of the Morrow Point Dam to the three components of the Taft ground motion (Fok and Chopra 1986a, 1987; Tan and Chopra 1996b). Patterned after the interpretation of response to gravity dams (Chapter 6), these investigations provided a structural-dynamics-based interpretation of how and why the various interaction mechanisms affect the response. However, these interpretations were more challenging because the three-dimensional dynamics of arch dams is much more complicated compared to the planar dynamics of a gravity dam monolith; and because, unlike gravity dams, the response of arch dams is not dominated by a single mode of vibration. Furthermore, the quantitative observations of the interaction effects cannot be generalized because they are strongly dependent on the three-dimensional geometry and thickness of the arch dam, properties that vary widely among dams.

For these reasons, here we present only the overall conclusions. First hydrodynamic and reservoir boundary absorption effects are more significant in the response of arch dams than for gravity dams because the added hydrodynamic mass, damping, and force are a bigger factor in the vibration properties of, and in the effective earthquake forces on, a slender arch dam compared to a massive gravity dam. Second, dam–foundation interaction effects are significant for arch dams but less than they were for massive gravity dams. Third, any analysis procedure that assumes the rock to be massless, i.e. considers only the flexibility of the foundation but ignores its mass and the damping – material and radiation – arising from dam–foundation interaction will significantly overestimate the earthquake-induced stresses in arch dams. Fourth, the effects of water compressibility will be significant in the response of most arch dams. The third and fourth conclusions will be further demonstrated in the next section based on earthquake responses of four arch dams.

9.2 BUREAU OF RECLAMATION ANALYSES

Starting in 1996, the US Bureau of Reclamation embarked upon a major program to evaluate the seismic safety of their dams. This investigation included earthquake analysis of several arch dams. Results presented herein are for four dams: Hoover Dam, a 221-m-high curved gravity dam

Figure 9.2.1 Hoover Dam: a 221-m-high curved gravity dam. Source: Arnkjell Løkke/

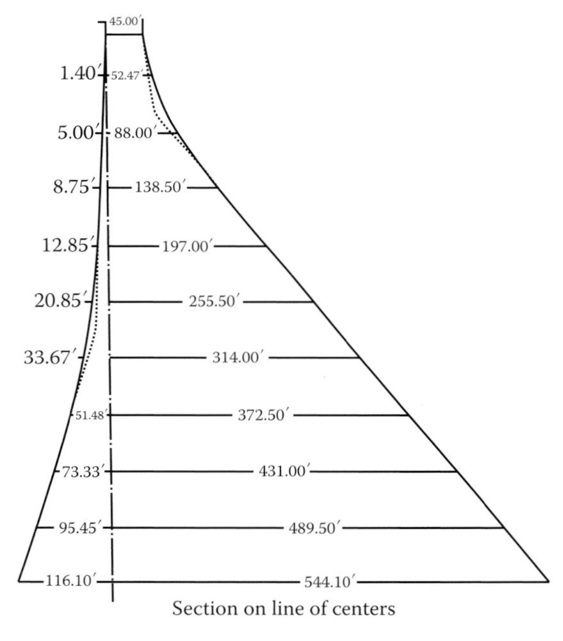

Figure 9.2.2 Cross section of Hoover Dam.

(Figures 9.2.1 and 9.2.2); Deadwood Dam, a 50-m-high single-curvature arch dam (Figure 9.2.3); Monticello Dam, a 93-m-high double-curvature arch dam (Figure 9.2.4); and Morrow Point Dam, a 142-m-high double-curvature arch dam (Figure 9.2.5). Ground motions consistent with the seismic hazard for each of the four dam sites were selected. All analyses were implemented by the EACD-3D-96 computer software (Bureau of Reclamation 1998a,b,c and 1999).

Figure 9.2.3 Deadwood Dam: a 50-m-high single-curvature dam.

Figure 9.2.4 Monticello Dam: a 93-m-high double-curvature arch dam.

Figure 9.2.5 Morrow Point Dam: a 142-m-high double-curvature dam.

9.2.1 Implications of Ignoring Foundation Mass

Compared are the earthquake-induced stresses in each of the four dams computed for two cases: (i) including all effects of dam–foundation interaction; and (ii) ignoring the mass of foundation rock, i.e. considering foundation flexibility only (Figures 9.2.6–9.2.9). For these two cases, the largest arch stress on the upstream or downstream face of the dam is 3.3 MPa (476 psi) versus 5.8 MPa (844 psi) for Deadwood Dam; 5.0 MPa (730 psi) versus 9.7 MPa (1410 psi) for Monticello Dam; 4.6 MPa (665 psi) versus 9.2 MPa (1336 psi) for Morrow Point Dam; and 5.2 MPa (758 psi) versus 15.2 MPa (2204 psi) for Hoover Dam. These results demonstrate that if the mass of foundation rock is ignored, stresses are overestimated by a factor of about 2 for the first three dams and by a factor of about 3 for Hoover Dam.

Dam–foundation interaction lowers the fundamental resonant frequency of the dam and increases the overall damping in the system. The lowering of the frequency is primarily due to

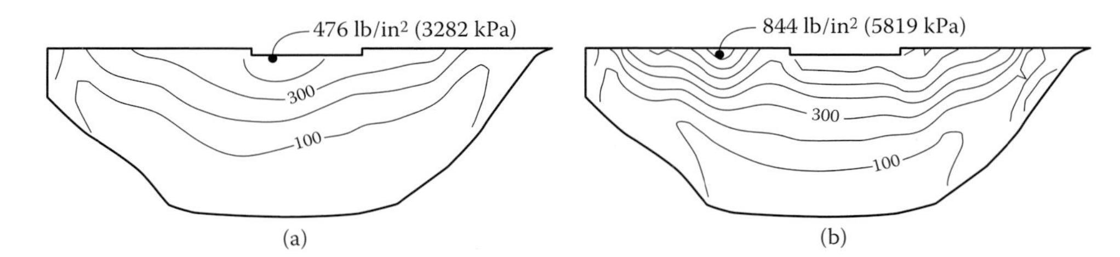

Figure 9.2.6 Peak values of tensile arch stresses in Deadwood Dam for two cases: (a) including all effects of dam–foundation interaction, and (b) ignoring the mass of foundation rock.

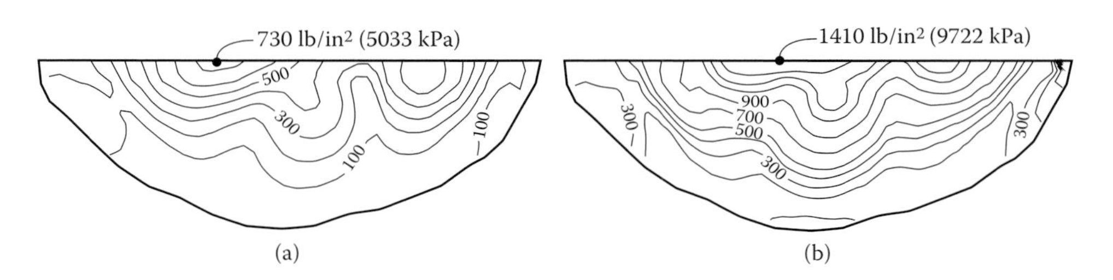

Figure 9.2.7 Peak values of tensile arch stresses in Monticello Dam for two cases: (a) including all effects of dam–foundation interaction, and (b) ignoring the mass of foundation rock.

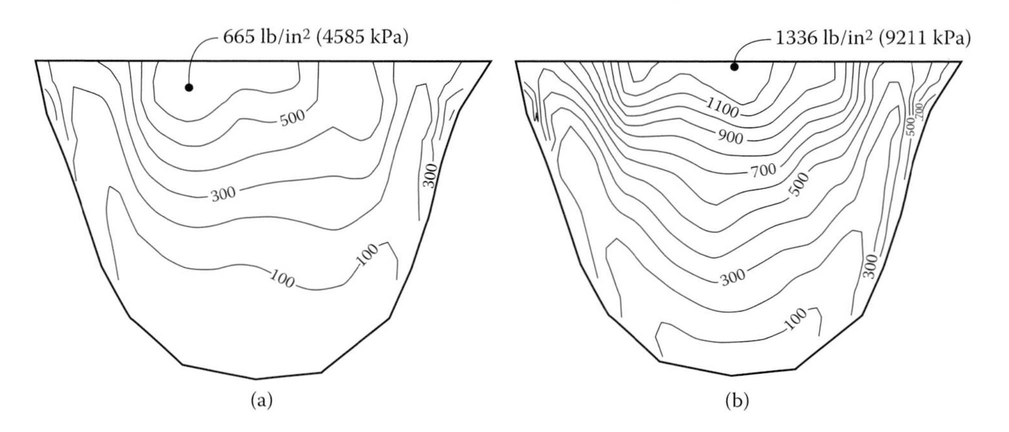

Figure 9.2.8 Peak values of tensile arch stresses in Morrow Point Dam for two cases: (a) including all effects of dam–foundation interaction, and (b) ignoring the mass of foundation rock.

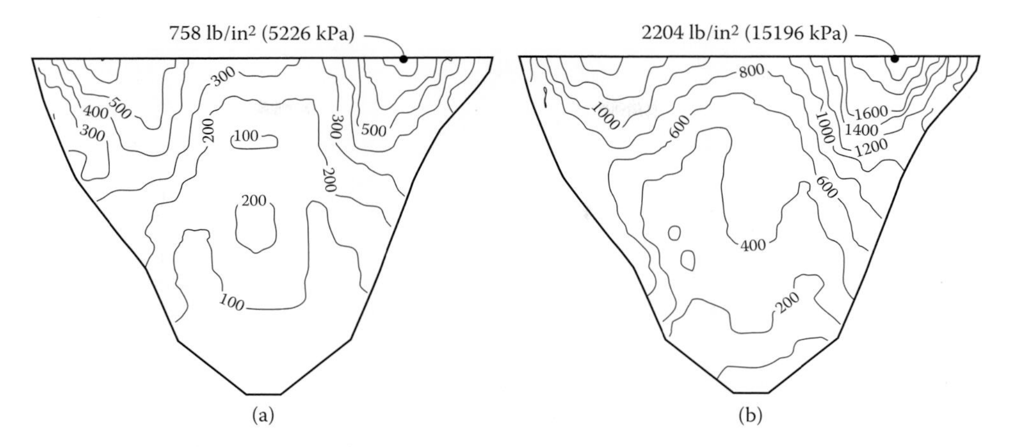

Figure 9.2.9 Peak values of tensile arch stresses in Hoover Dam for two cases: (a) including all effects of dam–foundation interaction, and (b) ignoring the mass of foundation rock.

foundation flexibility with secondary influence of foundation mass and damping (Tan and Chopra 1995c). However, if the mass of foundation rock is ignored, the increase in overall damping due to foundation damping – material and radiation – is implicitly ignored (Tan and Chopra 1995c), resulting in overestimation of earthquake response by factors of 2–3 for the dams considered.

As seen in Chapter 8, dam–foundation interaction introduces frequency-dependent (dynamic) stiffness matrix for the foundation, thus requiring analysis of the dam by a sub-structure method formulated in the frequency domain, implemented in specialized computer software, such as EACD-3D-2008. The temptation to use standard commercial finite-element analysis software instead has motivated the practice of assuming the foundation rock to be massless. It is evident from the results presented in Figures 9.2.6–9.2.9 that this assumption leads to excessive overestimation of stresses that may lead to over-conservative designs of new dams and to the erroneous conclusion that an existing dam is unsafe, thus requiring upgrading, which is invariably very expensive.

Independent confirmation of the preceding observations has been gleaned from earthquake records from M3.6 to M4.9 earthquakes centered 12–20 km away from three well-instrumented arch dams in Switzerland. Finite element models of the dam–foundation system, assuming the foundation to be massless, were developed and their properties calibrated against ambient vibration tests and forced vibration tests; such calibration led to damping ratios of 2–3% for the dams (Proulx et al. 2004). The response of the Mauvoisin and Punt-dal-Gall dams due to ground motion recorded in the free field away from the dam base, and of Emosson Dam due to base motion, were computed. Although the models were calibrated against data from ambient or forced vibration tests, the computed motions at the dam crest were much larger than the motions recorded during the earthquake.

Much better agreement between computed and recorded motions was achieved when mass of the rock was included – and consequently material and radiation damping was modeled – and the ground motion was defined as varying spatially, consistent with the earthquake records (see Section 10.1).

9.2.2 Implications of Ignoring Water Compressibility

Compared are the earthquake-induced stresses in two of the four Reclamation dams computed under two conditions: (i) including compressibility of water; and (ii) ignoring water compressibility; in the latter case, reservoir boundary absorption effects are implicitly ignored. By neglecting water compressibility and reservoir boundary absorption, the stresses may be significantly underestimated, as in the case of Monticello Dam (Figure 9.2.10), or considerably overestimated

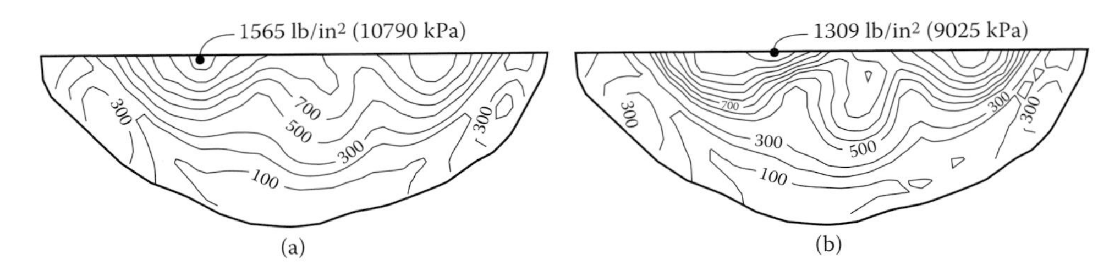

Figure 9.2.10 Peak values of tensile arch stresses in Monticello Dam computed under two conditions: (a) water compressibility included, and (b) water compressibility ignored.

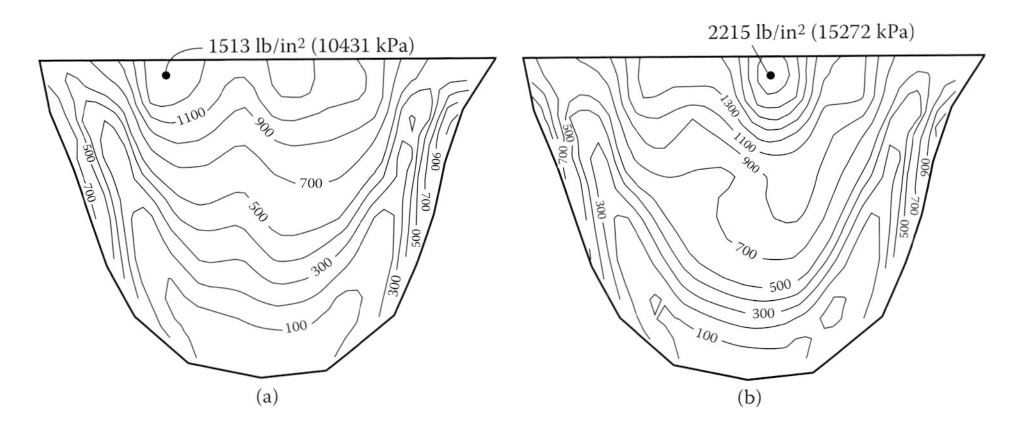

Figure 9.2.11 Peak values of tensile arch stresses in Morrow Point Dam computed under two conditions: (a) water compressibility included; and (b) water compressibility ignored.

as in the case of Morrow Point Dam (Figure 9.2.11); observe that these discrepancies vary with the location on the dam surface.

It has not been possible to identify the system parameters for which dam response will be overestimated (or underestimated) if water compressibility is ignored. This is because water compressibility modifies the resonant frequency, increases the overall damping (because of radiation of energy by waves traveling upstream or absorption of energy at the reservoir boundary), and significantly influences the shape of the frequency response curve (Fok and Chopra 1987). Thus, the influence of water compressibility on earthquake response of a dam would depend on the frequency characteristics of the ground motion as well as the vibration periods of the dam, increasing the response of some dams, but reducing it for others, as observed previously.

Water compressibility is expected to be significant in the response of most arch dams because the elastic modulus, E_s, for concrete is typically in the range of 2–5 million psi, and, for dams with close to full reservoir, Ω_r, would be much smaller than 2. Thus, we conclude that hydrodynamic effects should not be modeled by an added mass of water vibrating with the dam – a model that ignores water compressibility – in finite element analysis of dams.

9.3 INFLUENCE OF SPATIAL VARIATIONS IN GROUND MOTIONS

The response of Pacoima Dam (Figure 10.2.1a) to spatially varying ground motions recorded at the dam–foundation interface during the earthquakes of January 13, 2001, and January 17, 1994,[†] was determined by the analysis procedure presented in Chapter 8 (Chopra and

[†] The actual records obtained during this earthquake were incomplete; they were "completed" by Alves (2004) and are subsequently referred to as "recorded" motions.

Wang 2008). A detailed description of the dam, instrumentation, and recorded motions is presented in Chapter 10. Starting with the ground motions recorded at three locations near the dam–foundation interface (Figure 10.2.2), the motions at all nodes on the interface were determined by interpolating or extrapolating the records to define the spatially-varying input motions in the EACD-3D-2008 computer program.

9.3.1 January 13, 2001 Earthquake

The quasi-static component (Chopra 2017; Sections 9.7 and 13.5) is a significant but not a dominant part of the displacement response of Pacoima Dam to the January 13, 2001 earthquake records. Figure 9.3.1, which identifies the quasi-static component in the history of displacements at the center of the dam crest, shows that the quasi-static component is not a major part of the displacement in the radial direction (the direction of largest response), whereas it is dominant in the displacements in the tangential and vertical directions.

Therefore, the spatial variations in ground motions are expected to significantly influence, but not dominate, the stresses in the dam, as shown in Figure 9.3.2, where the peak values of the tensile stresses in the cantilever direction on the downstream face of the dam are presented; similar figures for arch and cantilever stresses on both faces of the dam are available in Chopra and Wang (2008). Presented are stresses due to four different excitations; the first three are spatially-uniform excitations defined by ground motions recorded at the base of the dam (Channels 9–11), the right (viewed from upstream) abutment (Channels 12–14), and the left abutment (Channels 15–17), respectively; see Figure 10.2.2. The fourth excitation is defined as the recorded (and interpolated or extrapolated) spatially-varying ground motions.

The stresses due to spatially-varying excitation may be smaller or larger than those due to spatially-uniform ground motion, depending on the intensity chosen for the latter excitation. They are larger when compared to the stresses due to the base motion – the least intense of the three spatially-uniform excitations – but are generally smaller than the stresses due to the abutment motions. Comparing the four parts of Figure 9.3.2 also reveals that spatial variations in ground motion significantly influence – but not dominate – the stresses in Pacoima Dam due to the January 13, 2001 earthquake.

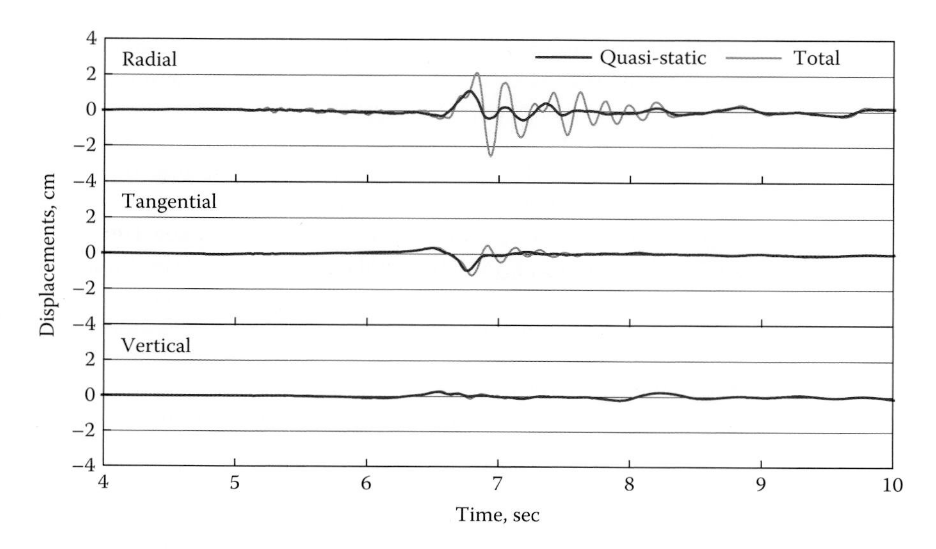

Figure 9.3.1 Quasi-static and total displacement histories at crest center of Pacoima Dam due to spatially-varying ground motion during the January 13, 2001, earthquake: (a) radial component; (b) tangential component; and (c) vertical component.

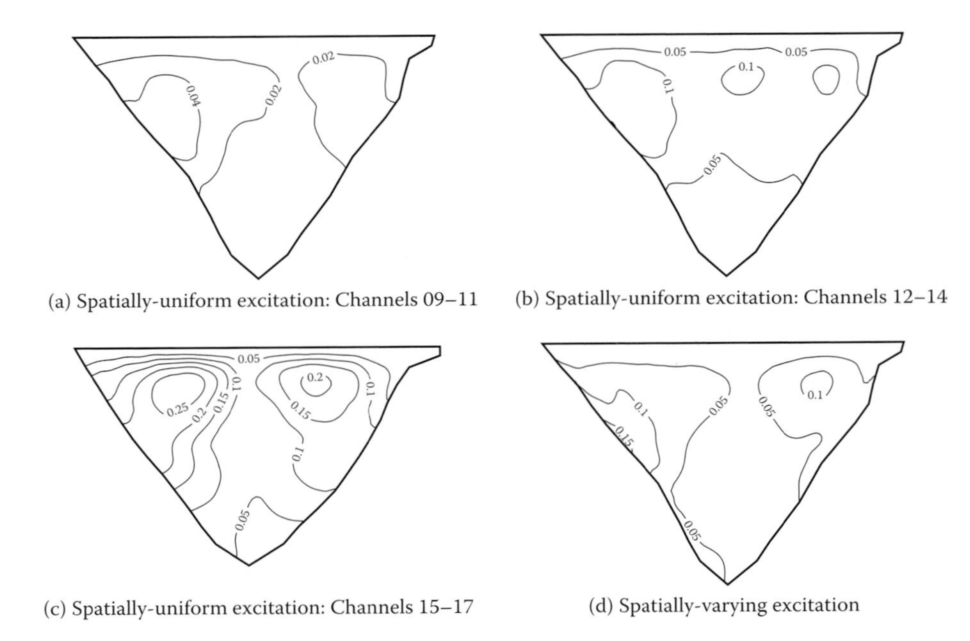

Figure 9.3.2 Peak values of tensile cantilever stress (MPa) on the downstream face of Pacoima Dam due to the January 13, 2001, earthquake: (a) spatially-uniform excitation defined by Channels 09–11; (b) spatially-uniform excitation defined by Channels 12–14; (c) spatially-uniform excitation defined by Channels 15–17; and (d) spatially-varying excitation.

9.3.2 January 17, 1994 Northridge Earthquake

The quasi-static component is dominant in the displacement response of Pacoima Dam to the January 17, 1994 Northridge earthquake records; therefore, the spatial variations in ground motion have profound influence on the computed stresses in the dam. Figure 9.3.3 identifies the quasi-static component in the history of displacements at the center of the dam crest. The ratio of the peak values of the quasi-static component and the total displacement at the crest center is 87%, 93%, and 97% for radial, tangential, and vertical directions, respectively. Consequently, the spatial variations in ground motion are expected to profoundly influence the stresses in the dam. Figure 9.3.4, where the peak value of the tensile stresses in the arch direction on the downstream face of the dam are presented, confirms this expectation; similar figures for arch and cantilever stresses on both faces of the dam are available in Chopra and Wang (2008). Presented are the stresses due to four different excitations. The first three are spatially uniform excitations defined by ground motion at the base of the dam (Channels 9–11), the right abutment (Channels 12–14), and the left abutment (Channels 15–17); see Figure 10.2.2. The fourth excitation is defined as the "recorded" (and interpolated or extrapolated) spatially-varying ground motion. The distribution pattern of stresses caused by the three spatially uniform excitations is similar, although the magnitude of stresses due to ground motions recorded at the base of the dam is much smaller than that due to motions recorded at the left or right abutment; the stresses due to the two abutment excitations are similar in magnitude. Comparing the stresses caused by spatially-varying versus spatially-uniform excitations shows that spatial

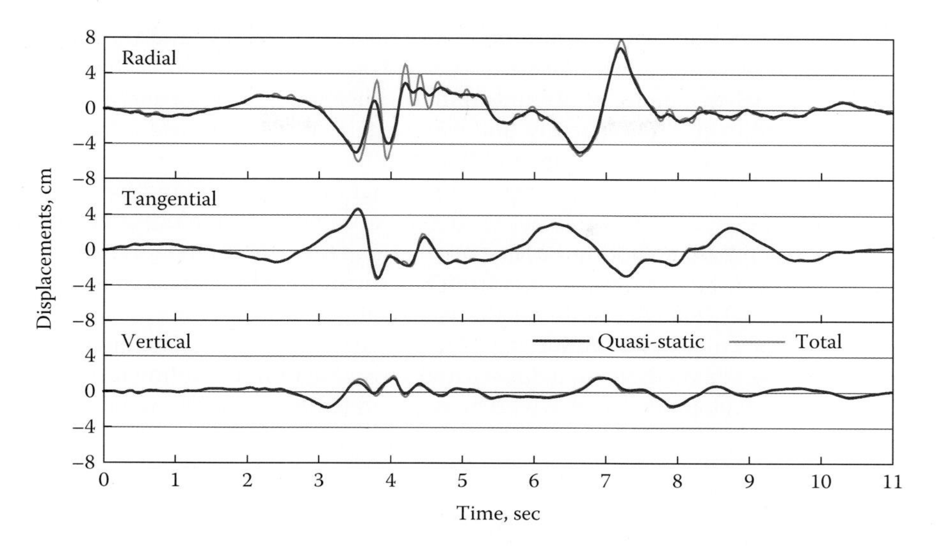

Figure 9.3.3 Quasi-static and total displacement histories at crest center of Pacoima Dam due to spatially-varying ground motion during the Northridge earthquake: (a) radial component; (b) tangential component; and (c) vertical component.

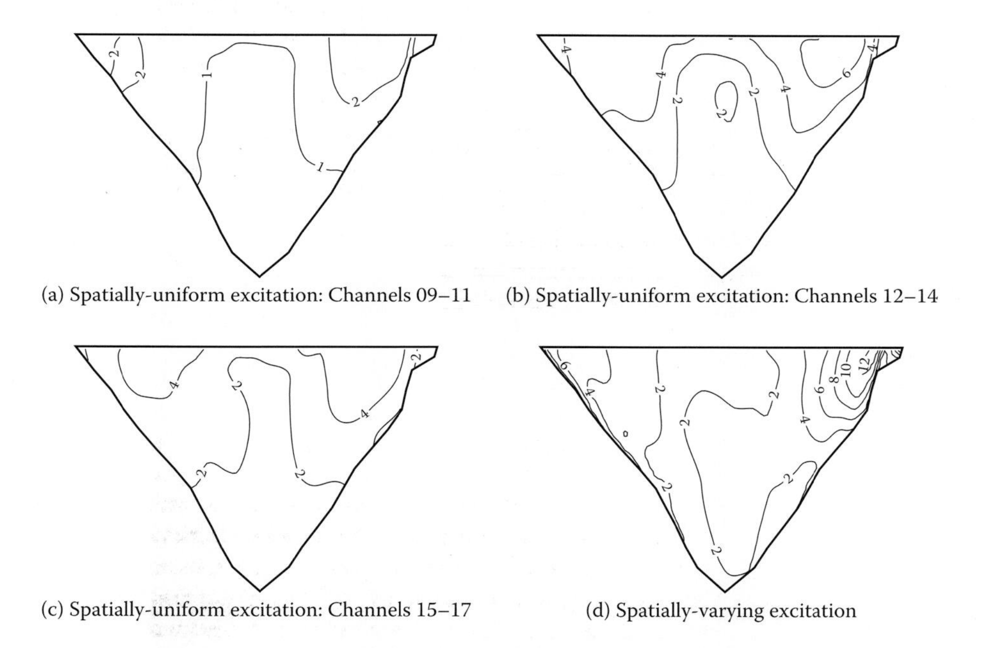

Figure 9.3.4 Peak values of tensile arch stress (MPa) on the downstream face of Pacoima due to the Northridge earthquake: (a) spatially-uniform excitation defined by Channels 09–11; (b) spatially-uniform excitation defined by Channels 12–14; (c) spatially-uniform excitation defined by Channels 15–17; and (d) spatially-varying excitation.

variations in ground motion had profound influence on the magnitude and the distribution of arch stresses (Figure 9.3.4). Observe the stress concentration near the top of the left abutment (Figure 9.3.4d). Spatial variations in ground motion cause much larger cantilever stresses on both faces (compared to all three spatially-uniform excitations) in portions of the dam adjacent to the dam–foundation interface (Chopra and Wang 2008).

The previous results demonstrate that spatial variations in ground motion can have profound influence on the earthquake-induced stresses in the dam. This influence obviously depends on the degree to which ground motion varies spatially along the dam–foundation interface. Thus, for the same dam, this influence would differ from one earthquake to the next, depending on the epicenter location and focal depth of the earthquake relative to the dam site.

Spatial variations in ground motion have typically been ignored in dam engineering practice for lack of well-tested methods to define spatially-varying ground motion around a canyon. Much research and development work would be required to develop such methods.

10
Comparison of Computed and Recorded Motions

PREVIEW

As mentioned in Chapter 9, ground motions recorded at arch dams vary spatially along the dam–foundation interface. Rarely are these spatial variations in ground motion considered in earthquake analysis of arch dams, and when they are included, dam–water–interaction is usually oversimplified: water compressibility, foundation mass, and foundation damping (material and radiation) are ignored, and the semi-unbounded extent of the foundation and impounded water domains is not recognized (Alves and Hall 2006a,b; Mojtahedi and Fenves 2000). Utilizing the linear analysis procedure presented in Chapter 8, which includes these interaction effects and recognizes the semi-unbounded extent of foundation and fluid domains, the response of two dams to spatially-varying ground motion recorded during past earthquakes is computed and compared with recorded motions to demonstrate the effectiveness of the analysis procedure.

10.1 EARTHQUAKE RESPONSE OF MAUVOISIN DAM

10.1.1 Mauvoisin Dam and Earthquake Records

Located in the Swiss Alps, Mauvoisin Dam is a 250-m-high double-curvature arch dam (Figure 10.1.1a). The dam was built to a height of 237 m during 1951–1957, and raised to its present height during 1989–1991. The base of the dam is at El. 1726 m above sea level and its crest at El. 1976 m. It is composed of 28 blocks for a total crest length of 520 m. The thickness of the crown cantilever decreases from 53.5 m at the base to 12 m at the crest (Figure 10.1.1b).

Figure 10.1.2 shows an array of 12 three-component strong-motion (SM) accelerographs operating at the dam since 1993. Accelerographs SM01–SM05 are located inside the upper gallery, 14 m below the crest; accelerographs SM06–SM08 are located at mid-height, and

Earthquake Engineering for Concrete Dams: Analysis, Design, and Evaluation, First Edition. Anil K. Chopra.

Figure 10.1.1 Mauvoisin Dam, Switzerland: (a) view from downstream; (b) cross section of crown cantilever.

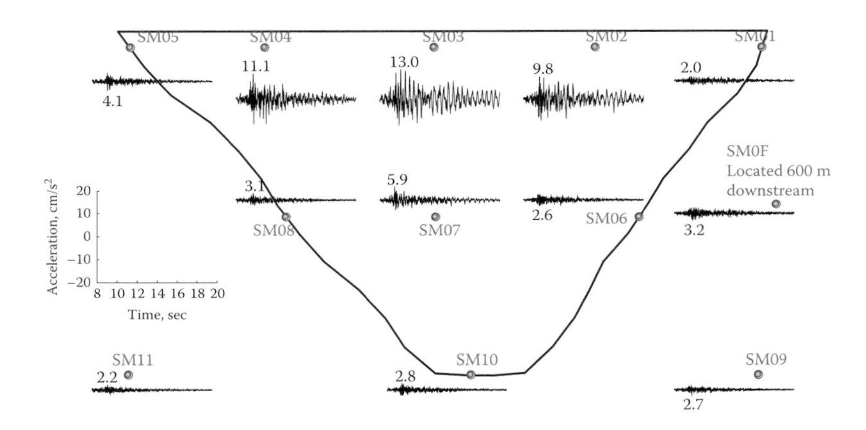

Figure 10.1.2 Recorded motions in stream direction; accelerations are in cm/sec^2; peak values are noted.

accelerographs SM09–SM11 are located at the base elevation. Installed in tunnels, accelerographs SM09 and SM11 are located essentially vertically below SM01 and SM05, respectively; SM10 is located at the base of the dam. Accelerographs SM01, SM06, SM08, and SM05 are located near the dam–foundation interface; SM06 and SM08 are mounted at the end of a gallery, 5 m away from rock. SM01 and SM06 are located on the left side (viewed from upstream) of the canyon and SM05 and SM08 on the right side of the canyon. An accelerograph SM0F is located in the free field 600 m downstream of the dam at El. 1840, i.e. 114 m above the dam base on the left side of the canyon.

Motions of Mauvoisin Dam during the magnitude 4.6 Valpelline earthquake of March 31, 1996, centered 13 km away from the dam, were recorded by the accelerograph array. At the time of the earthquake, the water level was at El. 1864, i.e. slightly more than half of the dam height. The stream components of motions recorded at the accelerograph locations are shown in Figure 10.1.2; similar figures for the cross-stream and vertical components are available in Chopra and Wang (2008).

10.1.2 System Analyzed

Figure 10.1.3 presents a finite-element model of the dam that includes 145 8-node thick shell elements and a total of 468 nodes, the finite-element mesh for the irregular region of the fluid domain, and the boundary element mesh at the dam–foundation interface. The foundation domain is treated as semi-unbounded and the fluid domain as unbounded in the upstream direction. To align with a horizontal surface of nodes in the finite-element model, the water surface is set at El. 1870 m compared to the actual El. 1863 m. Based on correlation of computed response with ambient vibration data and known concrete and rock properties, Proulx et al. (2004) selected parameter values for the dam–water–foundation system as follows: Concrete: elastic modulus = 36 GPa, Poisson's ratio = 0.2, and unit mass = 2400 kg/m^3; rock: elastic modulus = 72 Gpa, Poisson's ratio = 0.25, and unit mass = 2500 kg/m^3; water: unit mass = 1000 kg/m^3, and wave velocity = 1438 m/sec; and wave reflection coefficient at the reservoir boundary = 0.9. These material properties (elastic properties and unit weight) of the concrete and rock were retained here with one minor exception: a lower elastic modulus of 25 GPa was assigned to the uppermost 12.5 m of the dam, the part raised in 1991; this value comes from test data for the newer concrete (Proulx et al. 2004).

Data obtained from three concrete arch dams in the Swiss Alps provides a basis to choose damping values for earthquake analysis of these dams. Ambient vibration tests led to a viscous

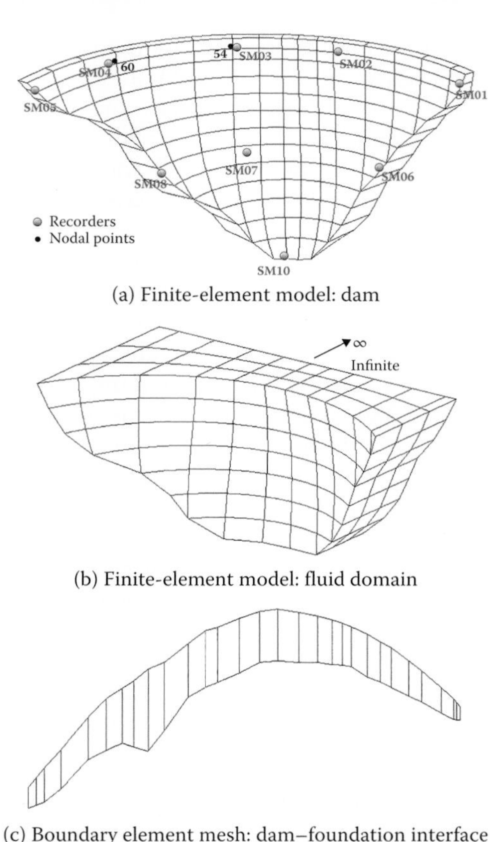

Figure 10.1.3 EACD-3D-2008 model for Mauvoisin Dam. (a) Finite-element model: dam. (b) Finite-element model: fluid domain. (c) Boundary element mesh: dam–foundation interface.

damping ratio in the range of 2–3% in the lower vibration modes of Mauvoisin Dam (Darbre et al. 2000), and similar values were obtained for the 130-m-high Punt-del-Gall Dam (Proulx et al. 2004). Forced vibration tests of Emosson Dam also resulted in 2–3% damping (Proulx et al. 2001), similar to Mauvoisin Dam. Although ambient vibration data for damping are usually considered less dependable than forced vibration data, the damping value for Mauvoisin Dam seems reliable, as the concrete and rock properties at the site are similar to those found at Emosson Dam.

Constant hysteretic damping factors for the dam alone and foundation rock separately were selected to achieve a viscous damping ratio for the overall dam–water–foundation system that is close to the "measured" value of 2–3%. For this purpose, frequency response functions for the response at the crest center due to spatially-uniform excitation at the dam–foundation interface in the stream and cross-stream directions were determined for several combinations of damping in the dam and in the foundation. In one such combination, the assumed values were $\eta_s = 0.02$ (i.e. a viscous damping ratio of 1% in all vibration modes of the dam alone) and $\eta_s = 0.06$ (i.e. viscous damping ratio of 3%) for the foundation, and determined from the resonance curve – by the half-power bandwidth method – the viscous damping ratio to be 2.2% in the first symmetric mode and 1.5% in the first anti-symmetric mode. Clearly, these chosen values for the η_s and η_f provide overall damping consistent with the aforementioned ambient vibration tests that led to viscous damping ratios of about 2% for the dam–water–foundation system, consistent with experimental data.

10.1.3 Spatially Varying Ground Motion

The five records – SM05, SM08, SM10, SM06, and SM01 – were obviously influenced by dam–water–foundation interaction, thus, they do not represent free-field motions around the canyon. Furthermore, they provide limited information on the motions at the dam–foundation interface; they describe the motions at only five locations that are not exactly at the dam–rock interface. In particular, SM01 and SM05 are mounted on bare rock approximately 7 m upstream of the interface; SM06 and SM08 are located at the end of a gallery, 4 m away from the interface (the gallery does not extend into the abutment rock), and SM10 is on the wall of the pendulum shaft located near the middle of the dam thickness. The motions at all other locations on the interface are unknown, thus no information is available about how the motions vary across the thickness of the dam. If all components of the motions at all locations on both edges of the dam–foundation contact zone had been recorded, the response of the dam to these motions, which obviously include dam–water–foundation interaction effects, could be determined by analyzing the dam (and impounded water) alone (without the foundation domain) subjected to the boundary motions. Such an analysis is not meaningful, however, if the information available on the boundary motions is incomplete, as is the case at Mauvoisin Dam, and will be the case at every dam.

The only avenue remaining is to treat records SM05, SM08, SM10, SM06, and SM01 as defining the free-field motions at locations on the interface that are closest to the recorders and fill in the missing information. Interpolating and extrapolating these records, the motions at all nodes (of the finite-element mesh) at the dam–foundation interface were determined (Chopra and Wang 2008) by a procedure developed by Alves and Hall (2006a), and the motions were assumed to be uniform over the thickness of the dam, thus defining the spatially-varying input motions in the EACD-3D-2008 computer program.

10.1.4 Comparison of Computed and Recorded Responses

With the spatially-varying excitation thus defined, the response of the dam is determined by the analysis procedure presented in Chapter 8. Fourier transforms of the computed acceleration at

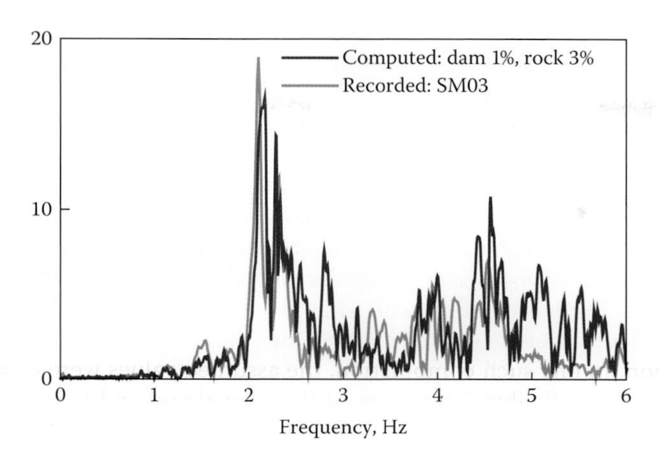

Figure 10.1.4 Comparison of Fourier transforms of recorded response at dam crest center (SM03) and computed response at node 54 to spatially-uniform excitation (SM0F record) for viscous damping values of 1% in the dam and 3% in the foundation, only the stream component of response is presented.

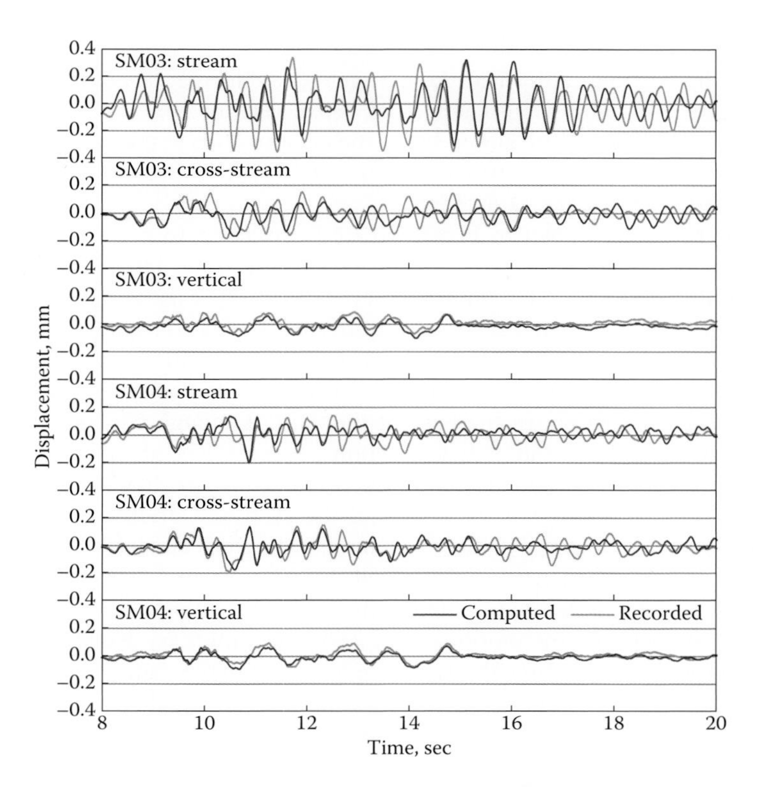

Figure 10.1.5 Comparison of recorded and computed displacements at crest center (SM03) and crest right quarter point (SM04); computed responses are for node 54 (near SM03) and node 60 (near SM04); stream, cross-stream, and vertical components of response at each location are included.

node 54 and the recorded acceleration at SM03 are compared in Figure 10.1.4, where we observe good agreement at the first two dominant peaks. The computed displacement responses of the dam are reasonably similar to the recorded displacements, but the agreement is far from perfect. The peak values of the computed displacements in the stream direction – the direction of the largest response – are very close to the recorded value (Figure 10.1.5). While the computed value at node 54 is 92% of the value recorded at SM03 (i.e. they differ by 8%), and the computed value at node 60 is 101% of the value recorded at SM04 (i.e. the two differ by 1%), the computed displacement histories do not agree as well. Although the computed displacement history is very close to that recorded over some time segments, the two differ significantly during other time segments (Figure 10.1.5), e.g. the computed displacement at node 54 in the stream direction is very close to the recorded motion at SM03 during 14–18 seconds, but the two differ significantly over other time segments.

Recognizing that the recorded ground motions provide an incomplete description of the earthquake excitation, and that no attempts were made to adjust the published data for parameter values for the mass and stiffness properties of the concrete and rock (Proulx et al. 2004), the agreement between computed and recorded motions is modestly good. Improved agreement could have been achieved by fine-tuning the system properties.

10.2 EARTHQUAKE RESPONSE OF PACOIMA DAM

10.2.1 Pacoima Dam and Earthquake Records

Located in the San Gabriel Mountains near Los Angeles, California, Pacoima Dam is a 113-m-high concrete arch dam, with a crest length of 180 m (Figure 10.2.1a). Completed in 1928, the dam varies in the thickness from about 3 m at the crest to 30 m at the base. A concrete thrust block supports the dam at the left abutment (Figure 10.2.1b). The 11 contraction joints in the dam body have beveled keys that are 30 cm deep.

The array of accelerographs shown in Figure 10.2.2 is designed to record 17 channels of motions. Three-component accelerographs are located at the base, left abutment, right abutment, and center of the crest, and one-component accelerographs exist at five locations on the dam body. Channels 12–17 are located at three stations near the dam–foundation interface on the downstream side. At each station, one channel is oriented in the east-west (stream) direction, one is vertical and one is north-south (cross-stream). Channels 9–11 are located on the downstream face about 10 ft above the base of the dam. Although Channels 9 and 11 are actually oriented in the radial and tangential directions, respectively, at this base location those directions are nearly east-west and north-south, respectively, so they are assumed to be equivalent. Channels 2–4 are located at the center of the crest and oriented in the radial, tangential, and vertical directions. Located on the crest at the right-third and left-quarter points, respectively, Channels 1 and 5 are oriented in the radial direction. Channels 6–8 on the dam body at about 80% height of the dam are all oriented in the radial direction.

This instrument array recorded the motions during the magnitude 6.7 Northridge earthquake of January 17, 1994, with its epicenter about 18 km southwest of the dam and a focal depth of 19 km. The water surface was 40 m below the crest during the earthquake. Most of the film recordings could not be fully digitized because the different records overlapped due to the large amplitudes and high frequencies. Portions of the digitized records were, therefore, missing, and Channels 7 and 14 did not record at all. Peak horizontal accelerations ranged from 0.5 g at the base of the dam to about 2.0 g on the canyon sidewalls near the crest. Spatial variations in the ground motions in both amplitude and phase from the bottom of the canyon to the top and from

Figure 10.2.1 Pacoima Dam: (a) dam and (b) left abutment.

one side of the canyon to the other were striking. Because of the intense ground shaking, the dam sustained significant damage that was repaired. The accelerometer array was also repaired and upgraded.

On January 13, 2001, a magnitude 4.3 earthquake occurred, with its epicenter about 6 km south of Pacoima Dam and focal depth of about 9 km. The water level was 41 m below the crest at the time of this earthquake. The stream (or radial) component of the recorded motions is presented in Figure 10.2.3; the cross-stream (or tangential) and vertical components are available in Chopra and Wang (2008). Spatial variation in ground motions along the dam–foundation interface is evident. In the stream direction, the peak acceleration of 13 cm/sec^2 at the base is amplified to 43 and 34 cm/sec^2 at the left and right abutments, respectively (Figure 10.2.3). In the cross-stream direction, the peak acceleration of 20 cm/sec^2 at the base is amplified to 95 cm/sec^2 at the left abutment and to 50 cm/sec^2 at the right abutment.

Alves (2004) conducted an ingenious and meticulous investigation of the spatial variations in ground motions recorded during the 2001 earthquake. Developing a method for generating

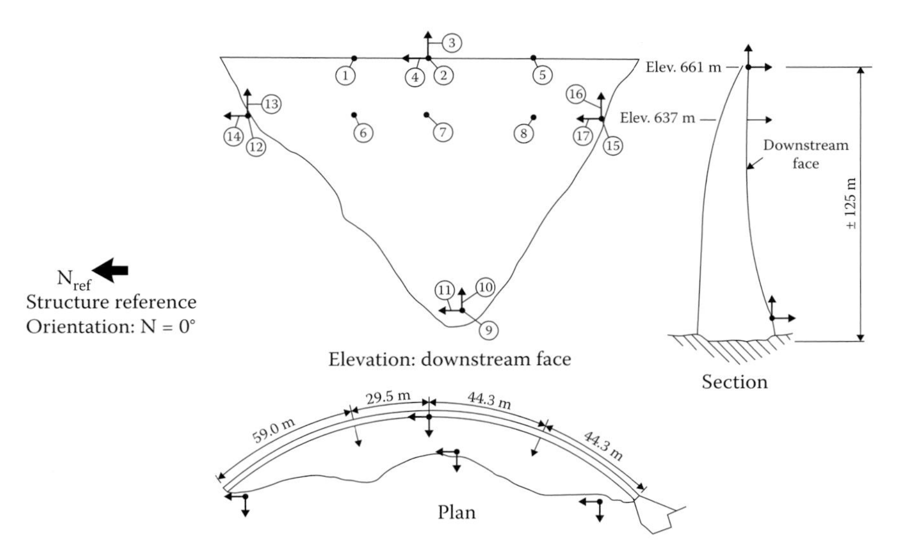

Figure 10.2.2 Accelerograph locations at Pacoima Dam. Source: CSMIP *Report OSMS 01–02.*

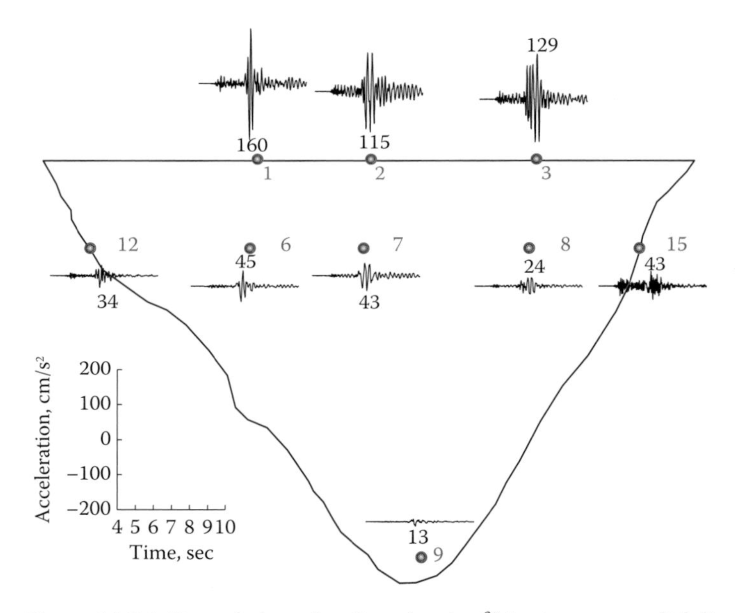

Figure 10.2.3 Recorded accelerations (cm/sec^2) in stream or radial direction at Channels 1–3, 6–9, 12, and 15 during the January 13, 2001 earthquake.

spatially-varying ground motion from a single three-component record at the base of the dam, he demonstrated its effectiveness for the 2001 earthquake by comparing generated motions with those recorded at the left and right abutments (Channels 12–17). After demonstrating the effectiveness of this method, he used it to generate abutment motions from the base motion recorded during the 1994 earthquake, where the actual records were incomplete (Alves and Hall 2006a); these generated motions are subsequently referred to as "recorded" motions.

The cross-stream (or tangential) component of the motions "recorded" during the 1994 earthquake are presented in Figure 10.2.4; the stream (or radial) and vertical components

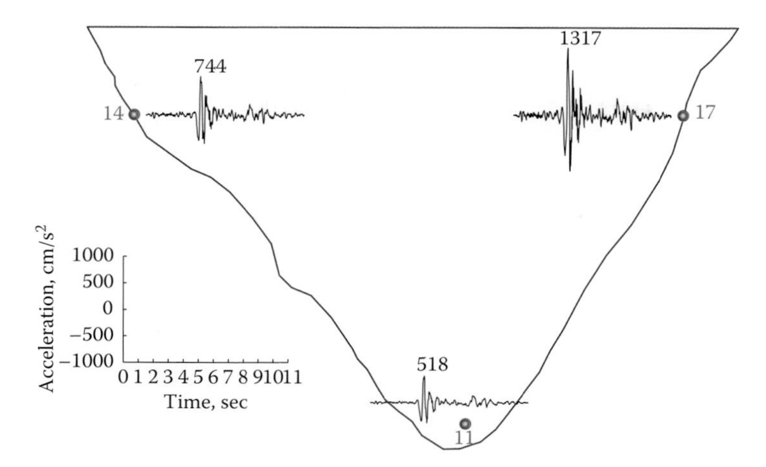

Figure 10.2.4 Accelerations (cm/sec^2) generated by Alves (2004) in cross-stream direction at Channels 11, 14, and 17 to represent motions during Northridge earthquake.

are available in Chopra and Wang (2008). Spatial variation in ground motions along the dam–foundation interface is evident. In the stream direction, the peak acceleration of 429 cm/sec^2 at the base is amplified by a factor of approximately two at the abutments. In the cross-stream direction, the peak acceleration of 518 cm/sec^2 at the base is amplified to 1317 and 744 cm/sec^2 at the left and right abutments. The time variation of motions at the two abutments are similar, but the difference in amplitude is striking; in contrast, the stream motions at the two abutments are similar.

10.2.2 System Analyzed

Figure 10.2.5 shows the finite-element model for Pacoima Dam, which includes 110 8-node thick shell elements with a total of 331 nodes. the finite-element mesh for the irregular region of the fluid domain. and the boundary element mesh at the dam–foundation interface; the foundation domain is treated as semi-unbounded, and the fluid domain as unbounded in the upstream direction.

The elastic properties were established after a comprehensive investigation that included calibration of Alves' finite-element model to approximate the modal properties determined by system identification using the earthquake records of January 13, 2001 (Alves and Hall 2006b). This investigation resulted in values of elastic moduli for the dam concrete and foundation rock as 21.9 GPa (3180 ksi) and 10.9 GPa (1575 ksi), respectively. The other properties were based in part on material tests: unit weight of concrete equal to 22.3 kN/m^3 (142 lb/ft^3), of rock equal to 25.9 kN/m^3 (165 lb/ft^3), and of water equal to 9.8 kN/m^3 (62.4 lb/ft^3); shear wave velocity in water = 1438 m/sec^1 (4720 ft/sec^1). Although Alves' model ignored the mass of foundation rock and compressibility of water, no attempt was made to refine his values for concrete and rock moduli for use in the EACD-3D-2008 model that includes foundation mass and water compressibility.

Constant hysteretic damping factors for the dam alone and foundation, separately, were selected to achieve a viscous damping ratio for the overall dam–water–foundation system that is close to the "measured" value of 6–7%. For this purpose, the frequency response functions at node 21 due to spatially-uniform excitation were computed for several combinations of damping in the dam and in the foundation. In one such combination, we assumed $\eta_s = 0.04$ (which corresponds to a viscous damping ratio of 2% in all vibration modes of the dam alone) and $\eta_f = 0.08$

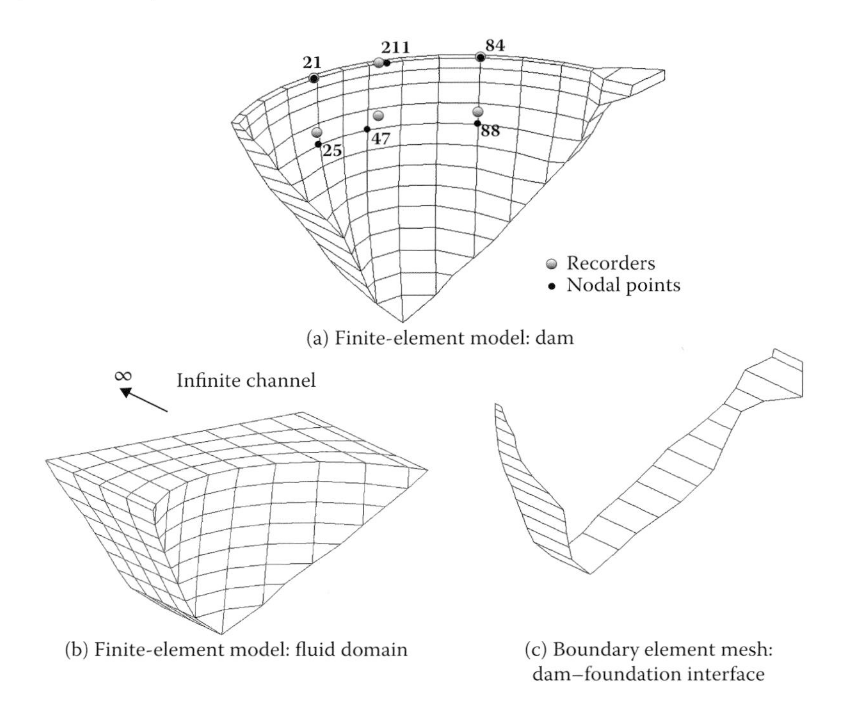

Figure 10.2.5 EACD-3D-2008 model for Pacoima Dam. (a) Finite-element model: dam. (b) Finite-element model: fluid domain. (c) Boundary element mesh: dam–foundation interface.

(i.e. viscous damping ratio of 4%) for the foundation. Determined from the resonance-curve – by the half-power bandwidth method – the viscous damping ratio was determined as 7.0% in the first symmetric mode and 6.7% in the first anti-symmetric mode. Clearly these chosen values of η_s and η_f provide overall damping consistent with the aforementioned system identification studies that had led to viscous damping ratios of 6.2% in the first symmetric vibration mode and 6.6% in the first anti-symmetric vibration mode of the dam (Alves 2004).

10.2.3 Comparison of Computed and Recorded Responses: January 13, 2001 Earthquake

Determined by the EACD-3D-2008 computer program, the response of Pacoima Dam to spatially-varying ground motions recorded during the 2001 earthquake is presented in Figure 10.2.6. The computed displacements compare well with the recorded displacements; the agreement is much better than was achieved earlier in the case of Mauvoisin Dam. The time-variation of the computed displacements is close to the recorded response over its entire duration; however, the peak displacement at Channels 1 and 2 is over-estimated. Recognizing that the recorded ground motions provide only an incomplete description of the earthquake excitation, and that no attempts were made to adjust the published data (Alves 2004) for the mass and stiffness parameters of concrete and rock, the agreement between computed and recorded displacements is judged to be satisfactory.

10.2.4 Comparison of Computed Responses and Observed Damage: Northridge Earthquake

A similar comparison of response computed by linear analysis against motions recorded during the Northridge earthquake of January 17, 1994 is not appropriate because the ground shaking was

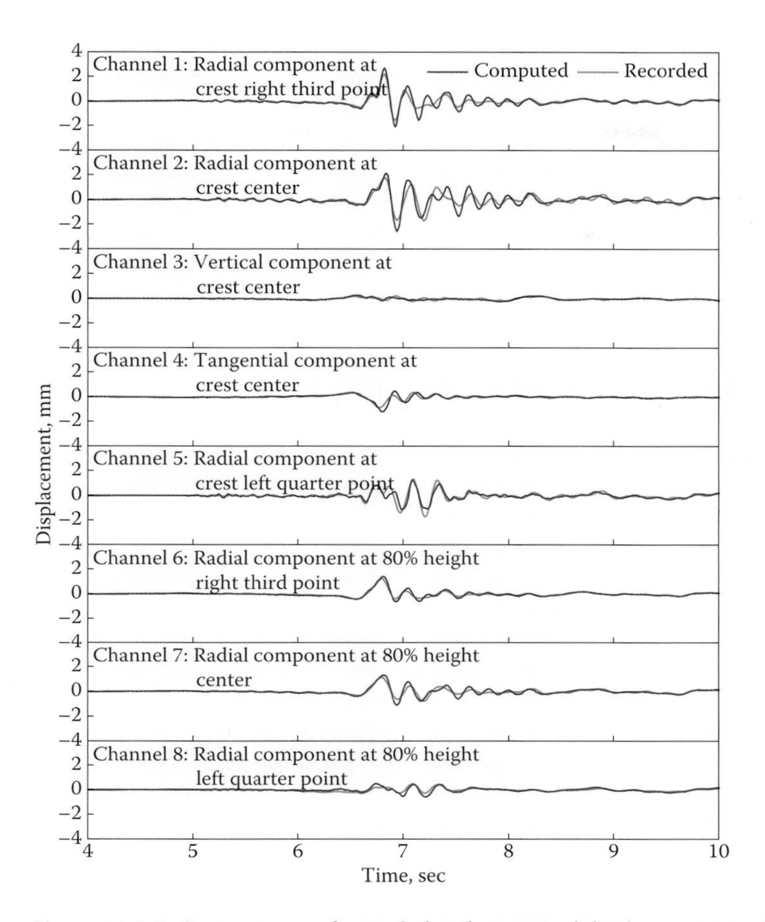

Figure 10.2.6 Comparisons of recorded and computed displacements at Channels 1–8 due to the January 13, 2001, earthquake. Computed responses is at the following nodal points; 21 (near Channel 1), 211 (near Channels 2–4), 84 (near Channel 5), 25 (near Channel 6), 47 (near Channel 7), and 88 (near Channel 8).

Figure 10.2.7 Joints opened and cracks occurred in the thrust block of Pacoima Dam during the Northridge earthquake.

intense enough to cause significantly nonlinear behavior of the dam. Obviously, stresses computed by linear analysis are not indicative of actual stresses that developed in the dam because vertical contraction joints opened and cracked occurred during the earthquake. However, a linear analysis by the EACD-3D-2008 computer program was able to identify the highly stressed zones in the dam. Such an analysis predicted large arch stresses in the thrust block between the dam and the left abutment and in the portion of the dam adjacent to the thrust block (see Figure 9.3.4d and additional figures in Chopra and Wang (2008)), suggesting that cracking would occur in these regions, which is what actually happened during the earthquake; cracking is visible in Figure 10.2.7.

10.3 CALIBRATION OF NUMERICAL MODEL: DAMPING

The numerical model for a concrete dam should be calibrated to match its actual vibration properties. Although the need to match vibration frequencies and modes is widely recognized, calibration of damping has not received as much attention.

Damping in the numerical model for the dam–water–foundation system should be consistent with measured values determined from low-amplitude motions – within the linear range of response – recorded during forced vibration tests, ambient vibration, or small earthquakes. Obviously, the measured values represent the overall damping in the system, including material damping, radiation ramping, and energy loss at reservoir boundaries; information on the contributions of individual sources of damping is generally not available.

Summarized in Figure 10.3.1[†] is the "measured" damping determined from forced vibration tests and ambient vibration data for 32 concrete dams (Hall 1998; Proulx and Paultre 1997; Proulx et al. 2004). Both gravity dams and arch dams covering a wide range of system geometry are included. The overall damping values measured at these dams are, but for a few exceptions, all

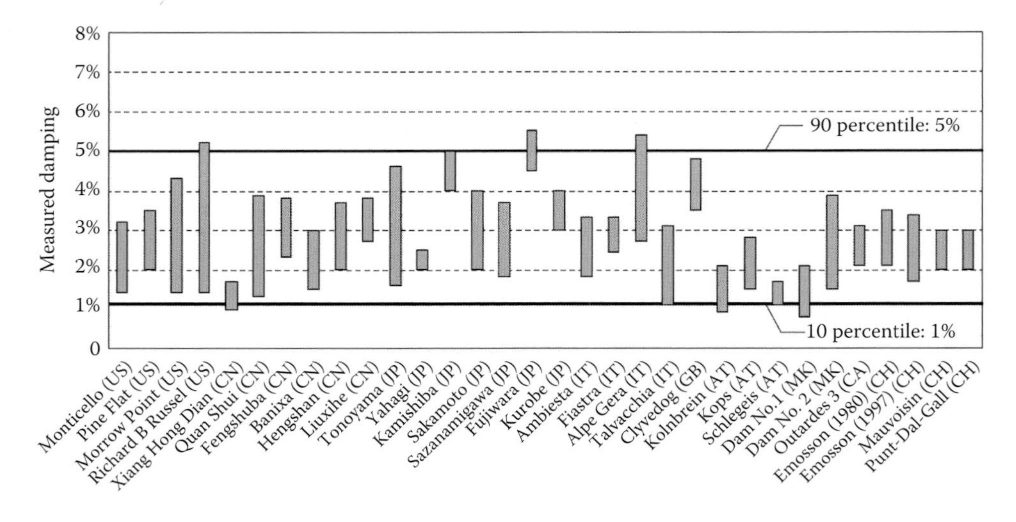

Figure 10.3.1 Measured damping at 32 concrete dams during forced vibration measurements compiled from Hall (1988), Proulx and Paultre (1997), and Proulx et al. (2004). The range for each dam shows the minimum and maximum damping values measured at the first few (1–5) resonant frequencies.

[†] Four data points that were judged as inaccurate due to excessive modal interference from the dam response (Hall 1988) are excluded from the dataset presented.

in the range of 1–5%. These comprehensive data lead to an important conclusion: overall damping in the numerical model should not exceed 5% unless a larger value was "measured" at the particular dam. In contrast, current practice of specifying a viscous damping ratio of 5% for the concrete dam alone and a similar value for the foundation domain separately will lead to damping in the range of 10–20% in the overall dam–water–foundation system; values up to 30% have been demonstrated in certain conditions (Bybordiani and Arici 2017). Thus, the current practice of choosing damping values should be abandoned because it will significantly underestimate the earthquake response of dams.

Material damping in the numerical model is usually specified separately for the two substructures: dam and foundation. In contrast, energy loss at reservoir boundaries and radiation damping are not known prior to earthquake analysis of the dam. Thus, it is not possible to know in advance the damping values that should be specified for the two substructures to achieve a target value of, say, 5%, for overall damping. Therefore, material damping in the two substructures must be determined by trial and error to achieve an overall damping consistent with the measured values – as described in Sections 10.1 and 10.2 for Mauvoisin and Pacoima dams. The overall damping in the numerical model can be determined from frequency response functions or system identification techniques applied to response histories recorded during earthquakes.

Researchers have demonstrated that damping in the range of 1–2% for the dam and 1–4% for the foundation is likely to lead to an overall damping in three-dimensional numerical models that is consistent with measured values. For example, viscous damping ratios of 1% and 3% were selected for the two substructures, respectively, in the case of Mauvoisin Dam; 2% and 4%, respectively, for Pacoima Dam; and 1% and 2%, respectively, for Morrow Point Dam. The resulting overall damping in the dam–water–foundation system was consistent with measured damping: 2–3% for Mauvoisin Dam; 6–7% for Pacoima Dam; and 1.5–4% for Morrow Point Dam. Responses computed from such numerical models were in good agreement with motions recorded during small earthquakes at Mauvoisin Dam and at Pacoima Dam.

Limiting the overall damping to less than 5% in 2D numerical models is very difficult because of the large amount of radiation damping associated with 2D homogeneous, semi-unbounded foundation models (Figure 10.3.2). Presented in Table 10.3.1 are data of total damping in the 2D model of Pine Flat Dam, computed by the substructure method (see Chapter 5), for several values of the parameters that characterize energy loss in the system: material damping in the dam, ζ_s; material damping in the foundation, ζ_f; the ratio of elastic moduli for foundation rock and dam concrete, E_f/E_s; and the reservoir bottom reflection coefficient α,

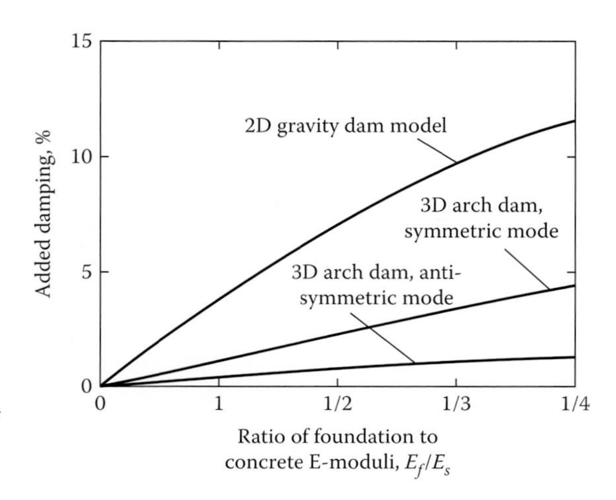

Figure 10.3.2 Additional damping in the fundamental mode of vibration due to dam–foundation interaction for 2D model of a gravity dam and 3D model of an arch dam supported on foundation modeled as homogenous half-space. Source: Data for gravity dams are from Fenves and Chopra (1984a) and for arch dams from Tan and Chopra (1995a).

Table 10.3.1 Overall damping in 2D numerical model of Pine Flat Dam computed by the substructure method for several values of the parameters that characterize energy loss in the system.

	Dam	Foundation		Reservoir bottom	Overall damping, %
Case	ζ_s, %	E_f/E_s	ζ_f, %	α	
1	5	1	5	0.68	13
2	2	1	2	0.68	10
3	0	1	0.5	0.68	8.5
4	0	2	0.5	0.77	5.0
5	0	3	0.5	0.80	3.7

which is computed from Eq. (5A.2.1) based on properties of rock (Appendix 5A.2). For example, 2% damping in the dam and 2% for the foundation domain leads to an overall damping of 8–10% in a 2D numerical model of Pine Flat Dam (Løkke and Chopra 2015), which is much higher than the 2–4% damping "measured" at the dam during forced vibration tests (Rea et al. 1975). Observe that it is not possible to achieve consistency with the measured damping values for this dam unless the foundation is much stiffer than concrete and essentially zero material damping is specified in the dam as well as the foundation. For response spectrum analysis of 2D models (Chapter 4), implemented in the computer program CADAM (Section 4.6), the damping ratio in the first mode can be limited to the desired level (e.g. 5%) if Eq. (4.1.5) leads to a larger value; limiting overall damping in response history analysis of 2D models in a similar manner is not feasible. However, site conditions, such as non-homogeneity in the rock, may reduce radiation damping. The last recourse is to use 3D models, which also permit realistic representation of the 3D geometries of gravity dams.

11

Nonlinear Response History Analysis of Dams

PREVIEW

Nonlinear dynamic analyses are necessary to estimate the performance of dams during ground motion intense enough to deform the structure beyond its linear range of behavior. Discussed in Part A are various nonlinear mechanisms that can develop. These include cracking of concrete, cyclic opening and closing of lift joints and contraction joints, sliding of rock blocks along fractures or weak planes, and cavitation in the impounded water. Modeling of all these nonlinear mechanisms, which is a vast subject, is discussed only briefly in Part A of this chapter.

This sets the stage for Part B, where the Direct Finite-Element Method (FEM) for response history analysis of a nonlinear model of the dam–water–foundation system is presented. Included are all effects known to be significant in earthquake response of dams: dam–water interaction including water compressibility and wave absorption at the reservoir bottom; dam–foundation interaction include mass, flexibility, and damping of the rock; water–foundation interaction; radiation damping associated with the semi-unbounded extent of the fluid and foundation domains; and spatial variation of ground motion. The procedure presented herein can be implemented in almost every commercial FE code without requiring modification of the source code. To achieve this goal, viscous dampers were selected to model wave-absorption boundaries to simulate the semi-unbounded domains, and a theory was developed to compute (in an auxiliary analysis) the effective earthquake forces that are applied at these boundaries. The Direct FEM is validated against the completely independent substructure method; results from the two methods are shown to be in close agreement.

Major simplifications of the Direct FEM, which in its rigorous form is computationally demanding, are presented toward the end of Part B. It is demonstrated that the effective earthquake forces at the side boundaries of the foundation domain may be computed by the much,

Earthquake Engineering for Concrete Dams: Analysis, Design, and Evaluation, First Edition. Anil K. Chopra.

much simpler one-dimensional analysis; those at the upstream boundary of the fluid domain may be neglected; and the sediments deposited at the reservoir boundary can be ignored.

The chapter closes with a section on major challenges in predicting the nonlinear response of dams, a topic that is discussed further in Chapter 12.

PART A: NONLINEAR MECHANISMS AND MODELING

11.1 LIMITATIONS OF LINEAR DYNAMIC ANALYSES

The analysis procedures presented in the preceding chapters are valid so long as the dam–water–foundation systems remain within its linear range of behavior. In particular, the computed tensile stresses in the dam are valid only if they are less than the tensile strength of concrete. Recall from Section 7.2.2, linear analysis of Koyna Dam resulted in stresses exceeding the tensile strength of concrete. Obviously, such computed stresses are unrealistic because well before these stresses are reached during the earthquake, the concrete will crack, thus limiting the development of stresses larger than the tensile strength. Despite these limitations, such linear analyses were useful in identifying the location of cracks in Koyna Dam (Section 7.2.2) and Pacoima Dam (Section 10.2.4).

However, linear analyses cannot predict the performance of dams during ground motions intense enough to cause extensive cracking in concrete or initiate other mechanisms of nonlinear behavior. Such is the case because the strength of materials, reduction in stiffness due to tensile cracking, redistribution of stiffness due to tensile cracking, redistribution of forces after cracking, or joint opening and closing, are not modeled in linear analyses. Thus, nonlinear analyses would be required to predict the distribution and extent of cracking of concrete or sliding along joints and interfaces.

11.2 NONLINEAR MECHANISMS

11.2.1 Concrete Dams

Mass concrete exhibits a nonlinear relationship between stress and strain that depends on loading rate and history. Experiments demonstrate that this relationship is essentially linear for compressive stresses up to approximately 50–60% of the compressive strength. However, nonlinear behavior of mass concrete becomes significant as stresses approach the compressive strength or the comparatively much smaller tensile strength when cracking is initiated. The tensile strength of concrete is an order of magnitude less than its compressive strength.

Nonlinear behavior of concrete is generally not relevant under pre-earthquake static loads because concrete dams are designed to resist gravity and hydrostatic loads through compressive stress fields; consequently, tensile stresses are essentially non-existent. Furthermore, the computed compressive stresses in well-designed dams are much less than the compressive strength of concrete, well within the aforementioned range of linear behavior.

However, intense ground motion can produce large cyclic strains in concrete dams that, combined with pre-earthquake strains, deform the material into the nonlinear range. Linear dynamic analyses of gravity and arch dams show that compressive stresses rarely exceed the linear limit. Thus, nonlinear behavior of concrete in compression is generally not an issue in earthquake response analysis of dams. In contrast, such analyses demonstrate that during an earthquake, the tensile limit may be exceeded at different locations within the dam. Although the dynamic stresses computed by linear analysis are comparable in compression and tension, as mentioned earlier, tensile strength of concrete is roughly an order of magnitude less than the compressive strength. For example, it was shown in Section 7.2.2 that the computed tensile stresses in Koyna Dam due to the 1967 Koyna earthquake exceeded the tensile strength of concrete by factors of 2 to 3 near the upstream and downstream faces in the upper part of the dam and at the heel. Thus, tensile cracking of mass concrete must be modeled in nonlinear dynamic analysis.

As stresses in a dam approach the tensile strength of the concrete, micro-cracks (which are always present in concrete) coalesce to form a crack surface that depends on the type of dam. Gravity dams resist loads primarily by cantilever action, implying that the stress state consists of essentially uniaxial vertical stresses. Thus, cracking in highly stressed areas of a gravity dam tends to occur on horizontal planes, first at weaker planes, such as lift joints and the concrete–rock interface at the base. A part of such a lift joint may open and undergo cyclic opening and closing during earthquake-induced vibrations of the dam.

Various crack and joint opening profiles that may develop in concrete gravity dams during earthquake excitation are presented schematically in Figure 11.2.1. Vibratory motion of the dam may initiate cracks on the upstream face or downstream face that propagate toward the interior of the cross section. The cracks may be horizontal along lift joints (3 and 4 in Figure 11.2.1) or the concrete–rock interface (5 in Figure 11.2.1); they could be inclined within the dam (6 and 8 in Figure 11.2.1) or in the foundation (7 and 9 in Figure 11.2.1).

On the other hand, arch dams resist loads by arch and cantilever actions, thus the stress state is essentially biaxial with stresses acting in the arch and cantilever directions. However, the vertical contraction joints between cantilever monoliths, which are grouted possibly with shear keys, limit the development of stresses in the arch direction because the joints are unable to resist any significant net tension. Thus, the joints may intermittently open during a part of a vibration cycle creating a tensile stress state at the joint, and then close later in the cycle when the stress state reverses to compression. This nonlinear response mechanism affects the dam response in two ways. First the opening of a joint temporarily reduces the resistance in the arch direction, causing transfer of load to cantilever action. The load picked up by cantilever bending would lead to increase in flexural stress and possible horizontal cracking of the monoliths. Second, repeated joint opening and closing may cause compression failure of the joint. Model tests have shown that as a joint opens on the tension side of the arch, the compressive stresses on the portion of the joint remaining in contact increase greatly, possibly crushing the concrete (Niwa and Clough 1982). These tests have also demonstrated that these nonlinear mechanisms have significant influence on the response of the dam. Horizontal cracking in cantilevers and failure of joints has the potential to create a failure surface of the type shown in Figure 11.2.2. After extensive cracking, it becomes extremely challenging to model the sliding and impact of sections.

During earthquake-induced vibration of a dam, lift joints and the concrete–rock interface – which are horizontal planes of weakness – may cyclically open and close and slide intermittently in the downstream direction.

The nonlinear behavior mechanisms mentioned above (tensile cracking of concrete and loss of joint integrity) and those related to the foundation rock to be mentioned later result from vibration of the dam due to earthquake excitation. The vibratory nature of dam response implies that the earthquake forces reverse direction during a vibration cycle. Thus, the forces usually do not act long enough in the downstream direction to initiate sliding or overturning of the monolith. Thus, overall sliding or overturning of the dam, which appeared prominently in traditional criteria for design of dams, are unlikely failure modes; however, the possibility of intermittent sliding and separation should be allowed in nonlinear models for analysis of dams.

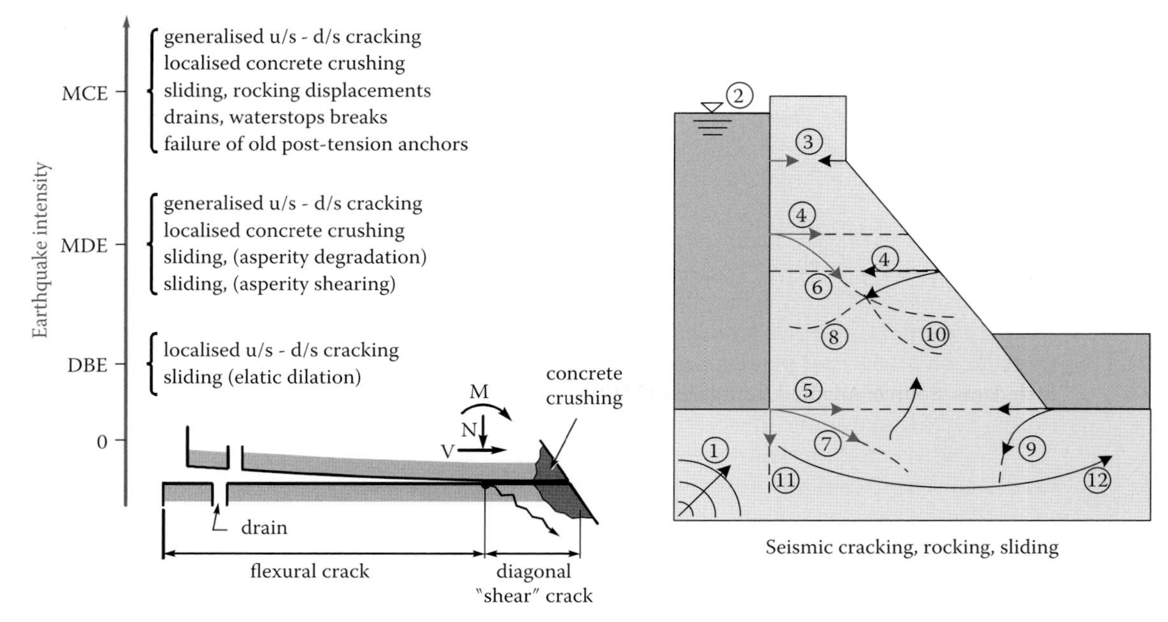

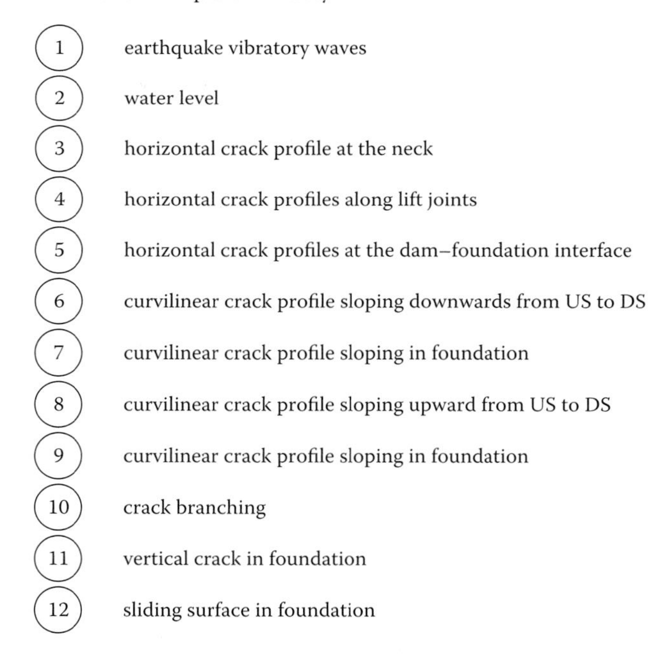

Figure 11.2.1 Seismic failure mechanisms of concrete gravity dams. Source: Léger et al. (2003), also see Léger (2019).

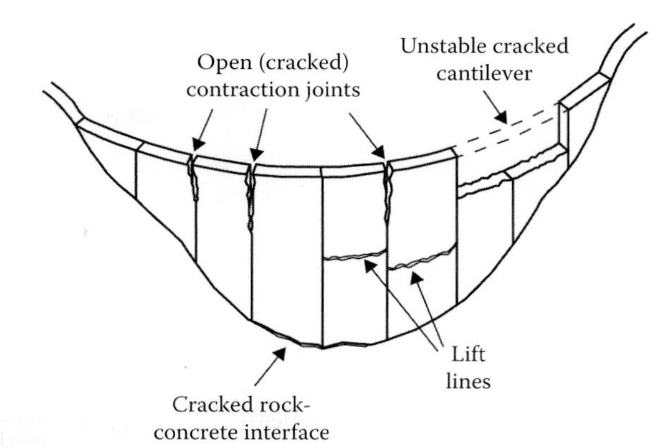

Figure 11.2.2 Failure mechanism of concrete arch dams. Source: Adapted from Léger (2019).

11.2.2 Foundation Rock

The foundation rock behaves nonlinearly because it is typically fractured and discontinuous, permitting sliding of rock blocks at these surfaces. During an earthquake, the forces transmitted to the foundation rock, can increase considerably beyond the pre-earthquake state. Such increase in forces transferred to the abutments of an arch dam may initiate shear failure along weak planes in the rock.

11.2.3 Impounded Water

Cavitation is the formation of vapor cavities in water – i.e. small liquid-free zones ("bubbles" or "voids") – when the local pressure declines to some point below the vapor pressure, followed by bubble implosion. Cavitation develops only during very intense ground motions. The extent of cavitation is typically confined to regions very close to the upstream face of the dam. The high-frequency spikes in water pressure caused by cavitation generally do not significantly affect the response of the dam (Zienkiewicz et al. 1983; Vargas-Loli and Fenves 1989; El-Aidi and Hall 1989b). Thus, water cavitation may generally be ignored in earthquake analysis of concrete dams.

11.2.4 Pre-Earthquake Static Analysis

Cracking of concrete predicted by nonlinear dynamic analysis of dams is likely to be sensitive to the static state of stress and strain that exists at the time the earthquake occurs. In determining this state, the sequence of dam construction and reservoir filling as well as the thermal history during construction should be modeled. Thermal stresses are especially significant in mass-concrete gravity dams if the vertical contraction joints between monoliths are grouted; in roller-compacted concrete dams, typically built without contraction joints or with a minimal number of such joints; and in arch dams after the joints between cantilevers have been grouted. The most severe temperature strains in a concrete dam occur during construction due to the heat of hydration of the concrete as it cures. In some dams, cracks developed during or soon after construction; such examples include Richard B. Russell Dam on the South Carolina/Georgia border and Dworshak Dam in Idaho.

11.3 NONLINEAR MATERIAL MODELS

11.3.1 Concrete Cracking

Formation and propagation of cracks in concrete structures have been modeled in finite element analysis as discrete cracks, "smeared," or the two combined. Since the late 1960s both models have been used because neither one is ideal; each model has its advantages and disadvantages. Here we comment on the salient aspects of both models; a comprehensive discussion is available elsewhere (Bhattacharjee and Léger 1992).

Discrete Crack Model. In the discrete crack model (Ngo and Scordelis 1967), a crack is represented by a gap at the boundary between finite elements. Growth (or propagation) of the crack is determined by constitutive models based on strength or on fracture mechanics (Ayari and Saouma 1990; Pekau et al. 1991). When a crack opens or propagates, the common nodal points are separated before proceeding to the next time step in the dynamic analysis (Figure 11.3.1a). Modeling of discrete cracks would appear to be a physically realistic approach that can explicitly consider penetration of water in the crack, uplift pressure on cracked-open surfaces, aggregate interlock at rough-crack surfaces, opening and closing of cracks, and impact and sliding of sections of the dam after extensive cracking. This approach can directly estimate the length and width of cracks. The discrete crack approach also permits the use of special finite elements to model weak planes and joints in the dam, such as the concrete–rock interface, horizontal lift planes, and vertical contraction joints. However, the discrete crack model is computationally challenging because the finite element mesh is redefined at each time-step during which a new crack is formed or the length of an existing crack changes, and it is difficult to determine the direction in which a crack will propagate during the next time-step. Furthermore in this approach, crack propagation is dependent on the size, shape, orientation, and order of finite elements in the mesh; thus the results are not unique.

The preferred way to model discrete propagation is the Extended Finite Element Method (XFEM) that allows cracks to propagate through finite elements, and hence does not require modifications to the FE mesh. This is achieved by adding enrichment functions to the finite element approximation to account for the presence of a crack that can propagate arbitrarily through elements (Belytschko and Black 1999; Moës et al. 1999). The XFEM has been employed to study the earthquake response of gravity dams and arch dams (Pan et al. 2011; Zhang et al. 2013; Goldgruber 2015). However, most implementations of XFEM do not allow for multiple cracks, which is a significant limitation for practical analysis of concrete dams.

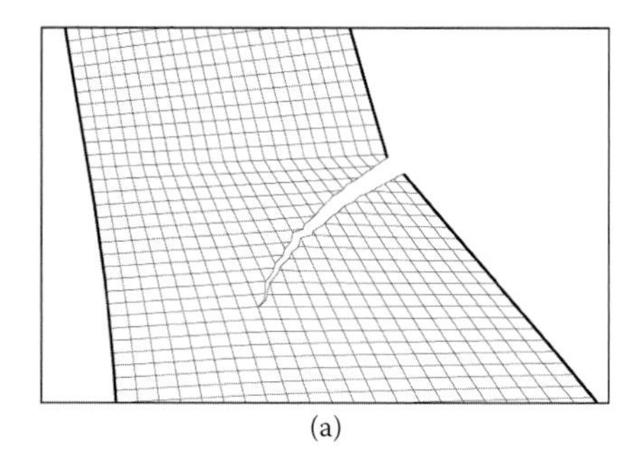

(a)

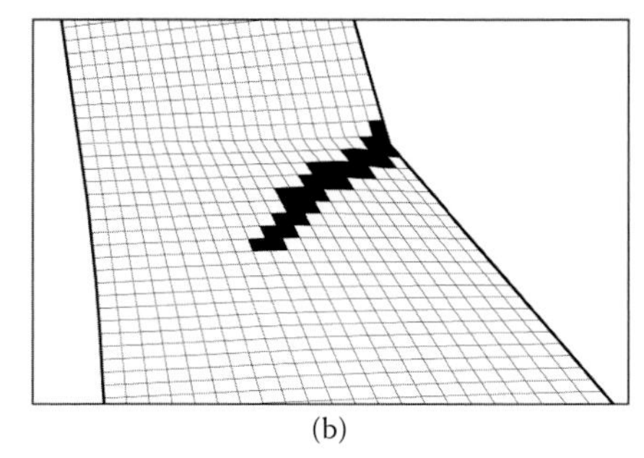

(b)

Figure 11.3.1 Two approaches to modeling crack propagation: (a) discrete crack model based on extended finite element method (XFEM); and (b) smeared crack model. Source: Adapted from Bhattacharjee and Léger (1992).

Smeared Crack Model. To overcome the aforementioned difficulties encountered in the discrete crack model, fracture can be idealized as "smeared" over an entire finite element (Figure 11.3.1b) or over a certain bandwidth of the element (Rashid 1968). The discontinuities in the displacement field across cracks is averaged over the finite element and approximated by continuous displacement functions that are then used to derive the element stiffness. Constitutive models describe the initiation of a crack and the softening response of concrete during crack propagation. After initiation of the fracture, the constitutive relation is changed from the original pre-crack condition to an updated damage relation, usually based on nonlinear fracture mechanics theory. This approach can include possible shear transfer across the crack, which can be significant in mass concrete because of the large-sized aggregate and the associated aggregate interlocking. In the "smeared" crack approach only the constitutive relation is updated, but the finite element mesh remains unchanged. This implies that only the tangent stiffness matrix is modified for the current state of cracking in an element. As a result, the smeared crack model can be easily incorporated into a nonlinear finite element analysis procedure.

However, the "smeared" crack model has its own share of disadvantages. First, the numerical results are not objective with respect to the finite element mesh; as the element size decreases, the fracture zone becomes smaller, and the force required to propagate the crack can decrease to a very small value. Second, in this approach it is difficult to model impact and sliding of sections after extensive cracking because the discontinuous displacements across the interacting surface are not well approximated. Third, the diffuse crack pattern computed by this approach does not provide any information on the length and width of the cracks or their location within elements.

The plastic damage model for cyclic loading of concrete developed by Lee and Fenves (1998) is a variation of the smeared crack model. Several features of this model have made it attractive for earthquake analysis of concrete dams. By keeping track of tensile damage and compressive damage through two damage variables d_t and d_c, the stiffness changes during cyclic loading – where the stiffness in compression is recovered when a crack closes, but the stiffness in tension is not recovered – can be simulated (Figure 11.3.2a). The model also enables the conceptual separation of the behavior in tension into two parts: linear stress–strain relation model for stresses less than the tensile strength, f_t (Figure 11.3.2b); and the softening behavior after initiation of cracking is described by fracture mechanics theory that relates stresses to crack-opening displacements u_r (Figure 11.3.2c).

This approach to defining the behavior after initiation of cracking offers two important advantages. First, the dissipated fracture energy per unit length of crack remains independent of the FE mesh; thus ensuring that the results are independent of mesh size (Lee and Fenves 1998; Hillerborg et al. 1976). Second, this approach introduces the concept of *specific fracture energy*, G_F, defined as the area under the stress–crack displacement curve (Figure 11.3.2c). The two significant parameters for this material model are the tensile strength f_t and specific fracture energy, G_F. They can be determined experimentally; f_t from a splitting-tensile or direct-tensile test (Raphael 1984) and G_F from a wedge-splitting test (Brühwiler 1990; Brühwiler and Wittmann 1990).

Various implementations of the smeared crack model have been used to study crack propagation in two-dimensional analysis of gravity dam monoliths (Bhattacharjee and Léger 1993; El-Aidi and Hall 1989a,b; Lee and Fenves 1998; Zhang et al. 2013), and in three-dimensional analysis of arch dams (Pan et al. 2011; Wang et al. 2013).

11.3.2 Contraction Joints: Opening, Closing, and Sliding

The nonlinear mechanisms associated with opening, closing, and sliding of contraction joints may be modeled by discrete joint elements (Dowling and Hall 1989; Fenves et al. 1992; Lau et al.

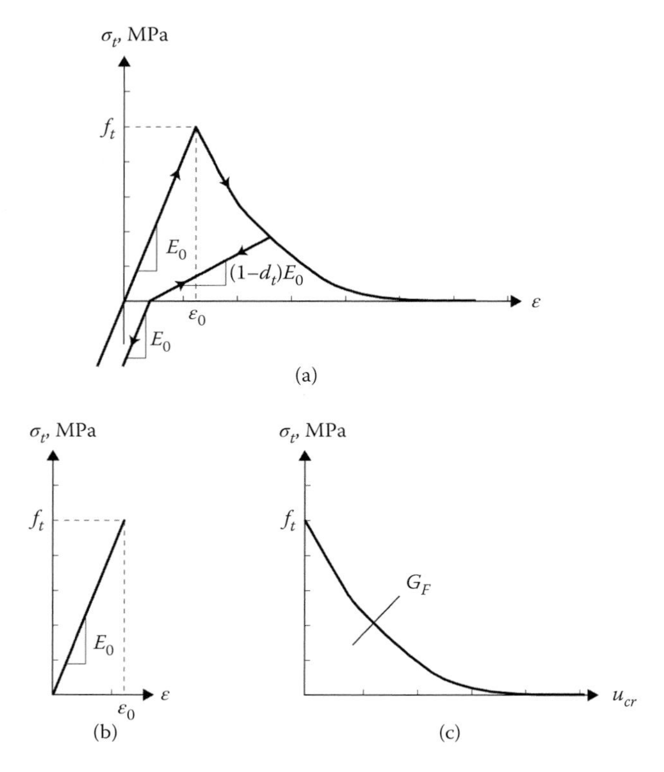

Figure 11.3.2 (a) Softening response of concrete under uniaxial cyclic loading; (b) linear stress–strain curve for pre-failure response; (c) post-failure stress-displacement curve, where the area under the curve is the fracture energy G_F. Source: Adapted from Brühwiler and Wittmann (1990).

1998) or by a contact formulation (Zhang et al. 2009; Goldgruber 2015). Opening and closing of the joint in the normal direction may be described by a zero-tensile-strength master–slave contact constraint with pressure overclosure relationship (Figure 11.3.3a) to prevent node penetration (see *ABAQUS User's Manual*). The behavior in the tangential direction is modeled by linear springs that simulate shear keys or by specifying a Mohr–Coulomb friction criterion (Figure 11.3.3b): $\tau = c + \sigma \tan \varphi$, where τ is the shear stress, σ is the normal stress, c represents cohesion, and φ denotes the friction angle.

11.3.3 Lift Joints and Concrete–Rock Interfaces: Sliding and Separation

The properties of lift joints are greatly influenced by how the joints were prepared before pouring the next lift of concrete. Even with good preparation, the strength and fracture energy at a joint are much lower than the values for mass concrete (Tinawi et al. 1998; Saouma et al. 1991; Fronteddu et al. 1998). Under the combined action of normal and shear forces, the failure of a joint is usually described by the Mohr–Coulomb criterion (Section 11.3.2); however, it is difficult to assign values to c and φ because of the roughness of lift joints and of the concrete-rock interface. Furthermore, understanding of the behavior of joints under cyclic displacements is limited because very few such experiments have been conducted (Puntel and Saouma 2008).

Special finite elements have been developed to model joints in mass concrete, rock, or at the concrete–rock contact. These include zero-thickness interface elements (Goodman et al. 1968), thin-layer interface elements (Zienkiewicz et al. 1970), and gap friction elements (Léger and

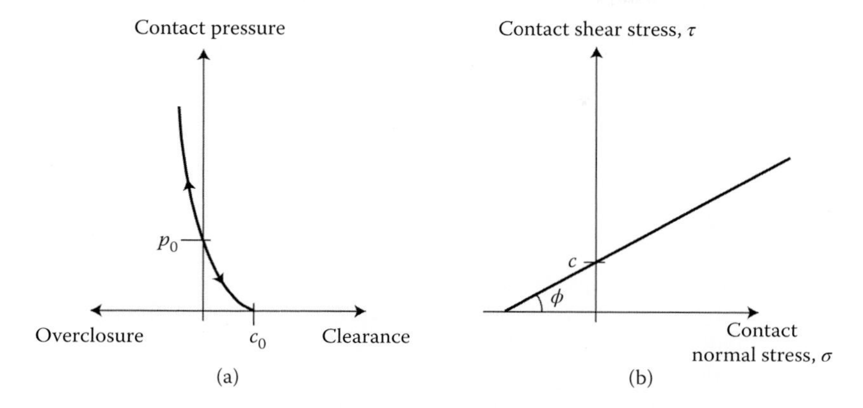

Figure 11.3.3 (a) Exponential pressure-overclosure relation for normal contact in contraction joint, where c_0 is the clearance at zero pressure and p_0 the pressure at zero opening; and (b) Mohr–Coulomb friction criterion. Source: Adapted from ABAQUS *Standard Users' Manual* (Dassault System n.d., Version 6.13).

Katsouli 1989). A comprehensive coverage of these elements is available in Fronteddu et al. (1998).

11.3.4 Discontinuities in Foundation Rock

Discontinuities and planes of weakness in the foundation rock can be modeled by nonlinear joint elements (Goodman et al. 1968; Léger and Katsouli 1989). Assemblages of rock wedges and blocks can be modeled by the Discrete Element Method (DEM) (Lemos 2008). Such modeling is a very challenging problem, particularly because detailed information about the foundation conditions is usually limited. Consequently, such detailed modeling is often excluded from dynamic analysis of the dam–water–foundation system. However, potential instability in the foundation must be evaluated by supplementary analyses.

11.4 MATERIAL MODELS IN COMMERCIAL FINITE-ELEMENT CODES

Modeling of the various nonlinear mechanisms mentioned in Section 11.2 differ among commercial FE and distinct element codes, e.g. ABAQUS, ADINA, DIANA, FLAC, LS-DYNA, NASTRAN, UDEC, and 3DEC. Most organizations – governmental or private – are predisposed to a particular code because they have used it for years for various types of analyses – static, thermal, and dynamic – they are familiar with user interfaces of the code, and have invested in developing in-house pre-processing and post-processing tools. Familiarity is often a dominant factor in selection of a commercial code; it tends to override other factors. However, it would obviously be better if choice of code was based on evaluating the relative capabilities of these codes to model nonlinear mechanisms that are important in the earthquake analysis of concrete dams. Irrespective of the computer code selected, extensive validations and verifications must be implemented at the material, element, and structural scales to develop confidence in the computed results.

No matter which code is selected, major issues arise in modeling the semi-unbounded physical system and in specifying earthquake excitation. These topics are addressed in Part B of this chapter where a Direct Finite-Element Method (FEM) for nonlinear RHA of the dam–water-foundation system is presented. The method is applicable to 2D analysis of gravity dam monoliths (Løkke and Chopra 2017) as well as 3D analysis of all types of dams, including gravity, arch, and buttress dams (Løkke and Chopra 2018, 2019a,b).

PART B: DIRECT FINITE-ELEMENT METHOD

11.5 CONCEPTS AND REQUIREMENTS

Earthquake analysis of arch dams requires three-dimensional models of dam–water–foundation systems that recognize the factors known to significantly influence the earthquake response of concrete dams (see Chapters 6 and 9): dam–water interaction including water compressibility and wave absorption at the reservoir bottom; dam–foundation interaction including mass, flexibility, and damping of the rock; radiation damping associated with the semi-unbounded extent of the fluid and foundation domains; and spatial variation of the ground motion. Furthermore, nonlinear mechanisms that become significant during intense ground motions should be modeled. These mechanisms include cracking of concrete; opening, closing, and sliding of contraction joints; sliding and separation at construction joints, lift joints, and concrete–rock interfaces; discontinuities in the rock due to local cracks and fissures; and cavitation in the fluid.

Analysis procedures based on the *substructure method* (Chapters 5 and 8) have long been available for frequency-domain analysis of three-dimensional dam–water–foundation systems. Restricted to linear systems with foundation and fluid domains that are simple in geometry and homogeneous in material properties, these procedures have played a central role in advancing our understanding of the profound influence of dam–water–foundation interaction and radiation damping on dam response (Chapters 6, 9, and 10).

In contrast, the Direct FEM of analysis is ideally suited for modeling of arbitrary geometry and inhomogeneous, nonlinear material properties of the dam, and foundation and fluid domains. In this method, the entire dam–water–foundation system is modeled by finite elements, and response history analysis (RHA) is implemented directly in the time domain. Facilitated by commercial FE software with their user-friendly interfaces, this method has been popular with professional engineers. While these programs are able to model nonlinear mechanisms, they often use simplistic models for dam–water–foundation interaction and the semi-unbounded domains.

Realistic modeling of dam–water–foundation systems requires a FE model that includes all the previously identified factors and truncated fluid and foundation domains with wave-absorbing boundaries (Wolf 1988) to simulate the semi-unbounded size of these domains. The seismic input is specified by *effective earthquake forces* applied directly to these boundaries (Wolf 1988; Zienkiewicz et al. 1989), or alternatively, in a single layer of elements interior of the boundaries (Bielak et al. 2003; Bielak and Christiano 1984). Utilizing the latter approach, Basu (2004) developed an advanced analysis procedure using Perfectly Matched Layer (PML) boundaries (Basu and Chopra 2004) and the Effective Seismic Input method (ESI) (Bielak and Christiano 1984) to specify effective earthquake forces. However, ESI and PML methodologies require modification of the FE source code, and the procedure is currently only available in LS-DYNA.

The goal here is to develop a methodology in the context of a Direct FEM that can be implemented in any commercial code without requiring modification of the source code. To achieve this goal, viscous dampers are selected to model wave-absorbing boundaries (Lysmer and Kuhlemeyer 1969), and a theory is presented to compute (in an auxiliary analysis) the effective earthquake forces that are applied at these boundaries. These features are available in almost every commercial code. The use of viscous dampers instead of the more sophisticated PML at

FE model truncations has the disadvantage that the sizes required for the truncated domains are much larger compared to the smaller domains that are effective in conjunction with the PML boundary. However, this disadvantage is outweighed by the attraction that the method can be implemented in any commercial FE program without modification of the source code.

11.6 SYSTEM AND GROUND MOTION

11.6.1 Semi-Unbounded Dam–Water–Foundation System

The idealized, three-dimensional semi-unbounded dam–water–foundation system considered consists of three subsystems (Figure 11.6.1): (i) the concrete dam with nonlinear properties; (ii) the foundation, consisting of a bounded region adjacent to the dam that may be constitutively nonlinear and inhomogeneous, and irregular in geometry; and the exterior, semi-unbounded, region with "regular" geometry that is homogeneous or horizontally layered and has linear constitutive properties; and (iii) the fluid domain, consisting of a bounded region of arbitrary geometry adjacent to the dam that may be constitutively nonlinear; and a uniform channel, unbounded in the upstream direction, that is restricted to be linear.

"Regular" geometry of the semi-unbounded foundation region implies that the canyon upstream of the bounded region has a uniform cross-section, and the canyon downstream of the bounded region has a uniform cross-section; however, the two cross-sections may be different. The assumption of homogeneous or horizontally layered properties in the exterior foundation region is introduced to permit use of a deconvolution method to define the seismic input for the system starting from a ground motion specified at the surface of the foundation rock (Section 11.6.2).

The semi-unbounded system in Figure 11.6.1 is modeled by a three-dimensional FE discretization of a bounded system with wave-absorbing boundaries – modeled by viscous dampers – at the bottom and side boundaries of the foundation domain, and at the upstream end of the fluid domain (Figure 11.6.2).

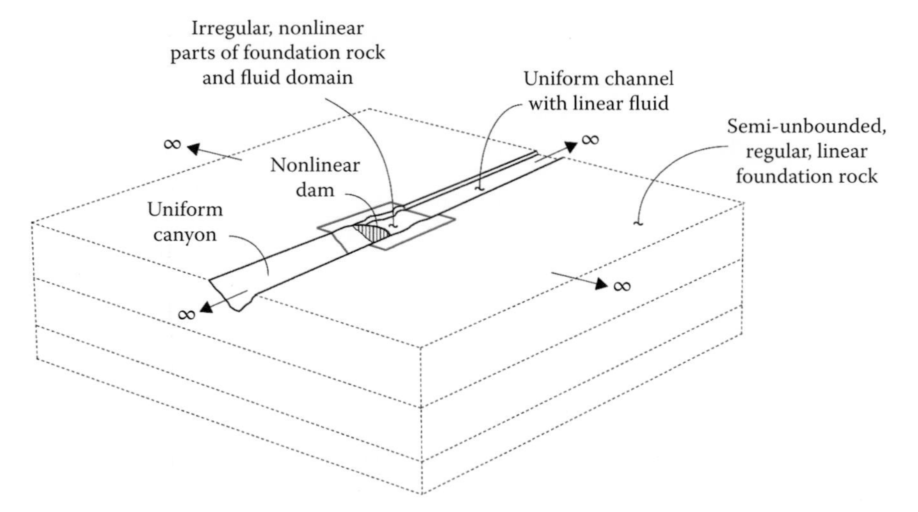

Figure 11.6.1 Three-dimensional semi-unbounded dam–water–foundation system showing three subsystems: (1) nonlinear dam; (2) foundation, consisting of an irregular, nonlinear region and a semi-unbounded linear region with "regular" geometry and properties; and (3) fluid domain, consisting of an irregular (possibly nonlinear) region, and a semi-unbounded uniform channel with linear fluid.

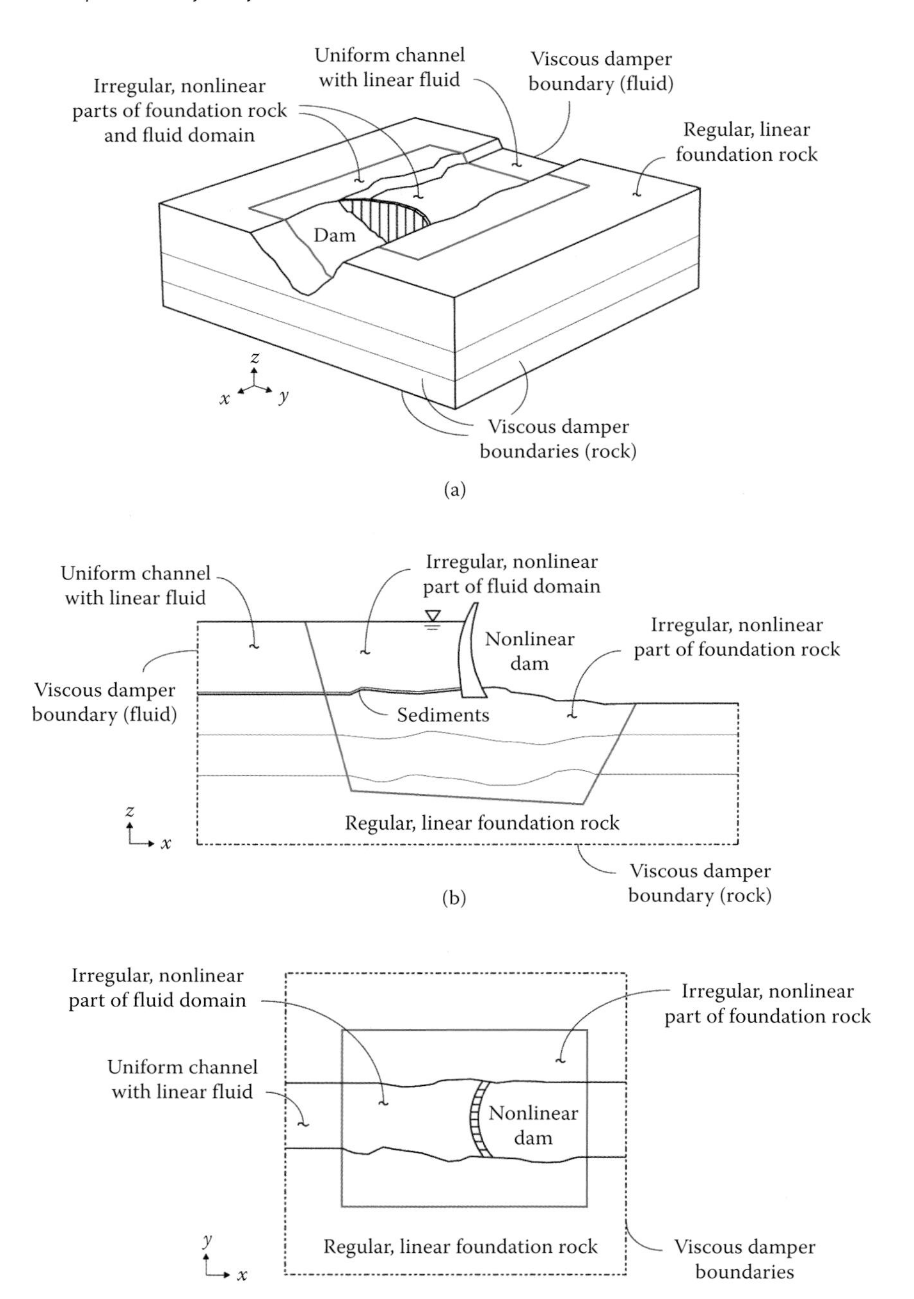

Figure 11.6.2 Dam–water–foundation system with truncated foundation and fluid domains: (a) 3D perspective view; (b) section view through center of canyon; and (c) plan view.

The linear, regular parts of the foundation and fluid domains included in the FE model (Figure 11.6.2) provide a transition from the subdomain adjacent to the dam with irregular geometry and nonlinear behavior to the exterior subdomain with regular geometry and linear behavior, a restriction required at the absorbing boundaries. The minimum sizes for these domains are determined by the ability of the viscous-damper boundaries to absorb outgoing (scattered) waves from the system.

The use of finite elements for the entire system permits modeling of arbitrary geometry and inhomogeneous, nonlinear material properties of the dam, canyon, foundation, and fluid domains adjacent to the dam. Furthermore, it allows for modeling of nonlinear mechanisms (Figures 11.6.1 and 11.6.2) such as cracking of concrete (Vargas-Loli and Fenves 1989; El-Aidi and Hall 1989a; Bhattacharjee and Léger 1993; Hall 1998; Lee and Fenves 1998; Zhang et al. 2009), sliding and separation at construction joints, lift joints, and at concrete-rock interfaces (Niwa and Clough 1982; El-Aidi and Hall 1989a,b; Léger and Katsouli 1989; Fenves et al. 1992; Hohlberg 1992; Hall 1998; Lau et al. 1998; Zhang et al. 2009), discontinuities in the rock due to local cracks and fissures (Lemos 2008), and cavitation in the fluid (Zienkiewicz et al. 1983).

11.6.2 Earthquake Excitation

Equations governing the motion of the system of Figure 11.6.2 subjected to the free-field ground motion – the motion that would occur in the foundation without the dam and water present – will be formulated in Section 11.7. Specifying such motions at all boundaries of the finite-element model remains a challenging problem.

The most general approach is to perform large-scale simulation of seismic wave propagation from the earthquake source to the dam site (Bao et al. 1998; Graves 1996; Moczo et al. 2007), as shown schematically in Figure 11.6.3a. Here, physics-based finite element or finite difference models of large regions subjected to a fault slip are analyzed. Although such regional simulations have been reported in the research literature, at present, they seem impractical for concrete dam analysis for two reasons: (i) the details of the earthquake fault rupture process and the properties of the geological materials in the source-to-site path are usually lacking; and (ii) simulation capability is currently limited to excitation frequencies lower than the natural vibration frequencies of concrete dams.

Another approach would be to use boundary element methods (BEMs) to compute the free-field motions resulting from incident plane waves propagating from infinity to the dam site at predefined angles; see Figure 11.6.3b. Such methods have been used to compute the free-field motions at the surface of canyons (Zhang and Chopra 1991a; Maeso et al. 2002) and to investigate the influence of assumed incident angles on the dam response (García et al. 2016). However, earthquake motion is composed of various types of waves – P, SV, SH, and surface – arriving from various directions that are not known in advance, thus limiting the application of this approach to practical problems.

Presently, a common approach starts with Probabilistic Seismic Hazard Analysis (PSHA) for the site to establish a design spectrum, and then select ground motions consistent with this spectrum (see Chapter 13). The earthquake excitation is defined by three components of free-field acceleration specified on the foundation (or ground) surface (Figure 11.6.3c): the stream component, $a_g^x(t)$, the cross-stream component $a_g^y(t)$ and the vertical component $a_g^z(t)$. This motion is usually specified at the elevation of the dam abutments, but other choices are possible, e.g. it could

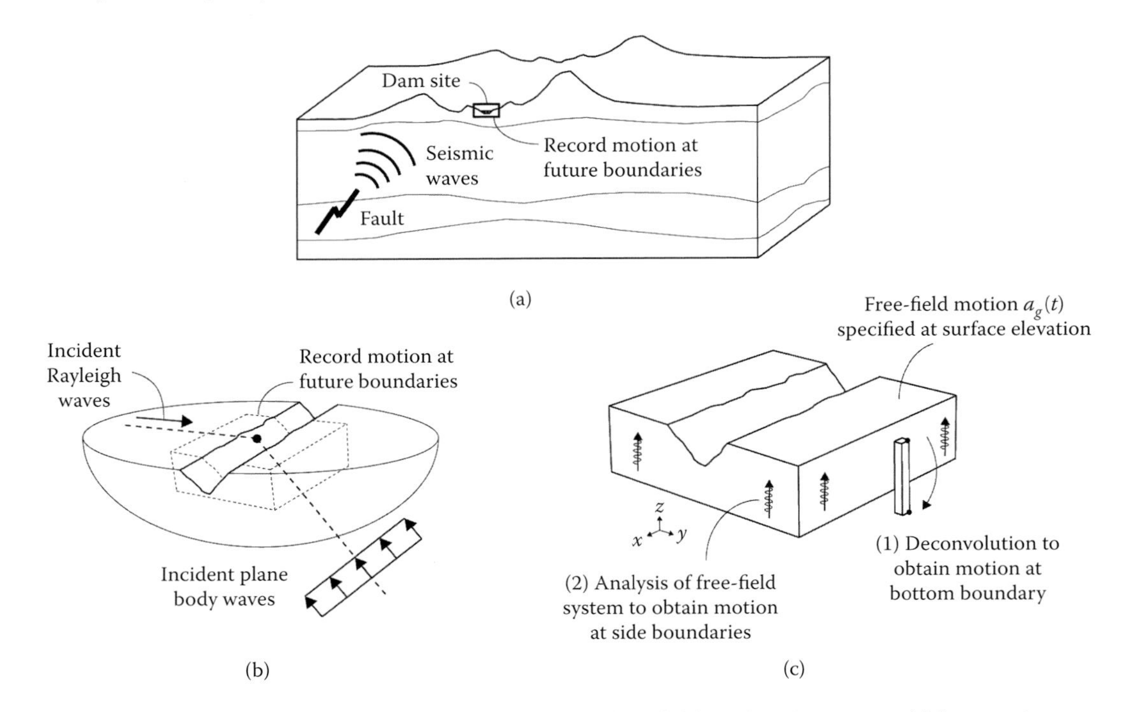

Figure 11.6.3 Schematic overview of methods to obtain free-field earthquake motion: (a) large scale fault-rupture simulation; (b) boundary element method with incident plane waves propagating from infinity at predefined angles; (c) deconvolution analysis starting with a free-field motion $a_g(t)$ specified at the foundation surface.

be the motion recorded near the base of an existing dam to be used an input to an analysis to compare computed and recorded motions at the crest of the dam.

Because PSHA typically does not consider topographic effects, the ground motions selected are intended for flat ground surface. Free-field motion at the bottom boundary of the truncated foundation domain is determined by deconvolution of the free-field motion $a_g(t)$ specified at the surface (Figure 11.6.3c). This one-dimensional deconvolution is based on the common assumption that the incident wave field consists solely of plane *SH*-, *SV*-, and *P*-waves propagating vertically upwards from the underlying semi-unbounded medium. This is clearly a major simplification but, at the present time, it seems to be a reasonable pragmatic choice. With the motion at the bottom boundary known, the free-field motion at the side boundaries is determined by analyzing a FE model of the foundation domain alone (without the dam or impounded water) by the Direct FEM, to be developed in Sections 11.7 and 11.8.

Results of such an analysis of the foundation domain of Figure 11.6.4, with the surface motion defined by the S69E component of the motion recorded at Taft Lincoln School Tunnel

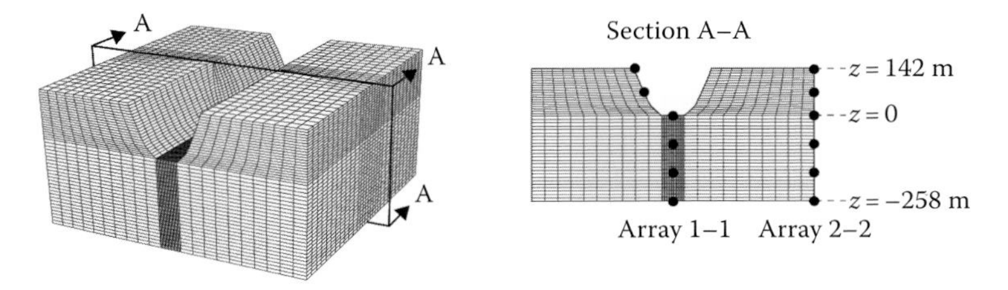

Figure 11.6.4 FE model of canyon showing location of two vertical node arrays: array 1–1 at the center of the model, and array 2–2 at the side boundary of the model.

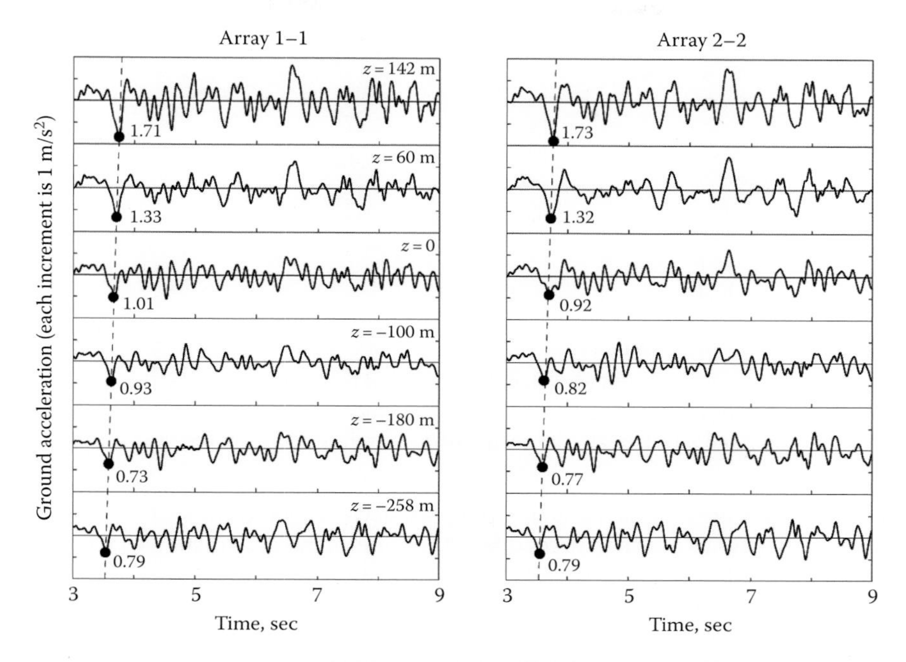

Figure 11.6.5 Stream component of free-field earthquake motion computed by the direct FE method at six different elevations at the two arrays. A specific peak in the acceleration history is identified and connected by a dashed line to demonstrate the amplitude change and time shift in the motions at higher elevations.

during the 1952 Kern County earthquake are presented in Figure 11.6.5. Observe from the results in Figure 11.6.5 that the amplitude and phase of the motion varies greatly with height. By comparing the motions at two locations at the same elevation (e.g. at $z = 0$ m), it is evident that scattering and diffraction of waves from the canyon results in some variation of motion in the horizontal direction, but this is relatively small because of the assumption of vertically incident waves.

11.7 EQUATIONS OF MOTION

The equations governing the motion of a three-dimensional dam–water–foundation system idealized as an ensemble of finite elements with viscous-damper absorbing boundaries were derived in Løkke and Chopra (2017, 2018) by interpreting dam–water–foundation interaction as a scattering problem in which the dam perturbs the "free-field" motion in an auxiliary state of the system. Presented here without derivation, the final equations of motion for the FE system (Figure 11.7.1) are

$$\begin{bmatrix} \mathbf{m} & \mathbf{0} \\ \rho(\mathbf{Q}_h^{\mathrm{T}} + \mathbf{Q}_b^{\mathrm{T}}) & \mathbf{s} \end{bmatrix} \begin{Bmatrix} \ddot{\mathbf{r}}^t \\ \ddot{\mathbf{p}}^t \end{Bmatrix} + \begin{bmatrix} \mathbf{c} + \mathbf{c}_f & \mathbf{0} \\ \mathbf{0} & \mathbf{b} + \mathbf{c}_r \end{bmatrix} \begin{Bmatrix} \dot{\mathbf{r}}^t \\ \dot{\mathbf{p}}^t \end{Bmatrix}$$
$$+ \begin{Bmatrix} \mathbf{f}(\mathbf{r}^t) \\ \mathbf{0} \end{Bmatrix} + \begin{bmatrix} \mathbf{0} & -(\mathbf{Q}_h + \mathbf{Q}_b) \\ \mathbf{0} & \mathbf{h} \end{bmatrix} \begin{Bmatrix} \mathbf{r}^t \\ \mathbf{p}^t \end{Bmatrix} = \begin{Bmatrix} \mathbf{R}^{\mathrm{st}} \\ \mathbf{0} \end{Bmatrix} + \begin{Bmatrix} \mathbf{P}_f^0 \\ \mathbf{P}_r^0 \end{Bmatrix} \tag{11.7.1}$$

where $\mathbf{r}^t$ is the vector of displacements of the dam and foundation domains; $\mathbf{p}^t$ is the vector of hydrodynamic pressures in the fluid idealized as a linear,† inviscid, irrotational, and compressible

† For convenience of notation, the fluid is assumed to be linear also in the irregular fluid subdomain; however, the formulation is applicable to nonlinear fluids with appropriate generalization.

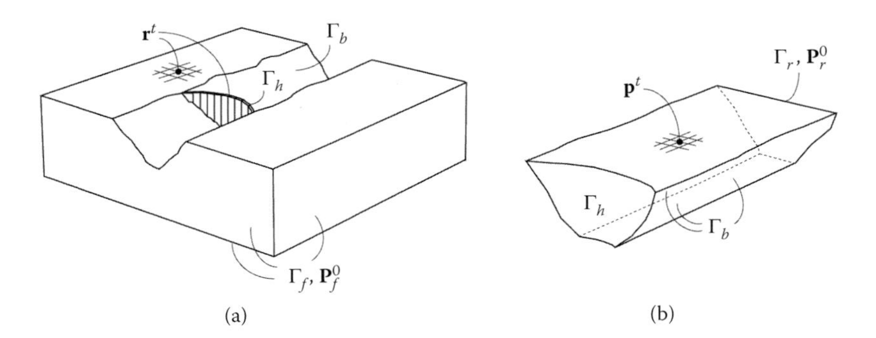

Figure 11.7.1 Schematic overview of FE model of (a) dam and foundation domain, (b) fluid domain. Highlighted are the viscous damper boundaries Γ_f and Γ_r at the truncation of the foundation and fluid domains, and the fluid-solid interfaces Γ_h at the upstream dam face and Γ_b at the reservoir bottom and sides.

acoustic fluid; **m** and **c** are the mass and damping matrices, respectively, for the dam–foundation system; $\mathbf{f}(\mathbf{r}^t)$ is the vector of internal forces due to (nonlinear) material response; **s**, **b**, and **h** are the "mass," "damping," and "stiffness" matrices for the acoustic fluid (Zienkiewicz and Bettess 1978), respectively; $\mathbf{c}_f$ is the matrix of damper coefficients for the viscous dampers acting in directions that are normal and tangential on the foundation boundaries Γ_f; $\mathbf{c}_r$ is the matrix of damper coefficients for the viscous dampers normal to the upstream fluid boundary Γ_r; ρ is the density of water; $\mathbf{Q}_h$ and $\mathbf{Q}_b$ are matrices that couple accelerations to hydrodynamic pressures at the dam–water interface Γ_h and water–foundation interface Γ_b, respectively; $\mathbf{R}^{st}$ is the vector of static forces that includes self-weight, hydrostatic pressures, and static foundation reactions at Γ_f.

Hydrodynamic wave energy is lost at Γ_b – the bottom and side boundaries of the reservoir – by means of two mechanisms. The first is wave absorption in sediments deposited at the reservoir boundaries, modeled by the reservoir bottom reflection coefficient α (Chapter 8), and its effects are included in Eq. (11.7.1) through the damping matrix **b**. The second mechanism, associated with water–foundation interaction, is explicitly considered in the FE model through the coupling matrix $\mathbf{Q}_b$. Because this mechanism automatically accounts for some absorption of hydrodynamic waves, care should be taken not to overestimate the total amount of energy lost at these boundaries when also including wave absorption in sediments deposited at the reservoir boundaries; see Section 10.3.

In the Direct FEM, sediments can be modeled explicitly by finite elements with appropriate constitutive properties, just like the fluid above and rock below are modeled (Wilson 2010; Section 23.15). Such a FE model would automatically account for absorption of fluid pressure waves at the reservoir boundary; for such a model, $\mathbf{b} = \mathbf{0}$ in Eq. (11.7.1).

The earthquake excitation appears in Eq. (11.7.1) as effective earthquake forces $\mathbf{P}_f^0$ and $\mathbf{P}_r^0$:

$$\mathbf{P}_f^0 = \mathbf{R}_f^0 + \mathbf{c}_f \dot{\mathbf{r}}_f^0 \text{ at the foundation boundaries } \Gamma_f \tag{11.7.2a}$$

$$\mathbf{P}_r^0 = \mathbf{c}_r\, \dot{\mathbf{p}}_r^0 \text{ at the upstream fluid boundary } \Gamma_r \tag{11.7.2b}$$

where the *free-field motion* $\mathbf{r}_f^0$ and *free-field boundary forces* $\mathbf{R}_f^0$ at Γ_f are to be computed from analysis of the free-field foundation domain, defined as the foundation domain without the dam or impounded water (Figure 11.7.2a); and the *free-field hydrodynamic pressures* $\mathbf{p}_r^0$ at Γ_r are to be computed from analysis of the free-field fluid domain, defined as the uniform channel upstream of Γ_r (Figure 11.7.2b). The free-field variables $\mathbf{r}_f^0$, $\mathbf{R}_f^0$, and $\mathbf{p}_r^0$ represent the minimal set of data required to specify earthquake excitation. Procedures for computing these variables at the different boundaries will be presented in the next section.

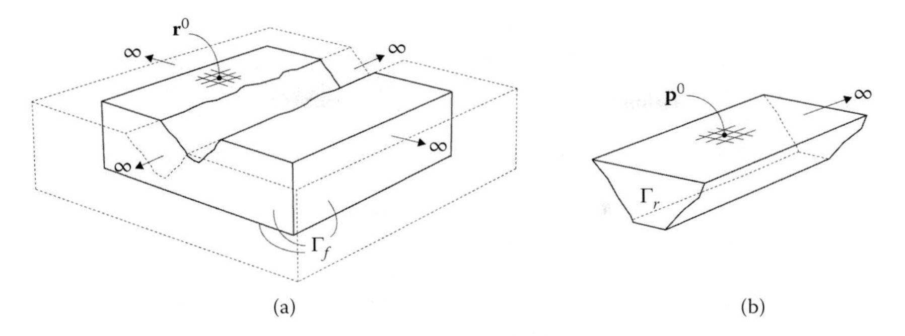

Figure 11.7.2 (a) Free-field foundation domain (without dam or impounded water) with displacements defined by $\mathbf{r}^0$; and (b) "free-field" fluid channel upstream of Γ_r with hydrodynamic pressures defined by $\mathbf{p}^0$.

Damping Matrix. Before closing this section, a brief discussion of how the standard damping matrix **c** should be modified for use in nonlinear RHA of dams is presented. For linear systems, the standard matrix **c** is constructed by assembling the damping matrices for two subsystems, the dam and foundation (Chopra 2017; Section 11.2.1). Although Rayleigh and Modal damping models are commonly used for several classes of structures, including buildings and bridges, they do not seem to be appropriate for continua in concrete and rock. Mass-proportional damping is inappropriate in the presence of sliding at the dam–rock interface or at a crack that extends through the thickness of the dam (El-Aidi and Hall 1989a,b; Hall 2006). Stiffness proportional damping is intuitively more appealing because the element damping forces are proportional to strain rates in the element. Damping matrices for the dam and foundation systems are defined as: $\mathbf{c}_c = a_{1c}\mathbf{k}_c$ and $\mathbf{c}_r = a_{1r}\mathbf{k}_r$ where $\mathbf{k}_c$ and $\mathbf{k}_r$ are the elastic stiffness matrices. The coefficient a_{1c} is determined from the damping ratio specified for the dam alone at a selected frequency. The coefficient a_{1r} is determined similarly but from the usually different damping ratio specified for the foundation domain alone at the same frequency. Usually, the frequency is selected as the fundamental natural frequency of the dam–foundation system. The damping ratios for the two substructures should be selected to be small enough to ensure that the overall damping in the dam–water–foundation system does not exceed the range of measured values (see Section 10.3).

The damping matrix, so constructed, should be modified to recognize that damping forces cannot exist in a finite element across a crack (El-Aidi and Hall 1989a,b; Bhattacharjee and Léger 1993; Hall 2006). This can be achieved in various ways: (i) removing the contribution of a cracked element to the overall damping matrix (El-Aidi and Hall 1989a,b); (ii) limiting the damping forces in each element (Hall 2006); (iii) defining the damping matrix to be proportional to the tangent stiffness matrix instead of the initial stiffness matrix (Bhattacharjee and Léger 1993); and (iv) replacing the initial stiffness matrix by a degraded elastic stiffness matrix that is a function of a damage variable computed in the plastic damage model for concrete (Lee and Fenves 1998).

11.8 EFFECTIVE EARTHQUAKE FORCES

11.8.1 Forces at Bottom Boundary of Foundation Domain

To facilitate implementation of the procedure, it is convenient to reformulate Eq. (11.7.2a) for the effective earthquake forces at the bottom boundary of the foundation domain

$$\mathbf{P}_f^0 = 2\mathbf{c}_f\,\dot{\mathbf{r}}_I^0 \tag{11.8.1}$$

where $\dot{\mathbf{r}}_I^0$ is the motion at the bottom boundary associated with the incident (upward propagating) seismic waves. Because this equation only requires the incident motion $\mathbf{r}_I^0$, it avoids computation of the free-field boundary tractions $\mathbf{R}_f^0$ that are required when using Eq. (11.7.2a) directly.

The incident motion $\mathbf{r}_I^0$ is obtained by 1D deconvolution of the surface free-field acceleration $a_g(t)$, assuming vertically propagating seismic waves and a homogeneous (or horizontally layered) half-space (Figure 11.6.3c). Deconvolution solves the inverse problem to determine the amplitude and frequency content of the incident motion that is consistent with the surface motion. It is most conveniently implemented in the frequency domain, either directly by computing the transfer function for a 1D half-space, or by utilizing 1D wave propagation software such as SHAKE (Schnabel et al. 1972) or DEEPSOIL (Hashash et al. 2011).

Although rather straightforward, deconvolution can be confusing because 1D wave propagation software operate with two possible motions at every depth; an *outcrop motion* and a *within motion*. By definition, the within motion is the superposition of the incident and reflected waves, i.e. it is the total (or "actual") motion at any given depth in the half-space. In contrast, the outcrop motion is the motion that would occur at a theoretical outcrop location at the same depth; this is equal to twice the amplitude of the incident motion. Thus, the incident motion $\mathbf{r}_I^0$ needed in Eq. (11.8.1) is one-half the outcrop motion at the bottom boundary determined from deconvolution analysis. The procedure to compute effective earthquake forces $\mathbf{P}_f^0$ is summarized elsewhere (Løkke and Chopra 2019a,b).

Some researchers have avoided deconvolution of the surface motion by idealizing the foundation as a homogeneous, undamped half-space. In this special case, a vertically propagating wave does not attenuate, implying that the incident motion $\mathbf{r}_I^0$ at the bottom boundary is equal to one-half the specified surface motion, except for a time shift. While this simplification may be appropriate for some cases (rock with high stiffness and very little inherent material damping), it may not be always valid. We will evaluate this approximation in Section 11.10.3.

11.8.2 Forces at Side Boundaries of Foundation Domain

The computation and application of the effective earthquake forces at the four side boundaries of the foundation domain [Eq. (11.7.2a)] can be automated with a FE code using a special free-field boundary element (Nielsen 2006, 2014). However, this element is available for 3D analysis only in a few commercial codes; examples are PLAXIS and FLAC. Consistent with the goal to develop a direct FE method that can be implemented in every commercial FE code without modification of the source code, an alternative procedure where the effective earthquake forces are computed in a separate auxiliary analysis of the free-field system is presented next.

The free-field motion $\mathbf{r}_f^0$ and boundary forces $\mathbf{R}_f^0$ required to compute the effective earthquake forces $\mathbf{P}_f^0$ at the four side boundaries of the foundation domain are determined from dynamic analysis of the foundation domain in its free-field state (Figure 11.8.1a). Although this system is much simpler than the actual dam–water–foundation system, it is still too complicated to analyze directly by the direct FE method because it contains the irregular canyon interior of the boundary Γ_f.

An alternative is to implement – for each component of ground motion – two sets of four simpler analyses: (i) four 1D corner columns subjected to forces of Eq. (11.8.1) at the base are analyzed first to provide $\dot{\mathbf{r}}_f^0$ and $\mathbf{R}_f^0$ for the nodes at the corners of the foundation domain, leading to $\mathbf{P}_f^0$; (ii) analyses of four 2D systems subjected to forces of Eq. (11.8.1) at its base and forces $\mathbf{P}_f^0$ on the sides determined from the first set of 1D analyses provides $\dot{\mathbf{r}}_f^0$ and $\mathbf{R}_f^0$ for nodes on all four side boundaries. The procedure is illustrated in Figure 11.8.1 and summarized in step-by-step form elsewhere (Løkke and Chopra 2018).

This procedure for computing $\mathbf{P}_f^0$ at the side boundaries of a system with arbitrary geometry is based on the assumption that the motion in each of the four 2D systems (Figure 11.8.1b) can

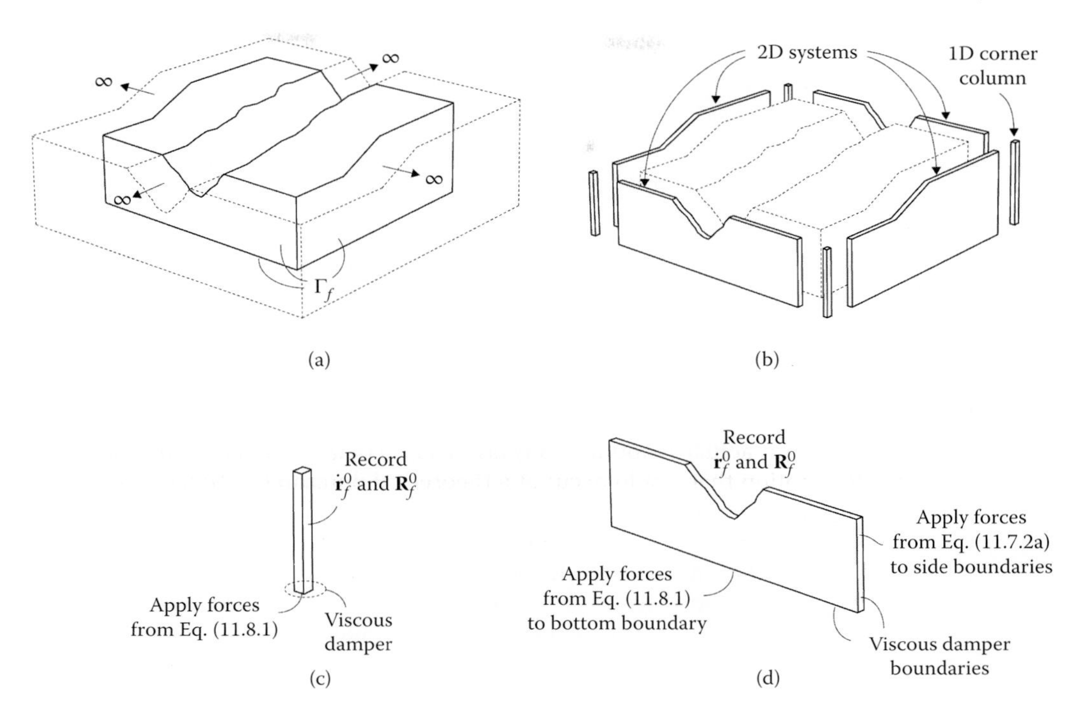

Figure 11.8.1 Computing $\mathbf{P}_f^0$ at side boundaries of foundation domain: (a) free-field system; (b) free-field system with corresponding 1D corner columns and 2D systems; (c) example analysis of 1D corner column to compute $\dot{\mathbf{r}}_f^0$ and $\mathbf{R}_f^0$ at corners; and (d) example analysis of 2D system to compute $\dot{\mathbf{r}}_f^0$ and $\mathbf{R}_f^0$ at the four side boundaries.

be determined independently of the other 2D systems. This assumption is reasonable as long as the foundation domain is large enough, which is generally the case because large domains are required to ensure that viscous-damper boundaries are effective in absorbing the outgoing waves and thus in simulating semi-unbounded domains. A more detailed discussion on the significance of this assumption can be found in Løkke and Chopra (2019a,b).

This procedure requires eight auxiliary analyses for each of the three components of ground motion. The computational effort required for these linear dynamic analyses is minimal, and the procedure can be automated in a pre-processing script that is set up and executed before the nonlinear dynamic analysis of the dam–water–foundation system is initiated. For this purpose, MATLAB was used to compute and store effective earthquake forces and the FE code OpenSees for analysis of the complete system. A similar methodology has been presented by Saouma et al. (2011) and implemented in the FE code MERLIN (Saouma et al. 2013).

This approach requires substantial management and transfer of large sets of data resulting from analysis of the free-field system to create the input to analysis of the actual system. This is a major disadvantage in analysis of 3D FE models that may easily have tens of thousands of boundary nodes. A simplified form of the free-field analysis of the foundation domain that drastically reduces these requirements is presented in Section 11.10.1.

11.8.3 Forces at Upstream Boundary of Fluid Domain

The free-field hydrodynamic pressures $\mathbf{p}_r^0$ required to compute $\mathbf{P}_r^0$ at Γ_r are to be determined by dynamic analysis of the fluid in its free-field state: a fluid channel that is unbounded in the upstream direction with uniform cross section (Figure 11.7.2b); the latter property permits

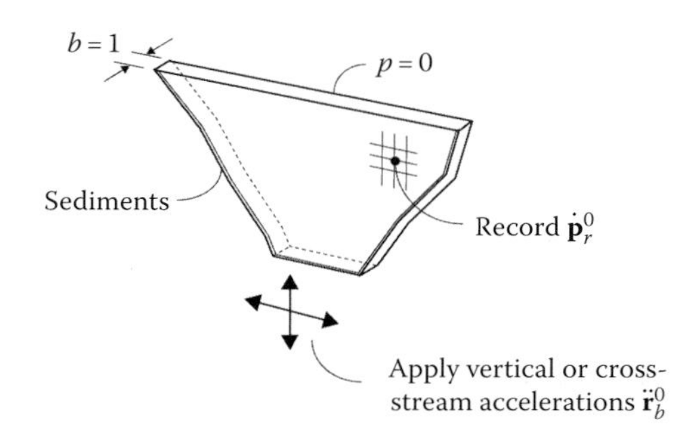

Figure 11.8.2 Computing $\mathbf{P}_r^0$ at upstream boundary of fluid domain: analysis of 2D fluid cross-section subjected to vertical and cross-stream excitation to compute $\dot{\mathbf{p}}_r^0$.

two-dimensional analysis of the system. The free-field pressures $\mathbf{p}_r^0$ on the boundary Γ_r may be computed by analyzing the 2D fluid-domain cross section (Figure 11.8.2); the stream component of ground motion will not generate any hydrodynamic pressures, implying $\mathbf{P}_r^0 = \mathbf{0}$. The procedure is summarized in step-by-step form elsewhere (Løkke and Chopra 2018).

Before closing this section, observe that the forces $\mathbf{P}_r^0$ are associated with earthquake-induced hydrodynamic pressures caused by vertical and cross-stream components of the earthquake excitation over the part of the fluid domain upstream of Γ_r that has been eliminated in the truncated model. These forces arise because the earthquake excitation is implicitly assumed to extend along the entire length of the unbounded fluid channel in the idealized system (Figures 11.6.1 and 11.8.2).

11.9 NUMERICAL VALIDATION OF THE DIRECT FINITE ELEMENT METHOD

The earthquake response of a concrete arch dam will be computed by the Direct FEM and compared against results from the completely independent substructure method (Chapter 8). Before presenting this comparison, we recall the important differences between the two methods:

1. The foundation and fluid domains are treated as semi-unbounded continua in the substructure method, whereas they are truncated at wave-absorbing (viscous damper) boundaries and modeled by finite elements in the Direct FEM;
2. The seismic excitation is specified as free-field ground motion at the dam–foundation interface in the substructure method, whereas it is input as effective earthquake forces at the wave-absorbing boundaries in the Direct FEM; and
3. The response history is determined via the frequency domain in the substructure method, but directly in the time domain in the Direct FEM.

11.9.1 System Considered and Validation Methodology

Chosen for analysis by both methods is Morrow Point Dam, a 142-m-high, approximately symmetric, single centered arch dam located on the Gunnison River in Colorado. The material properties and damping values selected for the dam concrete and foundation rock are based on the results from forced vibration tests of the dam and subsequent numerical studies performed to match the experimental data (Duron and Hall 1988; Nuss 2001; Nuss et al. 2003). The concrete and foundation rock are assumed to be homogeneous, isotropic, and linearly elastic. The concrete has a modulus of elasticity $E_s = 34.5$ GPa, density $\rho_s = 2403$ kg/m^3, and Poisson's ratio

$\nu_s = 0.20$. The foundation rock has a modulus of elasticity $E_f = 24.1$ GPa (i.e. $E_f/E_s = 0.70$), density $\rho_f = 2723\,\text{kg/m}^3$ and Poisson's ratio $\nu_f = 0.20$. The impounded water has the same depth as the height of the dam, density $\rho = 1000\,\text{kg/m}^3$, and pressure-wave velocity $C = 1440\,\text{m/s}^1$. For consistency with the substructure method, sediments are not modeled explicitly in the Direct FEM; instead they are modeled by a wave-absorbing reservoir bottom with $\alpha = 0.80$.

Material damping is modeled in the Direct FEM by Rayleigh damping with viscous damping ratios $\zeta_s = 1\%$ and $\zeta_f = 2\%$ for the dam and foundation, respectively, at two frequencies: $f_1 = 5$ Hz, the fundamental natural frequency of the dam on rigid foundation, and at three times this frequency. The damping matrix for the complete system is then constructed using standard procedures for assembling damping matrices for two subdomains (Chopra 2017; Chapter 11). Determined by the half-power bandwidth method applied to the resonance curve, the overall damping in the combined dam–water–foundation system is 3–5% for the first few modes of vibration, consistent with the range of measured damping values.

The FE model shown in Figure 11.9.1 is assembled using standard 8-node brick elements, with 800 full-integration solid elements for the dam, 42,000 reduced-integration solid elements

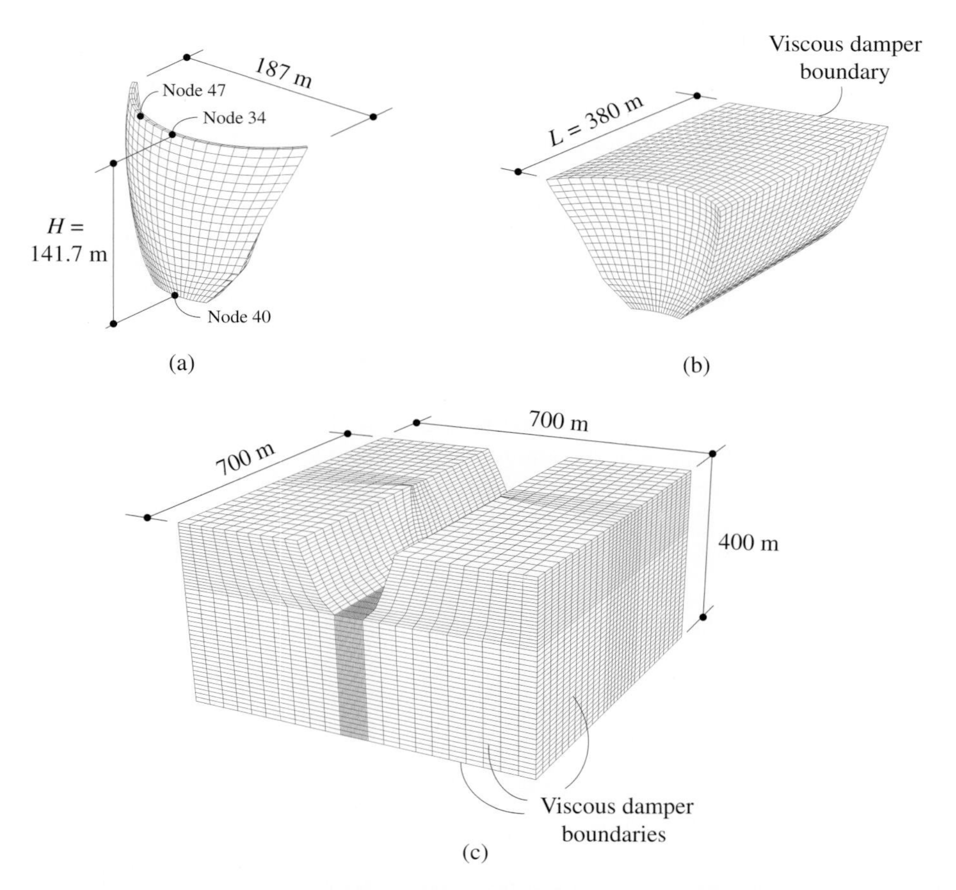

Figure 11.9.1 OpenSees FE model of Morrow Point Dam: (a) dam; (b) fluid domain; and (c) foundation domain.

for the foundation domain, and 9200 acoustic fluid elements for the fluid domain. Interface elements couple accelerations with hydrodynamic pressures at the fluid-solid interfaces, surface elements at the bottom and sides of the reservoir model a wave absorptive boundary, and viscous dampers at the truncation boundaries of the foundation domain and at the upstream end of the fluid domain model the semi-unbounded extent of these domains. The combined FE model consists of approx. 63,000 elements and 150,000 DOFs, and its overall dimensions are 700 m × 700 m × 400 m, corresponding to approximately $5H \times 5H \times 3H$, where H is the height of the dam. As mentioned in Chapter 8, the computer program EACD3D-08 was developed to implement the substructure method. The EACD3D-08 model, shown in Figure 11.9.2, includes 800 solid elements for the dam, the FE mesh for the irregular part of the fluid domain, and the boundary element mesh at the dam–foundation interface.

Material damping in the substructure method is modeled by rate-independent, constant hysteretic damping defined by damping factors $\eta_s = 0.02$ and $\eta_f = 0.04$ for the dam and foundation separately; these correspond to frequency-independent viscous damping ratios of $\zeta_s = 1\%$ and $\zeta_f = 2\%$. A numerical investigation confirmed that the damping in the Direct FEM, as defined earlier, is sufficiently close to this rate-independent damping over the frequency range of interest. Because EACD3D-08 ignores water–foundation interaction, this interaction mechanism is also excluded in the Direct FEM to ensure a meaningful comparison.

The earthquake excitation is specified as free-field ground motion on a rock outcrop at the top of the canyon (Figure 11.6.3c). Consistent with this specified motion, the seismic input is determined in the form required for each method of analysis. In the Direct FEM, the earthquake excitation appears in the form of effective earthquake forces $\mathbf{P}_f^0$ and $\mathbf{P}_r^0$ at the wave-absorbing boundaries; these forces are determined by the procedures described in Section 11.8. In the substructure method, the earthquake excitation is defined by the free-field motion over the canyon surface at the dam–foundation interface. To determine this motion consistent with the specified surface motion $a_g(t)$, we implement a direct FE analysis of the foundation domain without the dam or impounded water (Figure 11.9.3) subjected to the same boundary forces $\mathbf{P}_f^0$. The motion recorded at the dam–foundation interface is then the spatially varying excitation to be input into the EACD3D-08 analysis.

Results for the dam response are first presented in the form of dimensionless frequency response functions that represent the amplitude of radial acceleration at the crest of the dam† due to unit harmonic free-field surface motion, $a_g(t)$, at the control point. Implemented in the

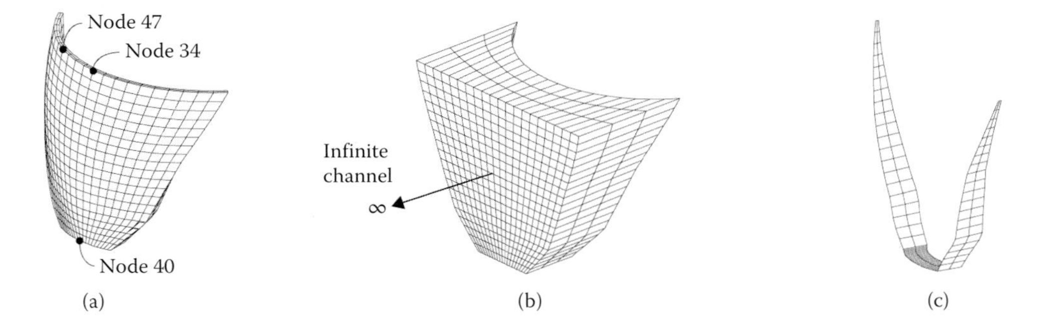

Figure 11.9.2 EACD3D-08 model for Morrow Point Dam: (a) FE model for dam; (b) FE model for semi-unbounded fluid domain; (c) boundary element mesh for the dam–foundation interface.

† The actual location at the dam crest is selected at node 34 – the center node – for upstream and vertical ground motions, and at node 47 for cross-stream motion.

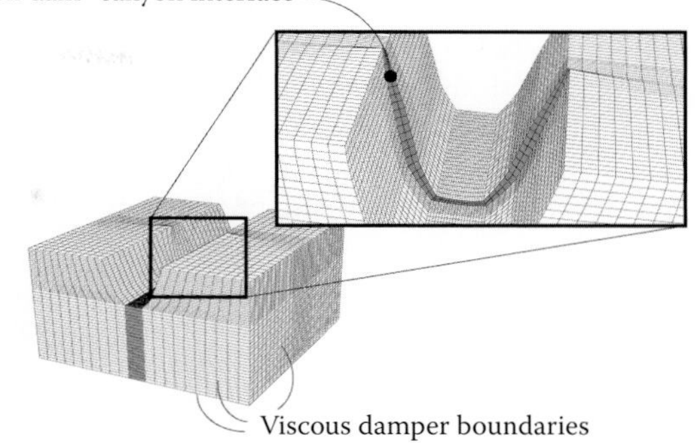

Figure 11.9.3 FE model of foundation domain to compute free-field motion at dam–foundation interface to be used as input to EACD3D-08 analysis.

frequency domain, the substructure method directly provides frequency response functions. In the direct FEM, these functions are determined by time-domain analysis of the system of Figure 11.8.1 subjected to a long sequence of unit harmonics with gradually increasing frequency; further details of this procedure are available in Løkke and Chopra (2019a,b).

11.9.2 Frequency Response Functions

Frequency response functions obtained by the Direct FEM and substructure methods for the dam on flexible foundation rock are compared for two cases: empty reservoir and full reservoir in Figures 11.9.4 and 11.9.5, respectively. Results for a full reservoir are presented here only for the stream component of ground motion, because limitations in the EACD3D-08 computer program do not allow for a meaningful comparison with the Direct FEM for cross-stream and vertical ground motions.

The response results obtained by the Direct FEM are very close to those from the substructure method. The small discrepancies near some of the resonant peaks (Figure 11.9.4), and at frequencies higher than 15 Hz (Figure 11.9.5), are primarily caused by reflections from the viscous-damper boundaries, which are incapable of perfectly absorbing all scattered waves. Such errors will generally decrease with larger domain sizes.

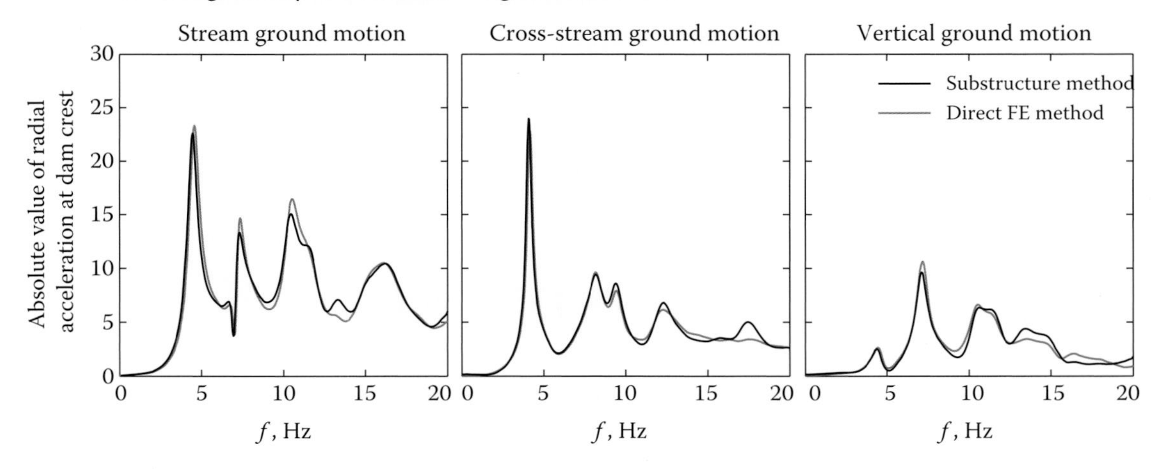

Figure 11.9.4 Frequency response functions for the amplitude of radial acceleration at the crest of Morrow Point Dam including dam–foundation interaction (empty reservoir) subjected to stream, cross-stream and vertical ground motions. Results are computed by Direct FE and substructure methods.

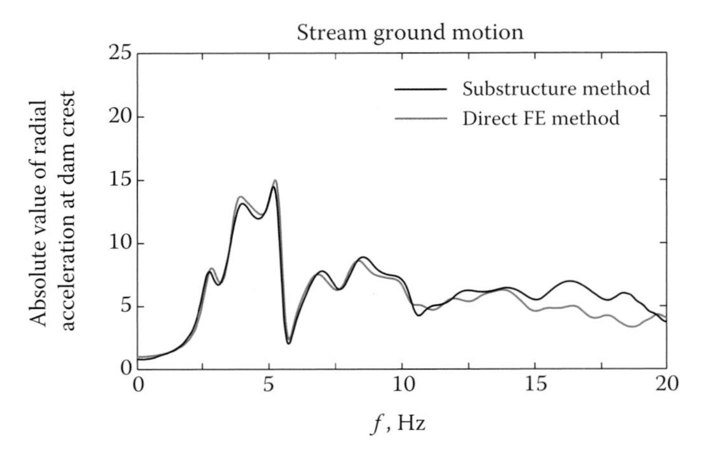

Figure 11.9.5 Frequency response functions for the amplitude of radial acceleration at the crest of Morrow Point Dam including dam–water–foundation interaction (full reservoir) subjected to stream ground motion. Results are computed by Direct FE and substructure methods.

11.9.3 Earthquake Response History

The response of the dam to the specified free-field motion in the stream direction, $a_g^x(t)$, defined by the S69E component of the Taft ground motion, determined by the Direct FE and substructure methods are compared next. The radial accelerations and displacements at the crest of the dam relative to the base of the dam (node 40) are presented in Figure 11.9.6, and envelope values of maximum tensile arch and cantilever stresses on the upstream face of the dam in Figure 11.9.7. The results computed by the Direct FEM closely match those from the substructure method: the displacements and accelerations at the crest show a near perfect match, and the envelope stress values are also close. The slight discrepancies in the stress contour plots were found to be caused by differences between the FE stress recovery algorithms in the two computer programs.

The excellent agreement demonstrates the ability of the Direct FEM to (i) model the factors important for earthquake analysis of arch dams: dam–water–foundation interaction including

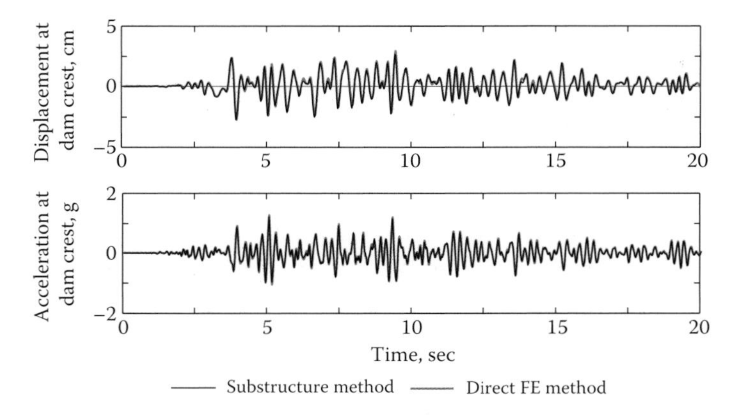

Figure 11.9.6 Radial displacements and accelerations at the crest of Morrow Point Dam including dam–water–foundation interaction subjected to S69E component of Taft ground motion applied in the stream direction. Results are computed by Direct FE and substructure methods.

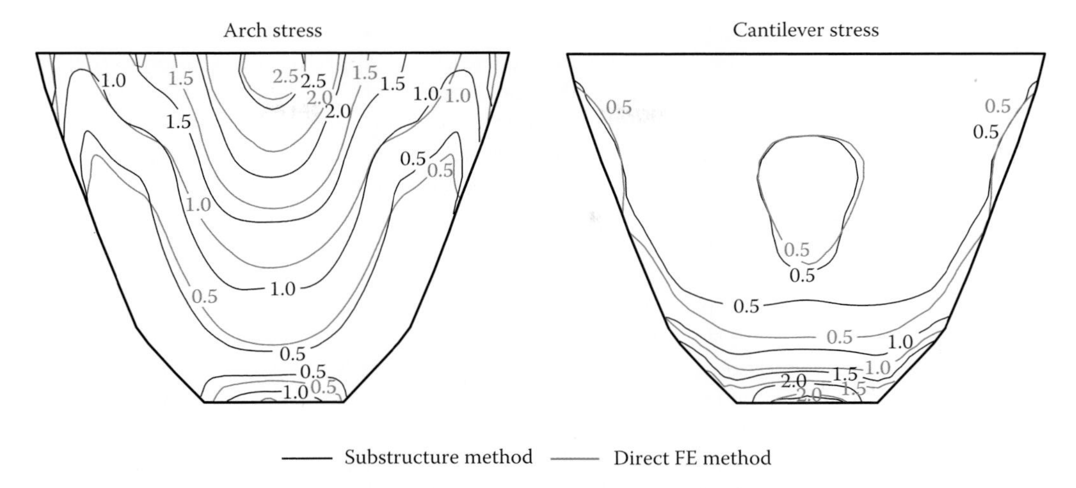

Figure 11.9.7 Envelope values of maximum tensile stresses, in MPa, on the upstream face of Morrow Point Dam including dam–water–foundation interaction subjected to S69E component of Taft ground motion applied in the stream direction; static stresses are excluded. Results are computed by Direct FE and substructure methods.

water compressibility and wave absorption at the reservoir boundaries, radiation damping in the semi-unbounded foundation and fluid domains, and (ii) convert the earthquake excitation specified at the surface of the foundation domain to effective earthquake forces at wave-absorbing boundaries.

The effectiveness of the Direct FEM with viscous-damper boundaries is apparent from the fact that these excellent results (Figures 11.9.4–11.9.7) are achieved even with relatively moderate domain sizes: the overall dimensions of the FE model are approximately $5H \times 5H \times 3H$, where H is the height of the dam.

The direct FE analyses were implemented in OpenSees on a laptop computer (without parallel processing capabilities) using simple MATLAB scripts to perform the data management for computing and applying effective earthquake forces. The CPU-time for dynamic analysis of the FE model in Figure 11.9.1 with roughly 150,000 DOFs was 68 min for 2000 time-steps; approximately 13 min were required for the auxiliary analyses to set up the effective earthquake forces, and 55 min for dynamic analysis of the complete system. Observe that the computational effort required to determine the effective earthquake forces for the system is small compared to the time required for dynamic analysis of the overall system. Clearly, it will become negligible compared to the CPU-time required for nonlinear dynamic analysis of such systems.

11.10 SIMPLIFICATIONS OF ANALYSIS PROCEDURE

11.10.1 Using 1D Analysis to Compute Effective Earthquake Forces

The procedure presented in Section 11.8.2 for computing effective earthquake forces $\mathbf{P}_f^0$ at the side boundaries of the foundation domain is rigorous because it satisfies the general requirement that the free-field system (Figure 11.7.2) is identical to the actual system (Figure 11.6.2a) in the region exterior to the absorbing boundary. However, it requires 24 auxiliary analyses (Figure 11.8.1) followed by management and transfer of large sets of data to the analysis of the system of Figure 11.6.2a.

This procedure may be greatly simplified by ignoring the effects of the canyon on the free-field motions at the absorbing boundaries. This approximation implies that the actual free-field foundation system (Figure 11.7.2a) is replaced by a much simpler system: a flat box with homogeneous (or horizontally layered) material properties (Figure 11.10.1a). Analysis of this three-dimensional flat box to vertically propagating seismic waves reduces to analysis of a one-dimensional column of foundation elements shown in Figure 11.10.1b.

Analysis of this 1D system, discretized to match the elevations of the boundary nodes in the main model, subjected to the forces of Eq. (11.8.1) at the base provides the motion $\mathbf{r}_f^0$ at every node along the height. Alternatively, $\mathbf{r}_f^0$ may be extracted at every elevation directly from deconvolution of the specified surface motion. Boundary tractions are computed from $\mathbf{r}_f^0$ using stress-strain relationships for a 1D system and converted to nodal forces $\mathbf{R}_f^0$. A step-by-step summary of the procedure and details on its implementation is available in Løkke and Chopra (2018). Such a single, one-dimensional free-field analysis is obviously much simpler compared to the rigorous procedure developed in Section 11.8.2. However, the free-field system of Figure 11.10.1a violates the general requirement that any free-field system must be identical to the actual system in the region exterior to the absorbing boundary and hence will result in error.

Frequency response functions for the amplitude of radial acceleration at the crest of Morrow Point Dam including dam–foundation interaction (with the reservoir empty) are presented in Figure 11.10.2, where results obtained using 3D and 1D free-field analyses to compute effective

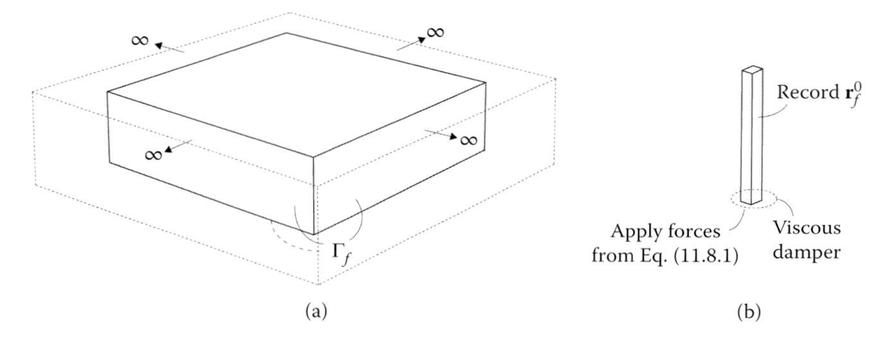

Figure 11.10.1 (a) Free-field foundation system without canyon; and (b) analysis of single column of foundation elements to compute $\mathbf{r}_f^0$.

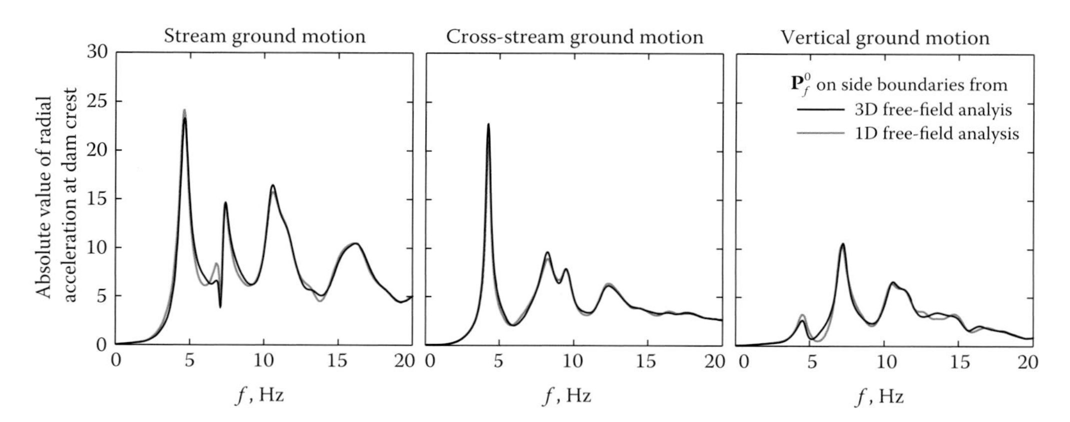

Figure 11.10.2 Errors due to use of 1D free-field analysis to determine effective earthquake forces $\mathbf{P}_f^0$ at the side boundaries on the response of Morrow Point Dam including dam–foundation interaction.

earthquake forces are compared. The closeness of the two sets of results justifies the use of 1D free-field analysis to compute effective earthquake forces.

The preceding results demonstrate that the errors introduced by ignoring the canyon in the free-field analysis are insignificant as long as the foundation domain is sufficiently large. This requirement is normally satisfied when absorbing boundaries are modeled by viscous dampers, which always require large domain sizes to ensure acceptable modeling of the semi-unbounded foundation domain. Thus, the use of 1D free-field analysis is appropriate for practical analyses where the foundation domain can be idealized as homogeneous or horizontally layered.

11.10.2 Ignoring Effective Earthquake Forces at Side Boundaries

Some dam engineers have been using a variation of the Direct FEM wherein effective earthquake forces are applied only to the bottom boundary of the foundation domain, and not to the side boundaries. This approximation is attractive because it eliminates the need for analysis of the free-field foundation domain, but as will be demonstrated, the resulting errors in the dam response can be unacceptably large.

Frequency response functions for the amplitude of radial acceleration at the crest of Morrow Point Dam (Figure 11.9.1) including dam–foundation interaction (with the reservoir empty) are shown in Figure 11.10.3. Results computed by the Direct FEM with and without effective earthquake forces applied to the side boundaries are compared. Excluding $\mathbf{P}_f^0$ at the side boundaries causes significant error in the dam response to all components of ground motion. Consequently, it is not prudent to use this gross simplification that may have been introduced to reduce computation. These forces can be readily determined using a 1D free-field analysis, which requires very little computation and still provides excellent results as demonstrated in Section 11.10.1.

11.10.3 Avoiding Deconvolution of the Surface Free-Field Motion

Some researchers have avoided deconvolution of the free-field motion $a_g(t)$ specified at the foundation surface by idealizing the foundation as a homogeneous, undamped, half-space (Zhang et al. 2009; Robbe et al. 2017). In this special case, a vertically propagating plane wave does not attenuate, implying that the incident earthquake motion at the bottom boundary, $\mathbf{r}_I^0$, that

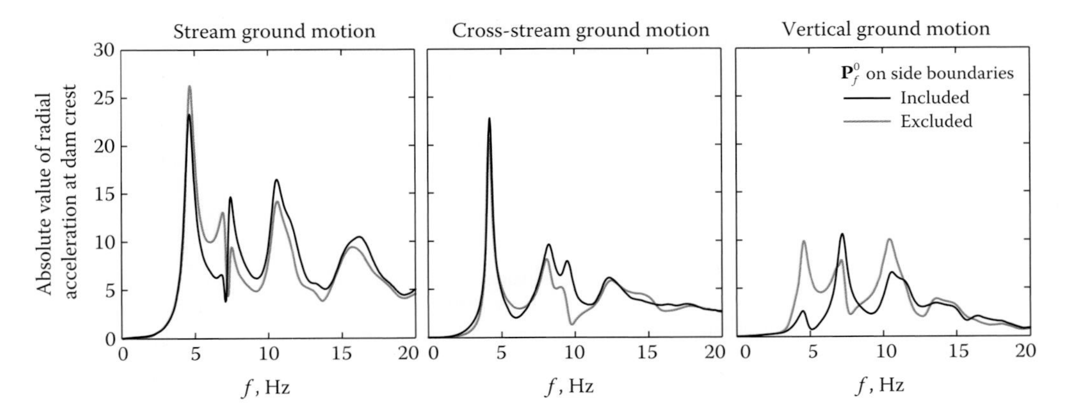

Figure 11.10.3 Discrepancies introduced by excluding effective earthquake forces $\mathbf{P}_f^0$ on the side boundaries of the foundation domain in the response of Morrow Point Dam including dam–foundation interaction.

enters into Eq. (11.8.1) is equal to one-half the free-field motion specified at the surface, except for a time shift. The validity of this approximation is investigated next for different foundation idealizations.

For this purpose, we first analyze a flat foundation box (Figure 11.10.4) by the Direct FEM and compare the computed motion at the surface of the foundation against the specified free-field motion at the surface. Effective earthquake forces at the bottom and side boundaries are computed from Eq. (11.8.1) and the methodology described in Section 11.10.1 using two methods for obtaining the incident motion $\mathbf{r}_I^0$ at the bottom boundary: (i) deconvolution of the specified free-field surface motion; and (ii) assumed as one-half of this surface motion. Such a comparison is presented in Figure. 11.10.5 for three different foundation idealizations†: homogeneous foundation with zero material damping; homogeneous foundation with 4% material damping; and horizontally layered foundation with zero material damping. Results are presented for the S69E, S21W, and vertical components of the Taft ground motion.

Observe from the results presented in Figure 11.10.5 that when the incident motion $\mathbf{r}_I^0$ is determined by deconvolution, the computed surface motion is essentially identical (to within FE discretization error) to the specified motion for all three foundation idealizations. In contrast, assuming $\mathbf{r}_I^0$ as one-half the specified surface motion gives essentially the exact results only if the foundation is homogeneous and undamped (Figure 11.10.5a), but leads to significant underestimation of the surface motion for a damped foundation (Figure 11.10.5b) and overestimation for a layered foundation (Figure 11.10.5c). These results suggest that dam response would be accurately computed with $\mathbf{r}_I^0$ assumed as one-half the specified surface motion only if the foundation domain is homogeneous and undamped, but that errors will be introduced in other cases.

This expectation is confirmed by the frequency response functions for the amplitude of radial acceleration at the crest of Morrow Point Dam including dam–foundation interaction (empty reservoir) presented in Figure 11.10.6 for (a) homogeneous, undamped foundation and (b) homogeneous foundation with 4% material damping. Assuming $\mathbf{r}_I^0$ as one-half the specified

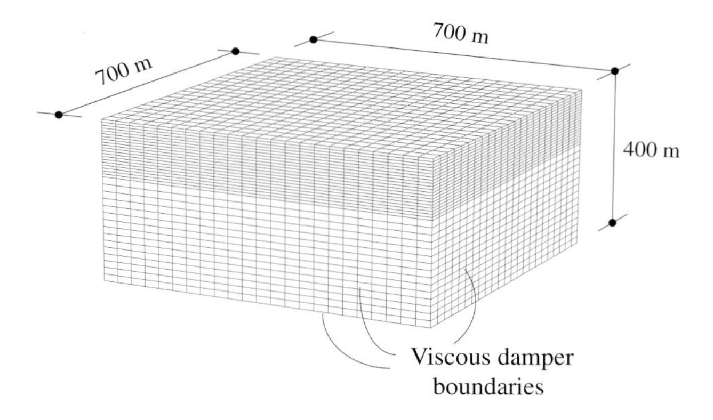

Figure 11.10.4 Flat foundation box.

† The material properties for the foundation are: density = 2723 kg/m^3 and Poisson's ratio = 0.20; the homogeneous foundation (Cases a and b) has shear wave velocity V_s = 2000 m/sec^1; the layered foundation (Case c) consists of three layers of equal thickness 133 m on top of homogenous bedrock, with shear wave velocities that increase with depth: $V_{s,1}$ = 1500 m/sec^1, $V_{s,2}$ = 2000 m/sec^1, $V_{s,3}$ = 2500 m/sec^1, and $V_{s,bedrock}$ = 3000 m/sec^1.

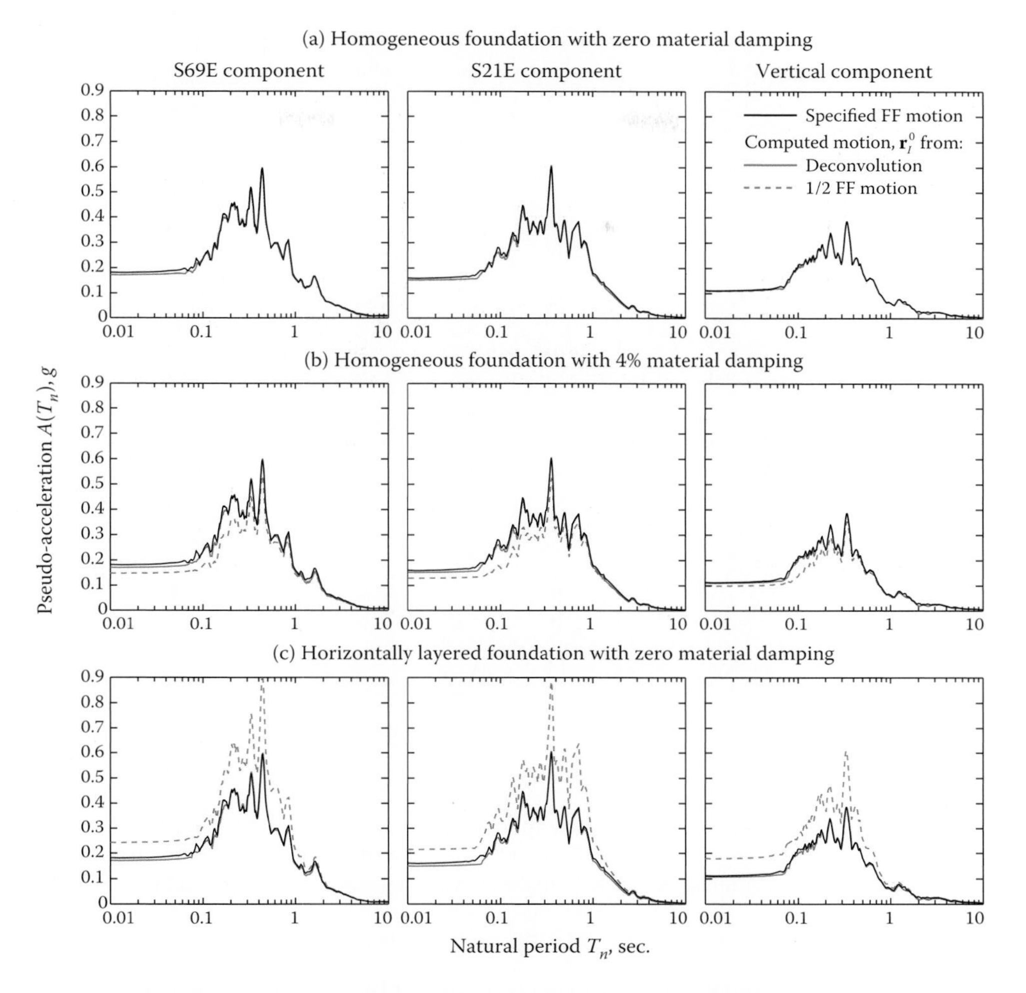

Figure 11.10.5 Comparison of pseudo-acceleration response spectra (5% damping) for free-field motion specified and computed at the surface of flat foundation box with the incident motion $\mathbf{r}_I^0$ determined by two methods: (1) deconvolution of the surface free-field motion; and (2) assumed as one-half the free-field motion. Results are presented for three cases: (a) homogeneous foundation with zero material damping; (b) homogeneous foundation with 4% material damping; and (c) horizontally layered foundation with zero material damping.

surface motion does not introduce error in dam response when the foundation is undamped but leads to significant underestimation of dam response if the foundation is damped.

To eliminate such errors, the incident motion $\mathbf{r}_I^0$ at the bottom boundary should be computed by 1D deconvolution of the surface free-field motion. There is little justification to bypass such analysis, especially because it is straightforward and requires very little computational effort compared to 3D analysis of the dam–water–foundation system.

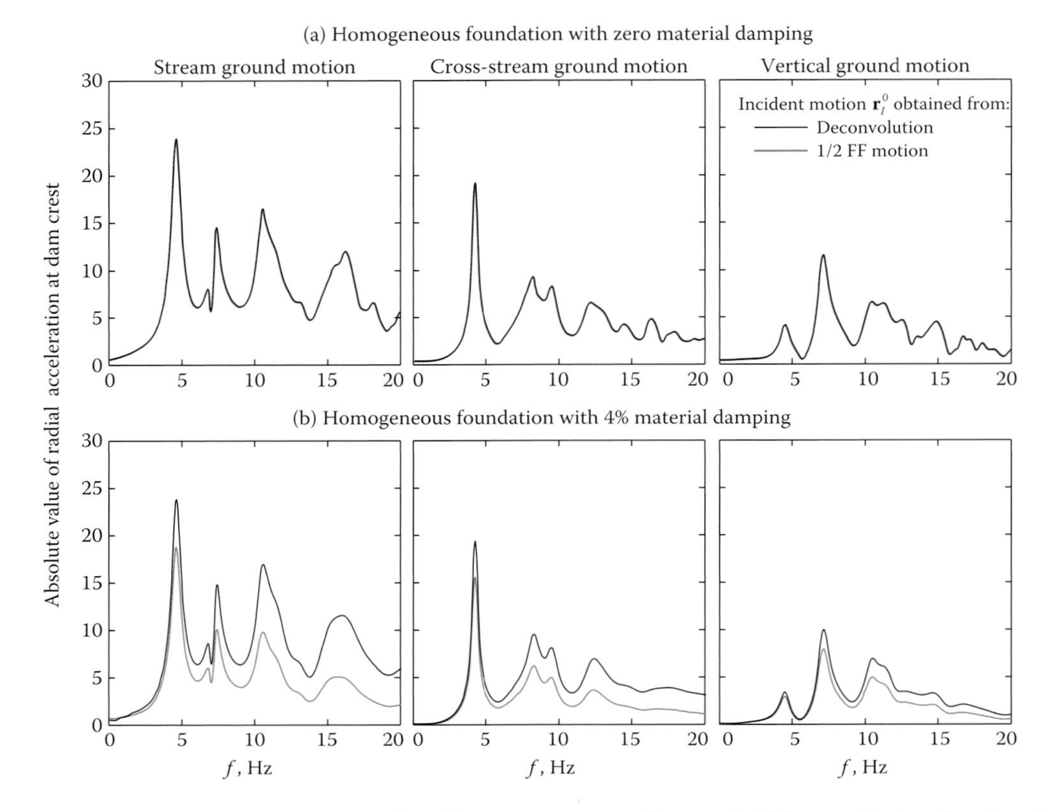

Figure 11.10.6 Discrepancies introduced by approximating $\mathbf{r}_l^0$ as one-half the specified surface free-field motion in the response of Morrow Point Dam including dam–foundation interaction with empty reservoir. Foundation is homogeneous with material damping equal to: (a) zero and (b) 4%.

11.10.4 Ignoring Effective Earthquake Forces at Upstream Boundary of Fluid Domain

Implementation of the Direct FEM may be simplified by excluding effective earthquake forces $\mathbf{P}_r^0$ at the upstream fluid boundary Γ_r, which are the forces associated with earthquake-induced hydrodynamic pressures in the semi-unbounded fluid channel upstream of Γ_r. This simplification is attractive because it eliminates the need for the two-dimensional auxiliary analysis of the fluid cross section (Section 11.8.3).

Presented in Figure 11.10.7 are frequency response functions for the radial acceleration at the crest of Morrow Point Dam supported on rigid foundation with full reservoir subjected to spatially-uniform ground motions at the dam–foundation and water–foundation interfaces for two values of the wave reflection coefficient: $\alpha = 0.50$ and $\alpha = 0.80$. Results for three cases presented for each α-value: (i) fluid domain of length $L = 2.5H$ including $\mathbf{P}_r^0$ on the upstream fluid boundary, which will be treated as the "exact" solution since it essentially matched results of the substructure method; (ii) $L = 2.5H$ excluding $\mathbf{P}_r^0$; and (iii) $L = 5H$ excluding $\mathbf{P}_r^0$.

Excluding $\mathbf{P}_r^0$ has little influence on the response for cross-stream and vertical ground motions if $\alpha = 0.50$, but leads to errors for higher values (e.g. $\alpha = 0.80$) and short fluid domains ($L = 2.5H$). For the stream component of ground motion, $\mathbf{P}_r^0 = \mathbf{0}$ (Section 11.8.3), so all three cases give identical results. This discrepancy in Cases 2 and 3 occurs because of the idealization of the system analyzed (Figure 11.6.1): a uniform fluid channel of unbounded length in the upstream direction with the earthquake excitation implicitly assumed to extend along the entire

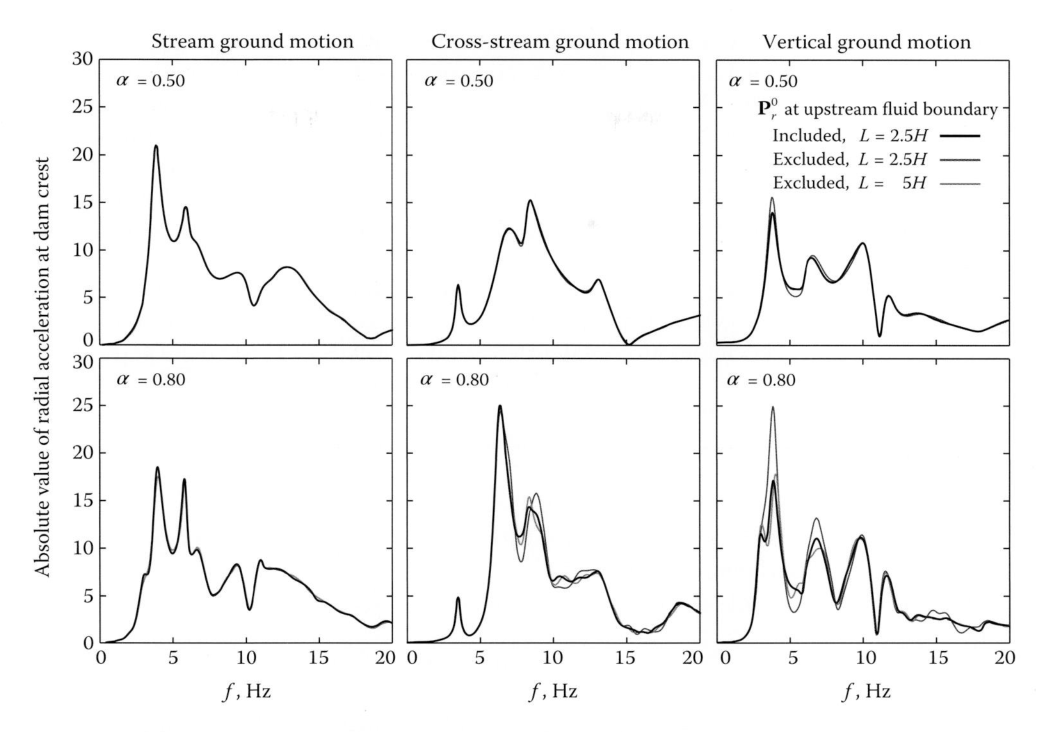

Figure 11.10.7 Discrepancies introduced by excluding effective earthquake forces $\mathbf{P}_r^0$ at the upstream fluid boundary in the response of Morrow Point Dam on rigid foundation with full reservoir. Results are presented for two values of the wave-reflection coefficient: $\alpha = 0.50$ and $\alpha = 0.80$. $\zeta_s = 3\%$ damping is specified for the dam alone. The results with $\mathbf{P}_r^0$ included and $L = 2.5H$ represent the "exact" solution.

length of this channel. In reality, neither the uniform fluid channel nor the ground motion would extend to infinity in the upstream direction, so excluding $\mathbf{P}_r^0$ from the analysis – implying that the excitation stops at the boundary Γ_r – seems to be a more appropriate idealization. However, the truncated fluid domain with viscous-damper boundaries should be long enough to accurately model dam–water interaction and radiation damping.

11.10.5 Ignoring Sediments at the Reservoir Boundary

We have seen that the sediments deposited at the reservoir bottom may influence the earthquake response of gravity dams greatly if the underlying foundation is rigid (Section 6.2.2); this influence is much less if the dam is supported on a flexible foundation and dam–foundation–interaction is included. In the latter case, the response to horizontal ground motions is affected relatively little (Figures 6.3.1a and 6.3.2a), but the response to vertical ground motions continues to be affected significantly (Figures 6.3.1b and 6.3.2b). These observations were gleaned from response of dams computed by the substructure method wherein the effects of water–foundation interaction were neglected and the sediments were modeled by the wave reflection coefficient, α. In the Direct FEM, however, water–foundation interaction is included automatically, and the sediments – with their thickness and extent – can be modeled explicitly.

Researchers have investigated two types of material models for sediments: (i) a two-phase fluid-saturated poroelastic model; and (ii) the viscoelastic model. The most sophisticated model is a two-phase fluid-saturated poroelastic material (Biot 1956). Researchers have demonstrated that sediments – modeled in this manner – have little influence on dam response if they are fully saturated but great influence if the sediments are partially – even slightly less than fully – saturated (Domínguez et al. 1997; Maeso et al. 2004). The two-phase poroelastic model is not ready for practical application for two reasons: (i) the dam response is extremely sensitive to the degree to which sediments are saturated, a property that cannot be determined precisely; and (ii) this material model requires detailed information on sediment properties – such as grain size, porosity, and hydraulic conductivity – for which data have not been available at reservoirs impounded behind dams.

A simpler model for sediments is a viscoelastic material, which is characterized by familiar parameters: modulus of elasticity, E_{sed}; Poisson's ratio, ν_{sed}; density, ρ_{sed}; and damping ratio. These properties have not been measured at dam sites, but data exists for river delta deposits and marine underwater sediments (Hamilton 1971). If needed, similar data could be developed at dam sites. Thus, the viscoelastic material model seems to be a pragmatic choice for sediments. However, it will be shown next, that sediments modeled in this manner have very little influence on dam response.

Gravity Dams. Determined by the Direct FEM, the frequency response function for the idealized dam described in Section 2.5.2[†] with full reservoir are presented in Figure 11.10.8 for two cases: no sediments at the reservoir bottom, and sediment layer modeled as a viscoelastic material.

These results demonstrate that sediments have small influence on dam response to horizontal ground motion; the first resonant peak, which is most significant in the earthquake response of dams, is unaffected, but the response at higher frequencies is noticeably affected. However, the overall influence of the sediments is not significant, as confirmed by the two sets of response histories presented in Figure 11.10.9, which are essentially identical. Similar conclusions have been reported by other researchers (Medina et al. 1990; Hatami 1997; Zhang et al. 2001).

Relatively speaking, sediments have more influence on complex frequency response functions for vertical ground motion (Figure 11.10.8). The discrepancy in response at the first two resonant frequencies is noticeable, although it is still small; however, the response at higher resonant frequencies is greatly influenced by sediments with the resonance peaks essentially eliminated. However, the earthquake response of the dam is essentially unaffected (Figure 11.10.9).

Arch Dams. Determined by the Direct FEM, the frequency response functions for Morrow Point Dam[‡] with full reservoir are presented in Figure 11.10.10 for two cases: no sediments at the reservoir boundaries and sediment layer modeled as viscoelastic material. These results demonstrate that, just as in the case of gravity dams, sediments have small influence on dam response; the first two resonant peaks due to ground motions in the stream and cross-stream directions, and the first resonant peak due to vertical ground motion are essentially unaffected, but responses at higher resonant frequencies are noticeably affected. However, the overall influence of sediments is not significant, as demonstrated by the two sets of response histories presented in Figure 11.10.11, which are nearly identical.

[†] System properties are as follows. Concrete: $E_s = 22.4$ GPa, density = 2483 kg/m^3, Poisson's ratio = 0.2, viscous damping = 2%. Rock: $E_f = 22.4$ GPa, density = 2643 kg/m^3, Poisson's ratio = 0.33, viscous damping = 10%. Height of dam = 120 m. Reservoir depth = 120 m. Sediments: $E_{\text{sed}} = 1.0$ GPa, density = 1600 kg/m^3, Poisson's ratio = 0.46; the associated pressure wave velocity is $V_{p,\text{sed}} = 1700$ m/sec^1. Rayleigh damping with 0% damping ratio specified at the first two resonant frequencies of the dam–water-foundation system, and sediment depth = 0.1 H = 12 m.

[‡] Properties of the dam–water–foundation system are the same as in Section 11.9.1. Sediments are modeled as a viscoelastic layer with uniform thickness, $H_{\text{sed}} = 0.1\,H$, where $H = 142$ m, $\nu_{\text{sed}} = 0.46$, and the two values for pressure wave velocity = 1400 m/sec^1 $\left(E_{\text{sed}} = 0.68\text{ GPa}\right)$ and 1800 m/sec^1 $\left(E_{\text{sed}} = 1.12\text{ GPa}\right)$.

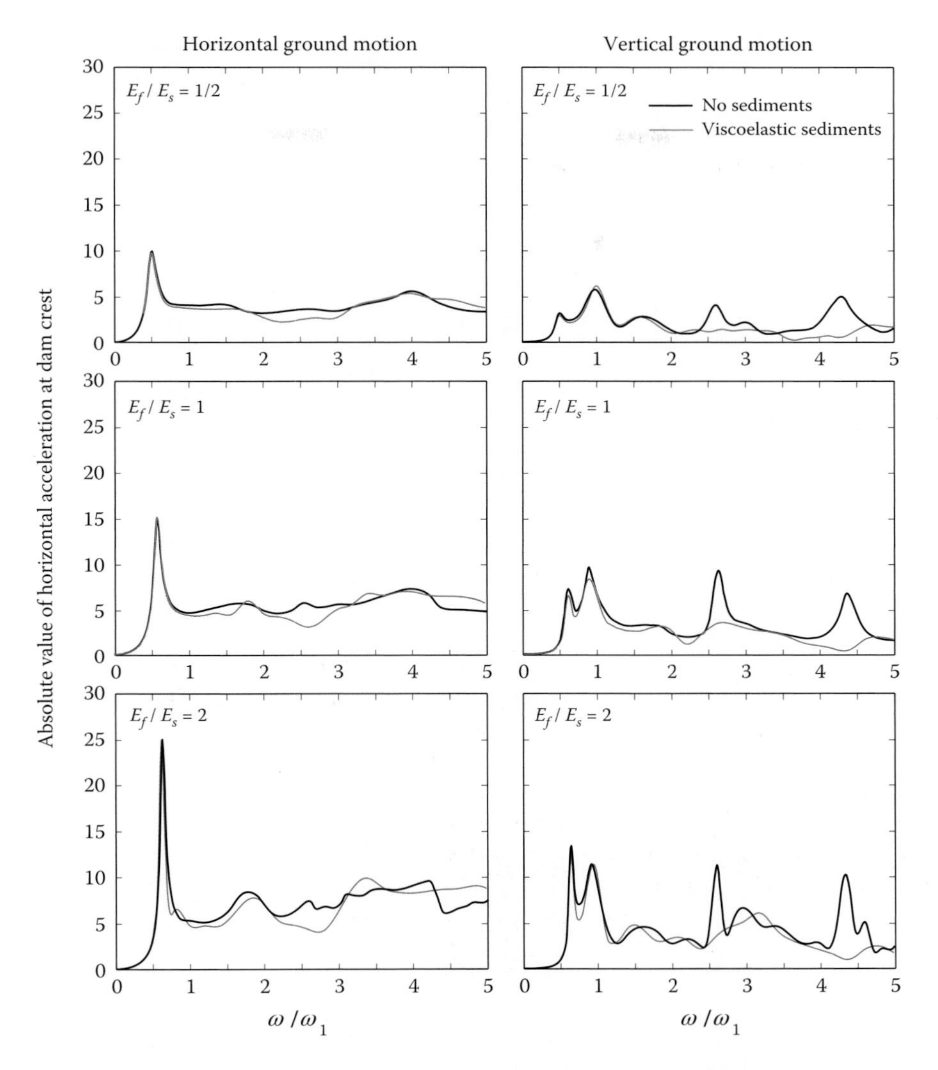

Figure 11.10.8 Influence of sediments on the frequency response functions for gravity dams with full reservoir due to horizontal and vertical ground motions. Results are presented for three values of $E_f/E_s = ½$, 1, and 2.

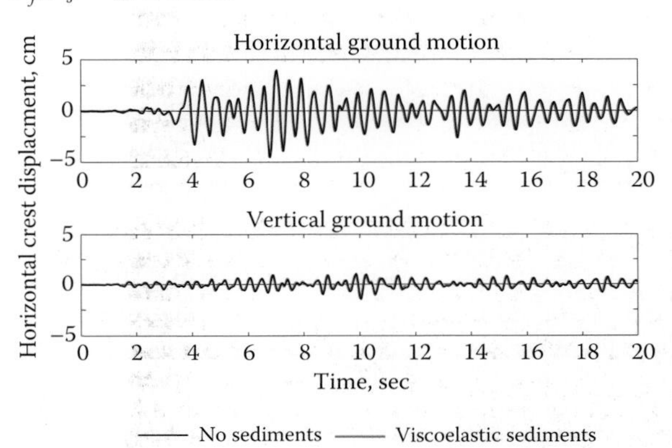

Figure 11.10.9 Influence of sediments on the earthquake response of a gravity dam on flexible foundation with full reservoir due to the S69E and vertical components, separately, of the Taft ground motion; $E_f/E_s = 1.0$.

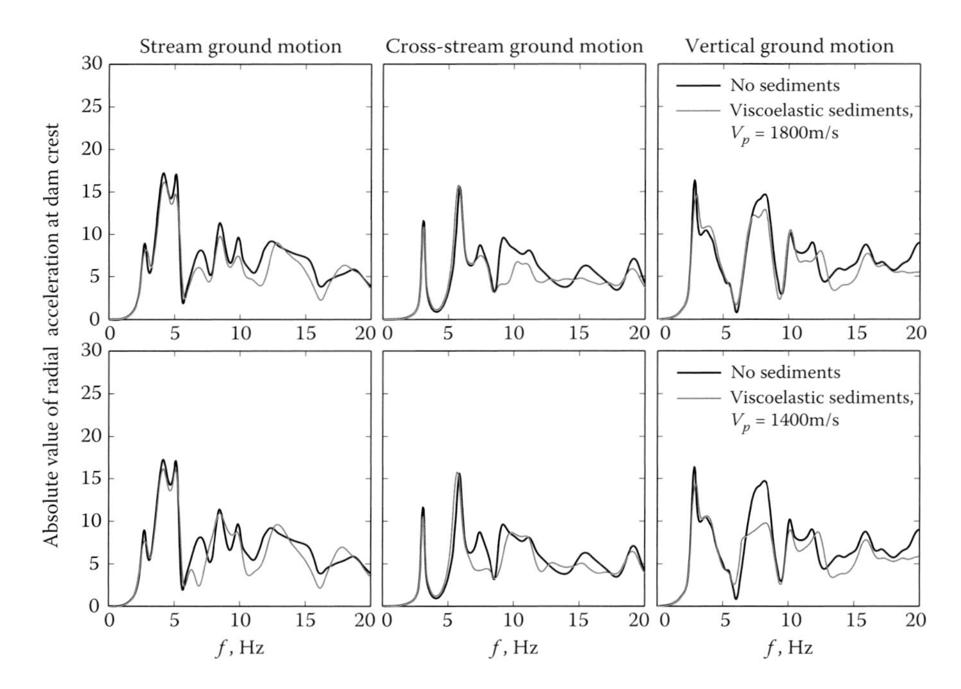

Figure 11.10.10 Influence of sediments on the frequency response functions for Morrow Point Dam with full reservoir subjected to stream, cross-stream, and vertical ground motions.

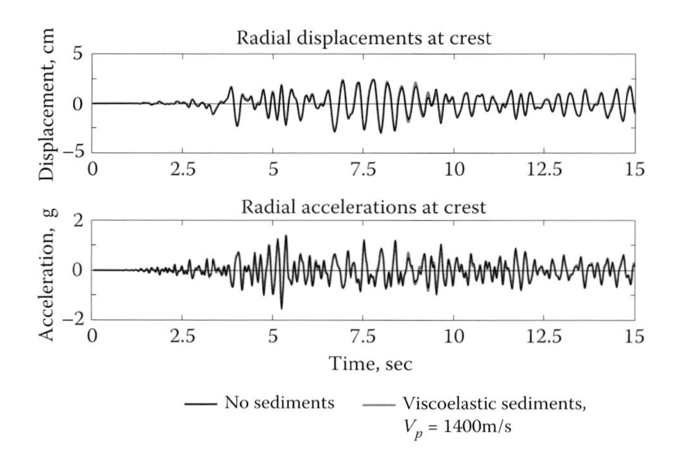

Figure 11.10.11 Influence of sediments on the earthquake response of Morrow Point Dam with full reservoir subjected to the S69E, S21W and vertical components (applied simultaneously) of the Taft ground motion.

Can Sediments Be Ignored?. The preceding results demonstrate that the influence of a layer of sediments at the reservoir bottom, modeled as a viscoelastic material, on the response of gravity dams as well as arch dams is negligible. Because water–foundation interaction and the associated radiation damping are included in the analysis, the additional loss of vibration energy due to sediments is apparently of little consequence. Thus, it seems reasonable to ignore sediments in analysis of the dam–water–foundation system by the Direct FEM.

This conclusion may seem to contradict earlier (1980s) results obtained by the substructure method, which indicated that sediments may influence dam response significantly. Because water–foundation interaction (and the associated radiation damping) is ignored in the substructure method, the loss of vibrational energy associated with wave absorption in the α-model apparently becomes significant. However, as demonstrated in Appendix 5, water–foundation interaction can be modeled satisfactorily by a wave reflection coefficient, α, computed from the properties of the underlying rock. Once such a model for water–foundation interaction is included; sediments may be ignored in the substructure method.

11.11 EXAMPLE NONLINEAR RESPONSE HISTORY ANALYSIS

Implementation of the Direct FEM for nonlinear RHA of an actual dam is summarized in this section; a detailed presentation is available in Løkke and Chopra (2019a,b).

11.11.1 System and Ground Motion

Chosen for this example analysis is Morrow Point Dam, with the properties of the dam–water–foundation system as described in Section 11.9.1 but for two exceptions: (i) sediments at the reservoir bottom or sides were not included, implying that they did not exist or they existed but were ignored because their effects on dynamic response of the dam are known to be relatively small once water–foundation interaction is properly modeled (Section 11.10.5); and (ii) material damping in the dam and foundation is modeled by stiffness proportional damping using the degraded elastic stiffness matrix instead of the initial stiffness (Lee and Fenves 1998). The initial damping matrices are $\mathbf{c}_c = a_{1c}\mathbf{k}_c$ and $\mathbf{c}_r = a_{ir}\mathbf{k}_r$; coefficients a_{1c} and a_{1r} are chosen to give 1% damping in the dam and 2% in the foundation domain at the fundamental frequency of the dam–water–foundation system. The overall damping in the system is consistent with the 3–5% measured at the first few resonant frequencies (Nuss 2001; Nuss et al. 2003).

The Direct FEM was developed in a form that can be implemented in any commercial FE code. Here, ABAQUS, a popular commercial FE code has been chosen. The FE model shown in Figure 11.11.1 contains 27,600 finite elements and 64,000 DOFs; 4196 solid elements for the dam (with four elements through the thickness of the dam), 14 175 solid elements for the foundation domain; and 9200 acoustic elements for the fluid domain. The overall dimensions of the FE model are 700 m × 700 m × 400 m corresponding to approximately $5H \times 5H \times 3H$, where H is the height of the dam. Viscous dampers at exterior boundaries were included to model the semi-unbounded extent of foundation and fluid domains.

Included in the model were several nonlinear mechanisms: cracking of concrete, opening and closing of contraction joints, opening and closing of the dam–foundation interface (but relative sliding at the interface is not permitted), tensile cracking in concrete, and effects of shear keys. Modeling of these nonlinear mechanisms is described in Løkke and Chopra (2019a,b).

Free-field ground motion at the horizontal surface of the foundation domain is specified as the S69E, S21W, and vertical components of the Taft ground motion – scaled by a factor of 2 – applied in the stream, cross stream, and vertical directions, respectively. From this specified motion, the effective earthquake forces to be applied at the absorbing boundaries are determined by the methods described in Section 11.8.

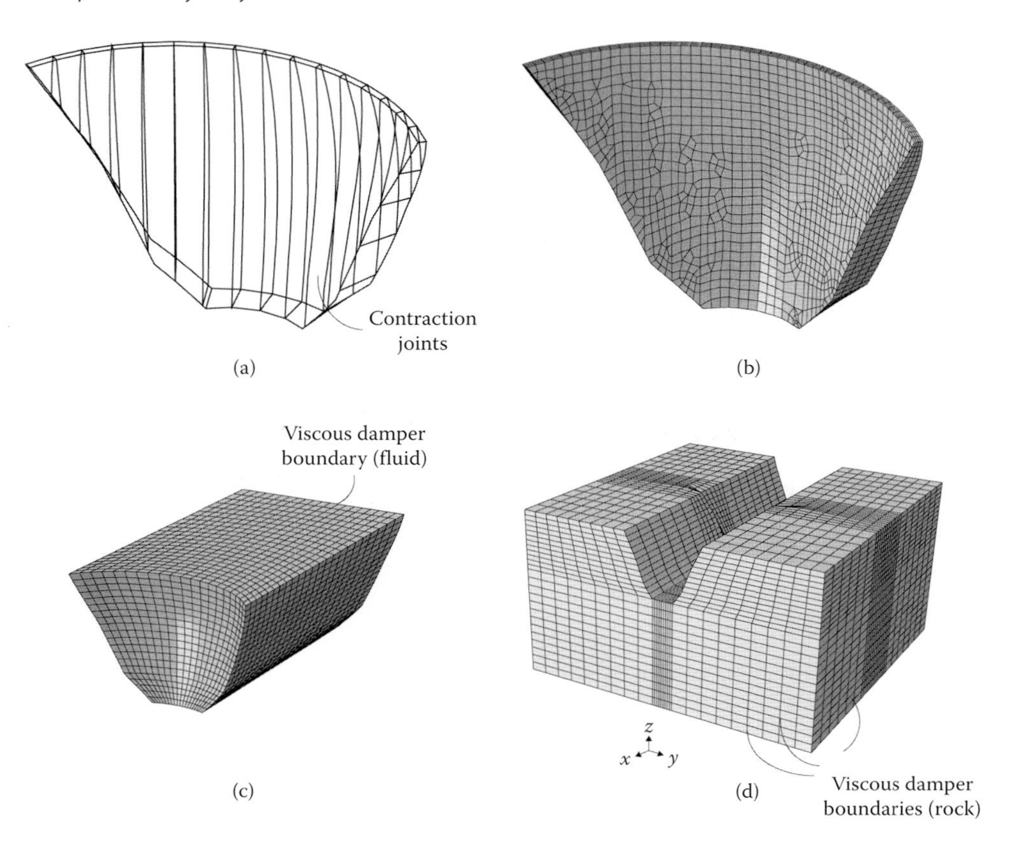

Figure 11.11.1 (a) Morrow Point Dam showing location of contraction joints; (b) FE model of dam; (c) FE model of fluid domain; and (d) FE model of foundation domain.

11.11.2 Computer Implementation

The Direct FEM is implemented in ABAQUS, a commercial FE program, using a pre-processing script in MATLAB that interacts with the ABAQUS input to compute and store the effective earthquake forces that define the excitation to this system. The analysis procedure is organized in three phases: (i) initial static analysis simulates the sequence of construction of the dam and filling of the reservoir; (ii) deconvolution analysis of a one-dimensional foundation column (Section 11.10.1) to determine the free-field motions that are in turn used to compute the effective earthquake forces at the bottom and side boundaries; (iii) nonlinear dynamic analysis of the FE model subjected to the effective earthquake forces determined in Step 2; the results of Step 1 provide the displacement and stress state of the system that becomes the initial conditions for the dynamic analysis. Detailed explanation of implementation of the Direct FEM in ABAQUS is available in Løkke and Chopra (2019a).

Before closing this section, note that Step 2 in the above procedure can be simplified by introducing "free-field boundary elements" at the side boundaries of the foundation domain that are processed in parallel with the main FE model as the analysis progresses in time (Nielsen 2006). However, such special elements are currently not available in most commercial FE codes used for analysis of concrete dams.

11.11.3 Earthquake Response Results

Commercial FE programs provide results for any response quantity of engineering interest. Examples of output are presented in Figures 11.11.2–11.11.6. Displacements in the stream, cross-stream, and vertical directions at the center of the crest of the dam are plotted as functions of time in Figure 11.11.2. Envelope values of the maximum (over time) displacements along the length of the crest of the dam are presented in Figure 11.11.3. Displacements and accelerations at the crest of the dam are of interest in checking the operability of appurtenant structures, such as mechanical equipment, gates, and roadway bridges over any spillway.

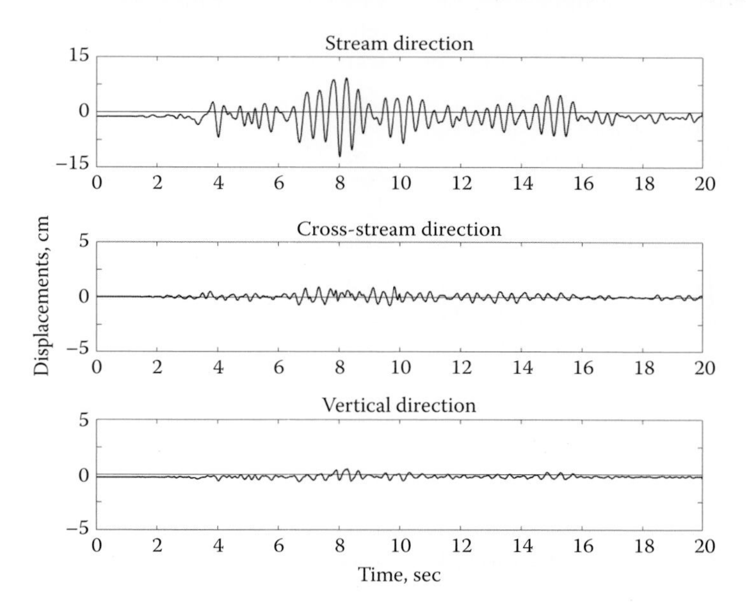

Figure 11.11.2 Displacement histories at center of dam crest in the stream, cross stream, and vertical directions.

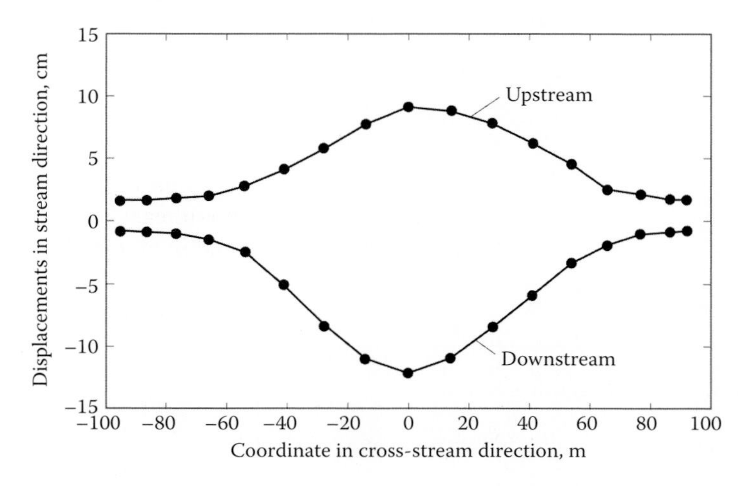

Figure 11.11.3 Envelope values of upstream and downstream displacements along the dam crest.

Opening of two contraction joints at the crest level are plotted as a function of time in Figure 11.11.4, and envelope values of openings of all joints are presented in Figure 11.11.5. The maximum opening of any contraction joint is approximately 25 mm, which is much less than the 150 mm depth of shear keys, implying that the shear keys remained interlocked during this ground motion.

The distribution of tensile damage over the two faces of the dam is presented in Figure 11.11.6. Such damage is greater on the downstream face and along one side of the dam–foundation interface (Figure 11.11.6a and b). The dam is beginning to show signs of a semi-circular crack pattern in the upper, central part of the dam, which has been observed as a potential failure mode during model studies of arch dams (Bureau of Reclamation 2002). However, no single crack has formed through the thickness of the dam (Figure. 11.11.6c) to develop such a failure mode.

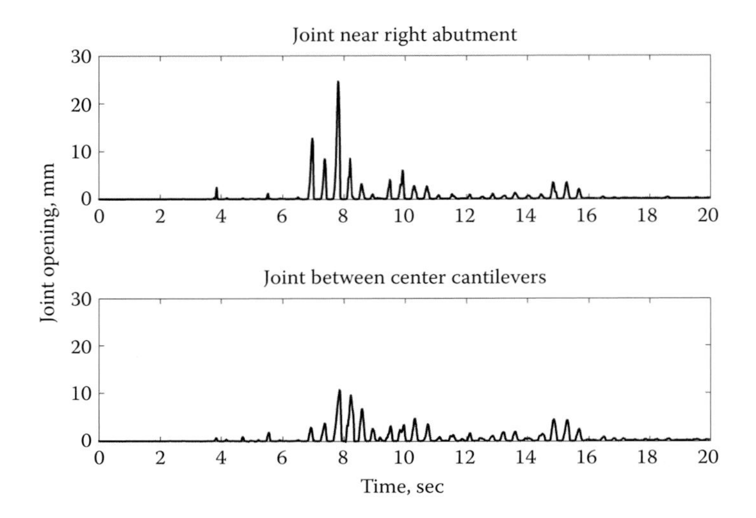

Figure 11.11.4 Opening of contraction joints at two locations: joint near the right abutment where maximum joint opening occurs, and the joint between the center cantilevers.

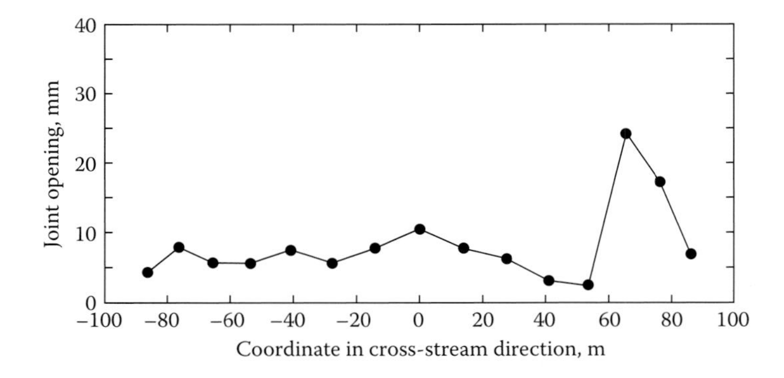

Figure 11.11.5 Envelope values of maximum contraction joint opening along the dam crest.

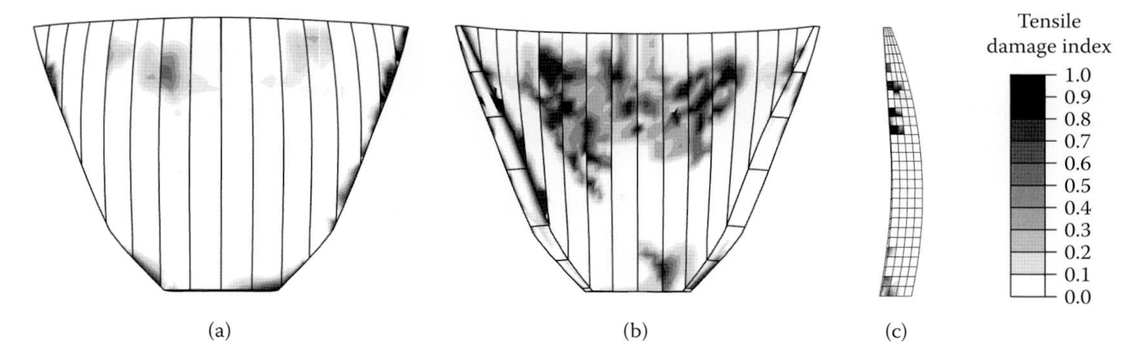

Figure 11.11.6 Distribution of tensile damage on (a) upstream face, (b) downstream face, and (c) section through center (crown) cantilever.

11.12 CHALLENGES IN PREDICTING NONLINEAR RESPONSE OF DAMS

Nonlinear response history analysis of dams is much, much more complicated relative to linear analysis. Computation times may be more than an order of magnitude longer in a nonlinear analysis, but this is not the biggest concern because enormous computing power is already available and will continue to improve in the future.

For many years, nonlinear FE analyses were limited by a major shortcoming: simplistic models for dam–water–foundation interaction and for semi-unbounded geometry of the foundation and fluid domains. These limitations were overcome in the Direct FEM presented earlier in this chapter where dam–water–foundation interaction and semi-unbounded fluid and foundation domains were modeled realistically. Because this method was developed in a form that can be implemented in commercial FE codes, the nonlinear material models available in these codes can be utilized to model the system.

The greatest impediment to reliable nonlinear analysis is the dearth of knowledge about the nonlinear constitutive properties of concrete and failure mechanisms under three-dimensional cyclic deformations at strain rates expected during vibration of the dam. These properties also vary over the life of the dam, depending on shrinkage of concrete, ambient temperature, and moisture content. To fill these knowledge gaps, a major systematic research program must be undertaken to test concrete cores of a size large enough to be representative of mass concrete. Typically, the maximum size of aggregate is 6 in. ASTM specifies that the concrete cores should be at least three times the size of aggregate, thus requiring 18-in.-diameter cores. However, rarely is this requirement enforced on existing dams because cores of such large size are very difficult to drill, extract, and test. Furthermore, very few laboratories possess the equipment required to test such large cores.

Another challenge of nonlinear RHA lies in developing mechanics-based nonlinear models for concrete and joints that are consistent with experimental data, and incorporating these models in FE codes. Existing models in commercial FE codes should be evaluated and modified as necessary, to be representative of mass concrete, lift joints, construction joints, and contraction joints between monoliths.

Incorporating these models in efficient numerical procedures is of great importance. Efficiency is critical for two reasons: (i) nonlinear RHA of large FE systems required to model practical dam–water–foundation systems is computationally very demanding; and (ii) a large number of such RHAs are required to consider the uncertainty in ground motion and in the material properties; both these uncertainties are known to significantly influence the nonlinear response of dams.

Although researchers and practitioners have investigated limited aspects of the nonlinear behavior of concrete dams, a great deal of innovative experimental, modeling, and analytical research must yet be done to develop practical nonlinear RHA procedures. When efficient analysis procedures have been developed and when the nonlinear behavior of mass concrete, rock, and joints can be modeled effectively, it will be necessary to perform extensive numerical parameter studies of concrete dams in order to understand the factors that significantly influence the seismic safety of dams. In addition, properly modeled shaking table tests and hybrid tests will be needed to verify the effectiveness of the safety evaluation procedures that have been developed.

Although many of these research needs were identified in a National Research Council Report (1990), they remain largely unaddressed, primarily for lack of research funding. However, limited investigations of various issues have occasionally been done for a specific dam, as part of its seismic safety evaluation.

Part III

DESIGN AND EVALUATION

12
Design and Evaluation Methodology

PREVIEW

Although detailed discussion of the seismic design and safety evaluation of concrete dams is beyond the scope of this book, this chapter presents a few important aspects of this broad subject. The two levels of ground motion (GM) and the corresponding requirements for satisfactory performance of the dam, specified by various professional and regulatory organizations, are presented in Section 12.1. Analysis of these established concepts brings out their implications that may not be widely recognized by dam engineers. As apparent from the preceding chapters, dynamic analysis of concrete dams is a highly complex problem, and interpretation of the results to predict the performance of a dam and estimate damage during intense GMs continues to be a major challenge. To gradually build confidence in any investigation, seismic demands on a dam should be determined by a series of dynamic analyses that become progressively more rigorous, and, similarly, the capacity of the structure to meet these demands should be systematically determined; these topics are presented in Sections 12.2 and 12.3. A discussion of the challenges in evaluating the seismic performance of dams subjected to GMs intense enough to cause damage, the limitations of current knowledge, and the research needed to overcome these limitations is presented in Sections 12.4 and 12.5.

12.1 DESIGN EARTHQUAKES AND GROUND MOTIONS

It is common to consider two levels of ground motion (GM) with corresponding performance requirements in the seismic design of a new dam. However, the terminology for these two GM levels and their definition has not been standardized. We first summarize the essentially identical terminology and definitions in publications of two major organizations: International

Commission on Large Dams (ICOLD) and Federal Emergency Management Agency (FEMA) in the United States (ICOLD 2016; FEMA 2005, 2014).

12.1.1 ICOLD and FEMA

The Operating Basis Earthquake (OBE) is the earthquake event that produces GM at the site that can reasonably be expected during the service life of the project. This statement has usually been interpreted as GM that has a 50% probability of exceedance (PE) in 100 years, the commonly assumed life of concrete dams. The corresponding mean return period is 144 years (calculated assuming a Poisson model for occurrence of events). At this level of ground shaking, the facility – dam, appurtenant structures, equipment, power house, etc. – should experience little or no damage and continue to function without interruption; this performance requirement implies that the dam remains essentially within the linear range of behavior. The OBE should be determined by Probabilistic Seismic Hazard Analysis (PSHA).

The Safety Evaluation Earthquake (SEE) or Maximum Design Earthquake (MDE) is the earthquake event that produces GM at the site that is rare. Factors to consider in selecting the intensity of this GM are the consequences of failure of the dam, criticality of project function (power generation, water supply, flood control, etc.), and turnaround time to restore the facility to be operational after the earthquake event. The MDE represents ground shaking at the site associated with a long mean return period: 10,000, 3000, or 1000 years for dams where the consequences of dam failure are high, moderate, or low, respectively. Mean return periods of 10,000 (precisely 9950) years and 1000 (precisely 949 years) represent ground shaking associated with a 1% and 10% PE in 100 years, respectively. The MDE should also be determined by PSHA. At this level of ground shaking, there should be no catastrophic failure, such as uncontrolled release of the impounded water, although significant damage or economic loss may be tolerated. This performance requirement implies that the dam is allowed to deform significantly into the nonlinear range.

The FEMA and ICOLD documents also define a Maximum Credible Earthquake (MCE). This represents the GM during the largest magnitude earthquake along a recognized fault or within a particular seismo-tectonic province. A deterministic approach is used to estimate the MCE-level GMs at the site. For each identified fault, the largest magnitude earthquake is used as input to a ground motion prediction model (GMPM) to provide the probability distribution of the GM intensity. The 84th percentile value is defined as the deterministic-based MCE-level motion.

At sites close to major faults with high-slip rates (e.g. the San Andreas and Hayward faults in California), the earthquake event that produces the MCE-level GM may have a relatively high annual rate of occurrence, e.g. 0.015[†] for the Hayward fault in California. Combining the annual rate of 0.015 with a GM with 16% probability of being exceeded results in a return period of 416[‡] years, a much, much shorter return period than the 10,000 years for the MDE. This implies that when the next large earthquake occurs on a major fault in California, the MCE-level GM would, on average, be exceeded at 16% of the dams. This does not seem to be prudent, suggesting that the MCE-level (84th percentile deterministic) GM is not intense enough. However, in other parts of the world where the slip rates on active faults are low, the MCE-level GM may be much more intense than the GM with a 10,000-year return period. Therefore, for safety evaluation of high-consequence dams the more intense of two GMs should be selected: (i) GM from MCE on known active faults; and (ii) GM associated with 1% PE in 100 years or return period of 10,000 years (ANCOLD 2017).

---

† The Hayward fault has a 31.7% chance of rupturing in a 6.7 or larger magnitude earthquake in the next 26 years; http://seismo.berkeley.edu/hayward/hayward_hazards.html. The associated annual rate of occurrence is 0.015.

‡ Return period = $(0.015 \times 0.16)^{-1}$ = 416 years.

12.1.2 U.S. Army Corps of Engineers (USACE)

In the USACE manuals (1999, 2016), the definition for the OBE is identical to the one stated in the last section, but the MDE is defined differently. For critical features of the project, the MDE is the same as the MCE. For all other features, the minimum MDE is an event with a 10% PE in 100 years, implying a mean return period of 950 (precisely 949) years. A shorter or longer return period for non-critical features may be appropriate, depending on the consequences associated with failure or the dam.

12.1.3 Division of Safety of Dams (DSOD), State of California

This influential agency continues to use deterministic methods to define seismic hazard. The MCE-level GM is defined as the 84th percentile estimate from the GMPMs.

12.1.4 U.S. Federal Energy Regulatory Commission (FERC)

Using deterministic methods to define seismic hazard, Federal Energy Regulatory Commission (FERC) requires dynamic analysis of the dam for the MCE-level GM followed by static analysis to evaluate the post-seismic stability of the damaged dam with reduced shear strengths and increased uplift pressures, resulting from damage to drains. FERC is concerned only about an uncontrolled release of the impounded water and not with operability of the facility at GMs of lower intensity.

12.1.5 Comments and Observations

The performance requirement that a facility should continue to function without interruption during the OBE seems to have been introduced to minimize economic losses during "frequent" earthquakes. However, it appears that the definition of the OBE has been based on tradition instead of extensive studies on life-cycle cost (economic) analysis. For a portfolio of 100 dams, the present definition of OBE-level GM and performance requirement implies that, on average, 50 dams would experience more intense GM during their lifetime, and, hence, may not remain functional. This number would, most likely, seem excessively large to many engineers and policy makers. To investigate this and related issues, systematic research on life-cycle cost analysis of concrete dams should be undertaken to develop a basis for making rational decisions on the choice of OBE-level GM. It is likely that such investigations would suggest that the GM should be associated with PE that is significantly lower than 50%, implying a return period much longer than 144 years. Some organizations have chosen a return period of 475 years (ANCOLD 2017).

Turning to the MDE, the performance requirement of no catastrophic failure during MDE-level GM – a rare, intense GM – seems appropriate. However, it is not clear that the definition of the MDE is appropriate. Consider a portfolio of 100 concrete dams and 100-year service life; the overall probability of at least one dam experiencing a MDE GM is

$$\begin{aligned} &1 - \text{Prob (no dam experiences MDE during service life)} \\ &= 1 - (1 - 0.01)^{100} \\ &= 0.63 \end{aligned}$$

where the exponent refers to the number of dams. This means that the probability of at least one dam experiencing this intense (or more intense) GM over a 100-year period is 63%. Will those dams all meet the performance requirement of no catastrophic failure? It is not possible to answer this question with confidence because of limitations in our current (2019) ability to determine the extent of cracking in concrete, sliding at joints and interfaces in the dam and in the foundation rock, or operability of equipment and appurtenant structures necessary to lower the level of impounded water in case of emergency after a major earthquake. Various reasons underlying these limitations were presented in Section 11.12. This unfortunate state of affairs suggests that until nonlinear modeling and dynamic analysis of concrete dams and their foundations improves to develop a high degree of confidence in the results, the design criteria should be made more stringent.

A case for this recommendation can be made by comparing the established performance criteria for concrete dams and nuclear power plants. For the GM corresponding to a return period of 10,000 years – known as the Safe Shutdown Earthquake (SSE) in the nuclear industry and MDE in the context of concrete dams – nuclear power plants are designed to remain linear, i.e. undamaged, but extensive damage is permitted in concrete dams. Major dams where the consequences of failure are unacceptable should be designed more conservatively. This conservatism can be achieved by two initiatives taken individually or together. First, the intensity of the OBE-level GM is increased by lowering the PE, i.e. lengthening the return period. Shown in Figure 12.1.1 are the Uniform Hazard Spectra (UHS) for several values of PE over 100 years – 50%, 10%, 5%, 2%, and 1% – for the Pine Flat Dam site. Obviously, the forces to be considered in linearly elastic design or evaluation of the dam, consistent with performance requirements for the OBE-level GM, increase as the PE decreases. Although the increase in design forces corresponding to lower values of PE may seem excessive to the point of being unacceptable, we should not rush to this judgment without considering several factors. First, modern nuclear power plants have been designed to remain linearly elastic during the 10,000-year GM. So this requirement may also be appropriate for large dams where consequences of failure can be catastrophic. Second, the additional cost of the dam designed for larger forces could be minimized by creative designs (e.g. see Section 7.2) that lead to improved earthquake performance. Based on life-cycle cost analysis, we should answer questions like: (i) Is it appropriate to follow the nuclear-industry tradition and design a high-consequence dam to remain undamaged during a 10,000-year return period GM?

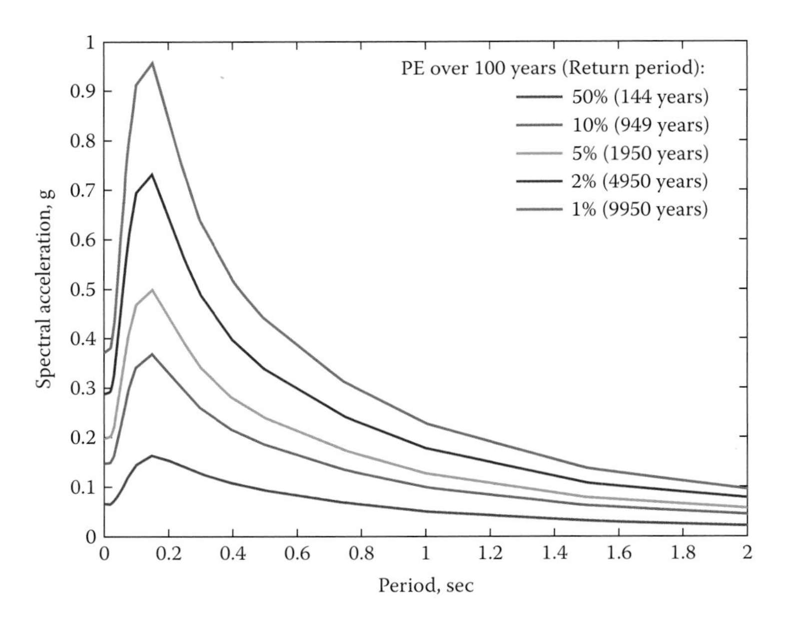

Figure 12.1.1 Uniform hazard spectra for several values of probability of exceedance (PE) over 100 years for the Pine Flat Dam site in California; corresponding return periods are noted.

Figure 12.1.2 Hoover Dam, a 221-m-high curved gravity dam on the Colorado River in the United States. Source: Arnkjell Løkke.

(ii) If this design goal is not economically prudent, what is an appropriate return period for which the dam should be designed to remain undamaged? Much research will be necessary to answer these questions. The second initiative could be to control more stringently the damage permitted during the MDE to a threshold much lower than what is envisaged in the current requirement of "no uncontrolled release of the impounded water."

Design earthquakes for concrete dams are based on specified PEs – 50% in the case of the OBE and 1% in the case of the MDE – over 100 years, the assumed service life of the project. However, the choice of 100 years may not be realistic. Consider two examples: it is difficult to imagine that the iconic Hoover Dam (Figure 12.1.2) built in 1933, will be abandoned in 2033, which is only 14 years from the time of this writing. The Aswan Low Dam (also known as Old

Figure 12.1.3 Aswan Low Dam, a 36-m-high masonry gravity dam with buttresses, on the Nile River in Egypt.

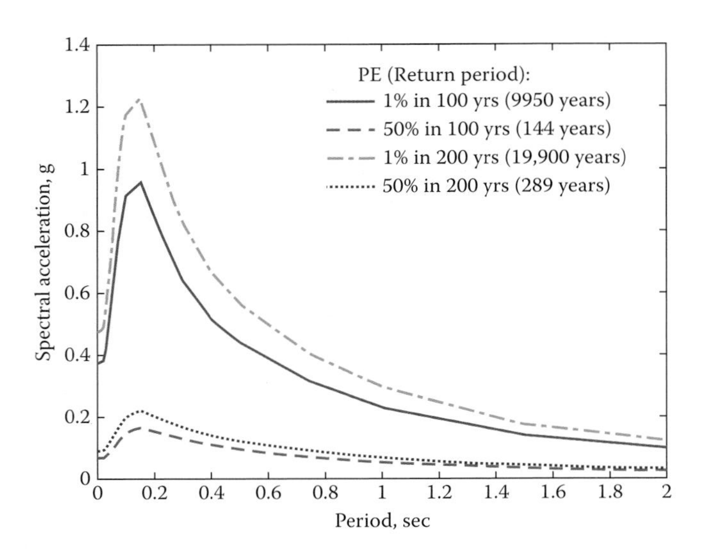

Figure 12.1.4 Uniform hazard spectra for the Pine Flat Dam site, California, corresponding to a probability of exceedance = 1% and 10% over 100 and 200 years; corresponding return periods are noted.

Aswan Dam) on the Nile River in Egypt (Figure 12.1.3), was constructed between 1899 and 1902, and continues (2019) to serve as a critically important facility in support of the power generation needs of Egypt. Having already existed for 117 years, most likely this dam will continue to remain functional for many years to come.

Suppose we consider a service life of 200 years. For the same PE, the GM will now be more intense. This is illustrated in Figure 12.1.4 where the UHSs are plotted for PE = 1% and 10% over 100 and 200 years. It is apparent that the spectral ordinates and design forces are increased by about 30–40% for service life of 200 years over the values for 100-year service life.

12.2 PROGRESSIVE SEISMIC DEMAND ANALYSES

The seismic demands imposed on the dam should be determined by a series of dynamic analyses that become progressively more rigorous. Each of these analyses should include the effects of dam–water–foundation interaction and the semi-unbounded extent of foundation and fluid domains, demonstrated to be significant earlier in this book.

The seismic demands imposed by the OBE can be computed by linear analysis in two stages: (i) a simplified response spectrum analysis (RSA) in which the response is estimated directly from the earthquake design spectrum, considering only those factors that are most important in the earthquake response of dams and yet is simple enough not to require the use of elaborate commercial computer programs; and (ii) a refined response history analysis (RHA) procedure for finite-element idealizations of the dam–water–foundation system. The former is recommended for purposes of preliminary design, and the latter for accurate computation of the dynamic response necessary to check the adequacy of the structure designed for the preliminary design forces. The simplified analysis procedure in Chapter 4 was developed for a 2D analysis of gravity dams but a corresponding procedure remains elusive for arch dams. As mentioned earlier (Chapters 5, 8, and 11), computer programs are available to implement refined RHA procedures for gravity, arch, and buttress dams.

The preliminary design of the dam should provide against overstressing in compression and tension; that is, the compressive and tensile stresses should not exceed the compressive and tensile strengths of concrete, respectively. The concrete strength requirements will be controlled by the tensile stresses because they will be similar in magnitude to the compressive stresses, whereas the tensile strength of mass concrete is an order of magnitude smaller than the compressive strength. The overturning and sliding stability criteria that have been used in standard design procedures in the past have little meaning in the context of oscillatory response of dams due to earthquakes (Chopra and Zhang 1991). These criteria could be satisfied only because the lateral earthquake force was unrealistically small in traditional design (Sections 1.3–1.5). However, they cannot be satisfied if the peak lateral force is determined by dynamic analyses. Researchers have proposed reducing this force to 50–60% of its full value in stability analysis of the dam (Tinawi et al. 2000). The end result of this phase of the design process is a preliminary design of the dam.

The adequacy of the preliminary design of the dam should be checked with the aid of refined, rigorous analysis procedures, such as those presented in Chapters 5, 8, and 11. The response of the preliminary design of the dam to selected GMs should be determined, resulting in more accurate values for the stresses and internal forces. Based on these results, the preliminary design of the dam should be revised, if necessary, to satisfy the same design criteria as mentioned in this section. The design modification may involve reshaping an arch dam, increasing the thickness of a gravity dam, and/or increasing the concrete strength.

However, as demonstrated in Section 7.2, the stresses in gravity dams can be significantly reduced by modifying the usual designs to reduce the weight near the crest of the dam. Instead of the solid concrete block added near the crest in typical designs of dams to support the roadway and to resist the impact of floating objects, lightweight structural systems would be preferable. Similarly, the auxiliary structures usually appended on the top of dams should be located with discretion so that they have a minimum adverse effect on stresses in the dam. Possible modifications in the geometry and mass distribution of arch dams that might lead to reduction of earthquake-induced stresses remain to be investigated fully.

A dam designed to remain within the linear range of behavior during the OBE should be evaluated to determine its performance in the event of a MDE. Before embarking upon a nonlinear RHA accompanied by a multitude of challenges in developing a numerical model, defining nonlinear constitutive properties of the materials, and dealing with sensitivity of results to uncertainty in GMs and material properties (see Section 11.12), the most rigorous linear RHA should be implemented. The results of such linear analysis would provide an initial understanding of degree of nonlinearity to be expected during the MDE. Such results can also assist in identifying areas of the dam that are likely to be strained beyond the linear range and require carefully developed nonlinear models. If the results of linear analysis indicate that the tensile strength of concrete or a joint is exceeded repeatedly during the duration of shaking, the engineer should consider the possibility of modifying the design to ensure essentially linear response even during the MDE. This could very well be economically preferable over repairing the damage that the original design is expected to experience during an MDE.

Many of the preceding comments in this section, after obvious modification, carry over to seismic evaluation of existing dams. In particular, a rigorous linear RHA should still be the first step in computing seismic demands on the dam, and the same criteria should be employed to determine the need for nonlinear RHA. In past investigations (of actual projects) that ignored mass of foundation rock and compressibility of water, the seismic demands on the dam were overestimated by factors up to 2 or 3 (Chapter 9). Such results could have led to the erroneous conclusion that an existing dam is unsafe, thus requiring upgrading, which is invariably very expensive.

However, if rigorous linear RHA of the dam demonstrates the potential for damage, a nonlinear analysis would be required. Despite its aforementioned limitations, nonlinear RHA of the dam–water–foundation system will provide a rough estimate of cracking in concrete, the amount of opening and sliding at contraction joints, lift lines, and joints in the rock. If this damage is deemed to be unacceptable, a retrofit scheme for the dam should be designed and earthquake response of the retrofitted dam computed to ensure that it meets the performance criteria.

12.3 PROGRESSIVE CAPACITY EVALUATION

An important property that determines the capacity of concrete dams to withstand earthquakes is the tensile strength of concrete. Ideally, the tensile strength should be determined from appropriate tests on specimens of concrete for the particular dam. However, a preliminary estimate of the tensile strength can be obtained from Figure 12.3.1, which presents four plots of tensile strength as a function of compressive strength, to be used depending on application. The lowest two plots, $f_t = 1.7f_c^{2/3}$ and $f_t = 2.3f_c^{2/3}$, are for long-time or static loading. The lowest curve represents actual tensile strength, whereas the second represents “apparent” tensile strength. The latter is not a quantity that can be measured; it is simply the stress corresponding to tensile strain at failure under the assumption of a linear stress–strain curve; see Figure 12.3.2. The apparent tensile strength is to be used to interpret the stresses computed by linear finite-element analysis. Similarly, the third and fourth plots, $f_t = 2.6f_c^{2/3}$ and $f_t = 3.4f_c^{2/3}$, are the actual and “apparent” tensile strengths at strain rates expected during earthquake-induced vibration.

If the stresses computed from linear RHA exceed repeatedly the “apparent” tensile strength determined by the empirical methods mentioned above, this important property should be determined for the actual concrete in the dam. Tensile strength can be determined from three types of tests: direct tension, splitting tension, and flexural tests. Results of these tests differ, and results of tests on cores taken in the field differ compared to tests on laboratory specimens. The

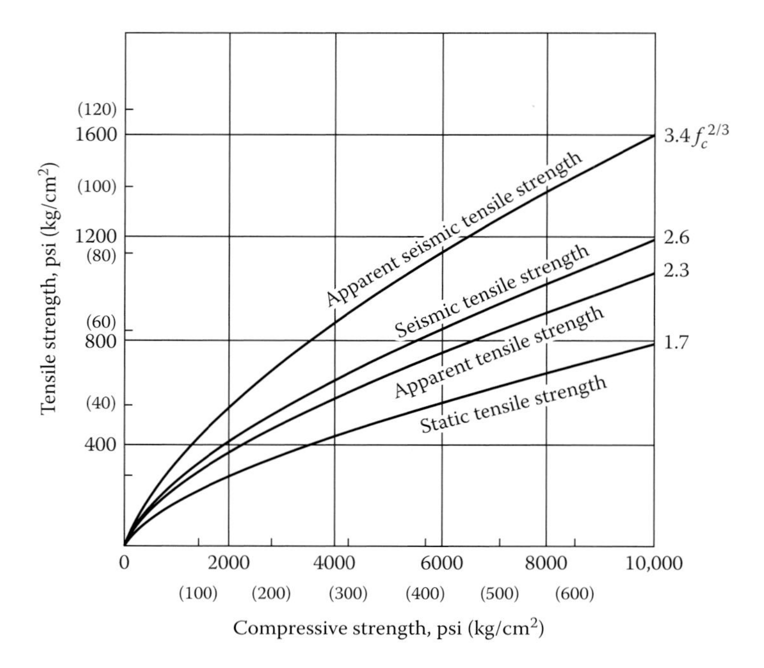

Figure 12.3.1 Design chart for tensile strength (Raphael 1984).

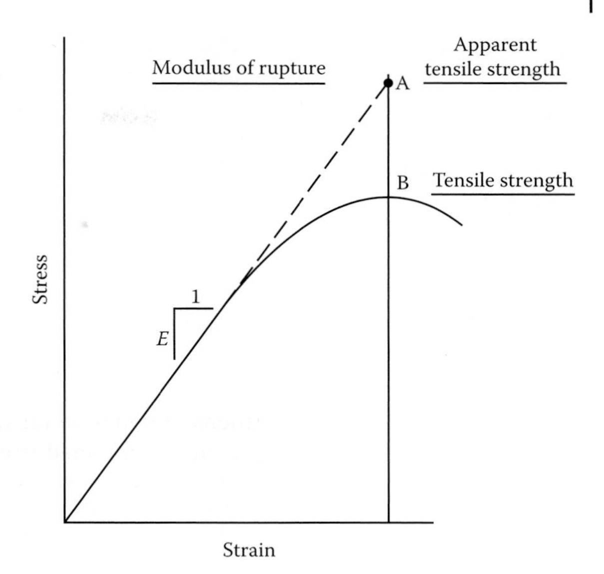

Figure 12.3.2 Apparent tensile strength (Raphael 1984).

direct tension test is difficult to accomplish and underestimates the tensile strength of concrete if the specimen is allowed to surface-dry. The flexural test, together with its usual linearly derived modulus of rupture, provides a basis to determine the tensile strength. The modulus of rupture should be multiplied by a factor that accounts for the nonlinear behavior of concrete and depends on the shape of the specimen. On the other hand, the splitting tension test is easiest to accomplish and gives the most reliable results. However, tensile strength obtained from the splitting tension test should be multiplied by about 4/3 to account for the nonlinear behavior of concrete near failure, before using it to interpret results of linear finite-element analysis (Raphael 1984).

Because the tensile strength of concrete increases with the rate of loading the aforementioned tests should be conducted at strain rates the concrete may experience during earthquake motions of the dam. In the absence of a facility to perform dynamic tests, Raphael (1984) recommended that the tensile strength of concrete for judging the seismic safety of a concrete dam be equal to the static value multiplied by 1.5. However, test data on concrete for some dams does not support this recommendation, nor is it appropriate in the presence of significant initial (static) tensile stresses in parts of arch dams. Thus, we recommend a smaller value, say, 1.25, unless evidence is available to justify a larger value.

These estimates of tensile strength are appropriate for mass concrete but not for weaker zones, e.g. horizontal lift joints in all types of dams and vertical contraction joints between cantilevers of arch dams. The tensile capacity of these joints is greatly influenced by the construction methods and details.

12.4 EVALUATING SEISMIC PERFORMANCE

Evaluating the seismic performance of concrete dams based on results of linear RHA is relatively straightforward. If computed tensile stresses do not exceed the tensile strength – which may be reduced by a factor to be conservative – we conclude that the dam will remain undamaged during the earthquake. As mentioned earlier, this is typically the performance requirement during the OBE.

Evaluating the seismic performance of dams subjected to GMs intense enough to cause damage is very challenging. Quantitative measures for the extent of damage – cracking in concrete, sliding at lift joints or at cracked interfaces, and opening of contraction joints – that dams can sustain and still retain the impounded water have not been developed for lack of research on sensitivity of computed response to uncertainty in material properties and GMs, and on experimental validation of results from dynamic analysis. Thus, performance evaluation of dams deforming beyond the linear range of behavior is open to interpretation and judgment, leaving open the possibility of different engineers arriving at contradictory conclusions.

After completing a nonlinear RHA of the dam, a post-earthquake analysis of the damaged dam is required to evaluate if the dam will remain stable and continue to contain the impounded water. Such analysis should model the dam in its damaged condition with uplift pressures modified to reflect the post-earthquake condition of the drains. Comprehensive but qualitative discussion of these topics is available in Chapter 6 of a National Research Council Report (1990).

Part of the difficulty in establishing quantitative criteria for evaluating results from nonlinear RHA is due to the dearth of definitive evidence – experimental or observational – on the evolution of failure mechanisms in concrete dams. There is a crying need for research on credible potential failure modes and how they could develop during an earthquake.

Evaluating foundation stability is also a very challenging problem. Results of nonlinear RHA by the direct finite-element method (FEM) (Chapter 11) provide time variation of forces acting on the foundation. Under these driving forces, the dam should remain stable against sliding along concrete-rock contact and the foundation blocks or wedges formed by intersecting rock discontinuities should also remain stable. Evaluating the performance of the dam and foundation against these criteria is challenging, especially because the driving forces vary with time. This is yet another subject where much research is necessary to develop methodologies and to demonstrate their reliability.

The need for research alluded to in this section was articulated almost 30 years ago by a panel of experts appointed by the National Research Council (1990). Progress since then has been meager for lack of research funding.

12.5 POTENTIAL FAILURE MODE ANALYSIS

It is useful to think of the various modes by which a dam can fail during an earthquake; failure is defined as uncontrolled release of the impounded water. Several potential failure modes could be identified, which in the context of gravity dams include (Figure 12.5.1):

1. Concentration of stress in the upper part of the dam, where the slope of the downstream face changes, resulting in cracking extending through the thickness of the monolith and finally resulting in excessive movement of the separated block above the crack.
2. Sliding along the dam–rock interface that may cause shearing of the vertical drains, making them ineffective, resulting in increased uplift pressures, and reduced frictional force. Failure may result from excessive sliding at the interface or by toppling of spillway piers (about the weak axis) or slender blocks of arch dams.
3. Sliding along an unbonded lift joint that may cause shearing of the vertical drains, making them ineffective, resulting in increased uplift pressures, and reduced frictional force. Failure may result from excessive sliding along the lift joint or by toppling of spillway piers or slender blocks of arch dams.
4. Sliding along a discontinuity in the foundation, resulting in excessive movement of the foundation blocks.

Some dam engineering organizations have now adopted a methodology for evaluating the seismic safety of existing dams or proposed new dams that consists of: (i) identifying potential failure modes; (ii) postulating the entire sequence of events leading to failure; and (iii) developing

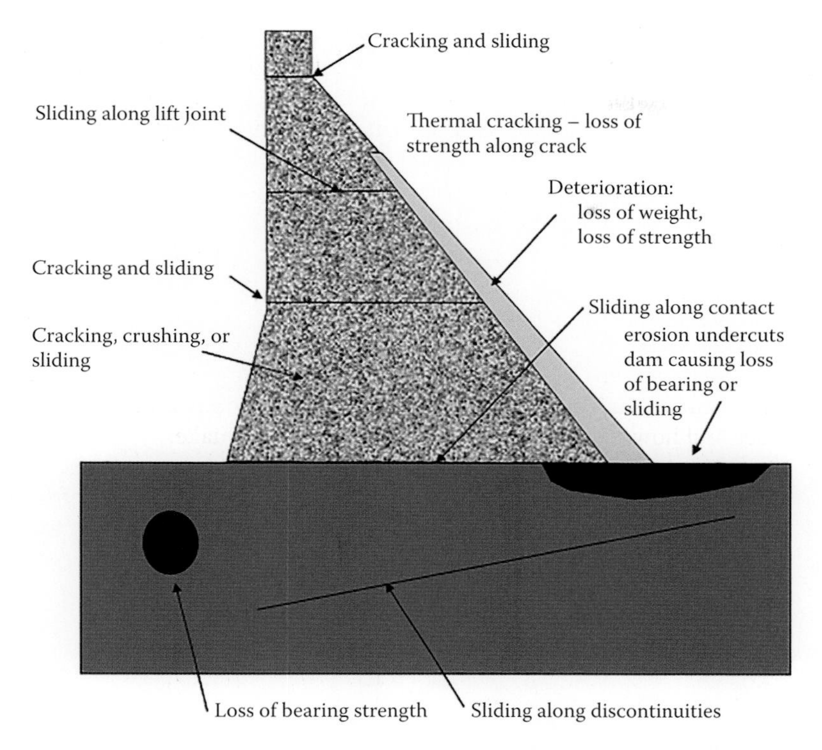

Figure 12.5.1 Potential failure modes given an initiating event (FEMA 2014).

logic (or event) trees, a graphical representation of all potential paths to failure. Documented in various publications (FEMA 2014; Hartford and Baecher 2004), this methodology is beyond the scope of this book. We simply observe that quantitative implementation of this methodology requires the ability to estimate the extent of cracking of concrete and sliding displacements along a crack, unbonded lift joint, or discontinuity within rock. Nonlinear RHA of the dam, developed in Chapter 11, provides such information.

13
Ground-Motion Selection and Modification

PREVIEW

The goal of nonlinear response history analysis (RHA) is to estimate seismic demands that may be imposed by unknown ground shaking at the site caused by future earthquakes. Because the ground motion (GM) cannot be defined uniquely, we are interested in the statistics of structural response to an ensemble of GMs. These GMs should, in some sense, be consistent with the seismic hazard at the site, usually characterized by the uniform hazard spectrum (UHS). Although the UHS is a candidate for the user-specified target spectrum, we will discuss in Section 13.1 the reasons why this is not the most appropriate choice, and present alternatives.

Although the number of GMs recorded during past earthquakes is large – now reaching several thousand – this database is still not large enough to enable selection of a subset of GMs consistent with the target spectrum, especially in highly seismic regions of the world, because of the paucity of records from large-magnitude earthquakes at short distances. Thus, it becomes necessary to modify selected GM records so that their response spectra are consistent with the target spectrum. Modification of GM records usually follows one of two approaches; amplitude scaling or spectral matching. In the first approach, a GM record that is initially selected because the shape of its response spectrum is generally consistent with that of the target spectrum is scaled (usually upwards) to achieve the desired intensity; thus, the scaled record $a(t) = \mathrm{SF}a_o(t)$ where SF is the scale factor and $a_o(t)$ is the original record. In the second approach, a GM record is modified such that its response spectrum matches very closely the target spectrum.

The subject of GM selection and modification (GMSM) has been a popular research topic since the early 2000s. However, much of the work has been restricted to a single horizontal component of GM. Within this restriction, the amplitude-scaling method is the subject of Sections 13.2–13.3, and the spectral-matching method is presented in Section 13.4. Thereafter, these GMSM procedures are extended to two horizontal components of GM in Sections 13.6 and 13.7, and subsequently to include the vertical component of GM in Section 13.8. These extensions are included in this presentation because all three components of GM should generally be considered in nonlinear response history analysis (RHA) of concrete dams, especially if the models are three-dimensional, as in the case of arch dams. This chapter is limited to GMs at sites where directivity and other near-field effects are of secondary importance.

PART A: SINGLE HORIZONTAL COMPONENT OF GROUND MOTION

13.1 TARGET SPECTRUM

13.1.1 Uniform Hazard Spectrum

A probabilistic seismic hazard analysis (PSHA) for a specific site determines the rate (or frequency) with which an earthquake characteristic, or intensity measure (IM), exceeds a selected threshold during some fixed exposure time in the future (McGuire 2004). The IM may be the peak ground acceleration (PGA); the ordinate of the pseudo-acceleration response spectrum at a vibration period of interest, say, the fundamental vibration period T_1 of the structure, for a selected value of damping; Modified Mercalli Intensity; the duration of ground shaking; or the displacement caused by a fault beneath a structure's foundation. For example, PSHA could answer the following types of questions: (i) What is the probability that the PGA at a site will exceed 0.5 g during 100 years? (ii) What is the PGA value that will be exceeded with 1% probability in 100 years? PSHA integrates the relative frequencies over all conceived earthquake occurrences (on all seismic sources in the region) and (GM) intensities to calculate a combined probability of an IM exceeding a specified threshold over a fixed exposure time.

The uniform hazard spectrum(UHS) is constructed by implementing PSHA for spectral acceleration at each vibration period (typically for 5% damping), independent of all other vibration periods. Figure 13.1.1 shows the UHS with a 1% probability of exceedance in 100 years for the Pine Flat Dam site in California (119.3°W and 36.8°N). This exceedance probability corresponds to a return period of 9950 years; this is the mean time between occurrences of the specified hazard, assuming that the exceedances follow a Poisson random process. A return period of 10,000 years is often selected for critical facilities such as major dams and nuclear power plants. The UHS was determined by OpenSHA, an open-source tool.[†]

13.1.2 Uniform Hazard Spectrum Versus Recorded Ground Motions

Probabilistic seismic hazard analysis (PSHA) also provides information about earthquake events contributing to the seismic hazard. Suppose that we are analyzing a structure with period $T_1 = 0.5$ sec for which the 1% in 100 years value of $A(0.5 \text{ sec})$ is 0.41 g, as shown in Figure 13.1.1. Disaggregation of the hazard shown in Figure 13.1.2, provides the percentage contribution of earthquakes of different magnitudes and distances to $A(0.5 \text{ sec}) = 0.41$ g; these contributions vary with the selected period. At the 0.5-sec period, the mean causal magnitude (M) is 6.24, the mean causal distance (R) is 21 km, and the mean ε is 1.02. The latter parameter is defined as the number of standard deviations by which a given $\ln A$ value differs from the mean predicted $\ln A$ value for a given M and R:

$$\varepsilon(T_n) = \frac{\ln A(T_n) - \mu_{\ln A}(M, R, T_n)}{\sigma_{\ln A}(T_n)} \tag{13.1.1}$$

[†] http://www.opensha.org/apps

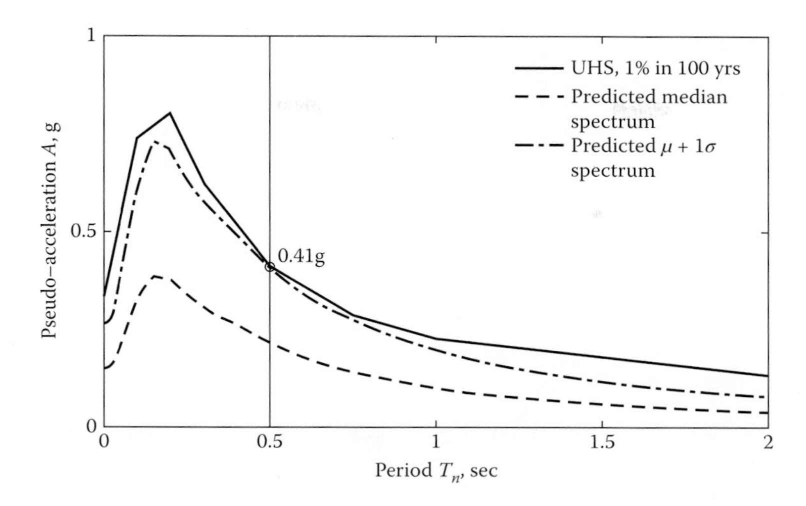

Figure 13.1.1 Uniform hazard spectrum (UHS) for the Pine Flat Dam site; also shown are the predicted median spectrum and $\mu + 1\sigma$ spectrum associated with an $M = 6.24$, $R = 21$ km event and a conditioning period T^* of 0.5 sec.

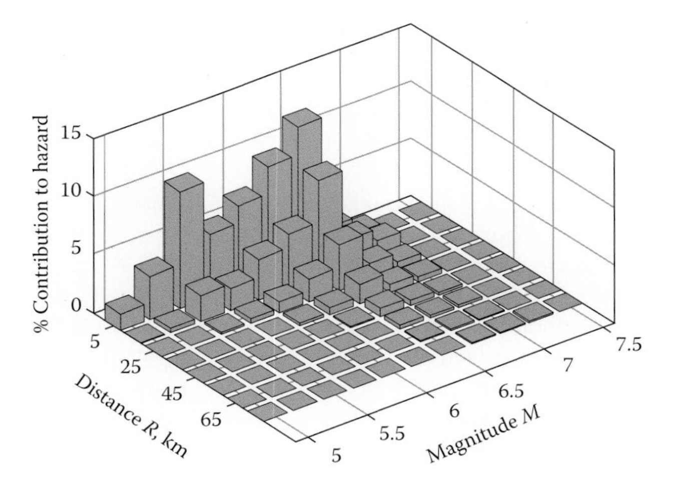

Figure 13.1.2 PSHA disaggregation for the Pine Flat Dam site, given 1% exceedance probability in 100 years, at a period of 0.5 sec.

where $\mu_{\ln A}(M, R, T_n)$ and $\sigma_{\ln A}(T_n)$ are the predicted mean and standard deviation, respectively, of $\ln A$ at period T_n, and $\ln A(T_n)$ is the natural logarithm of the actual spectral acceleration. Since $A(T_n)$ follows a lognormal distribution, its median and statistical dispersion (henceforth, dispersion for brevity) are identical to $\exp[\mu_{\ln A}(M, R, T_n)]$ and $\sigma_{\ln A}(T_n)$, respectively. Starting with essentially the entire database of recorded GMs, researchers have developed GM prediction models (GMPMs) to predict the median and dispersion of $A(T_n)$ for a given earthquake event (i.e. given M and R).

Computed from GMPMs, the predicted median spectrum associated with an earthquake of this magnitude ($M = 6.24$) at this distance ($R = 21$ km) is shown in Figure 13.1.1. The median $A(0.5\text{ sec})$ is clearly much smaller than the target value of $A(0.5\text{ sec}) = 0.41$ g, implying that this value of A is caused by GMs that are, on average, 1.02 standard deviations

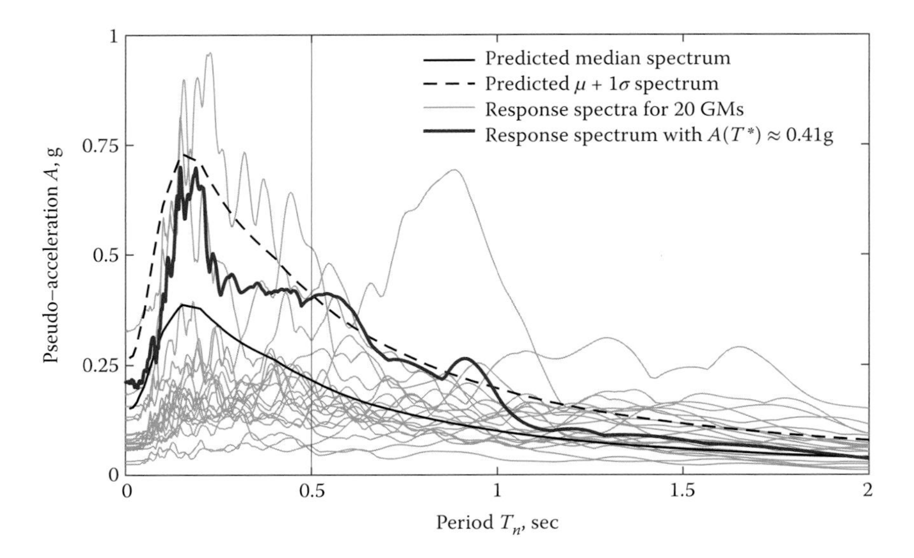

Figure 13.1.3 Response spectra of 20 ground motion records from earthquakes with approximate $M = 6.24$ and $R = 21$ km. The example spectrum shown with a heavier line is the L.A. Fletcher Drive station recording from the 1987 $M = 5.99$ Whittier Narrows LA, California, earthquake, with a distance to the fault rupture of 18.9 km.

larger than the predicted median GM associated with the causal earthquake event. This can be seen in Figure 13.1.1, where the predicted $\mu + 1\sigma$ spectrum is close to the target value of $A(0.5 \text{ sec}) = 0.41$ g.

This observation suggests that, for a given earthquake defined by M and R, there is considerable variability in the response spectra. Figure 13.1.3 shows the response spectra of 20 GMs due to earthquakes with approximately $M = 6.24$ and $R = 21$ km (more precisely, $5.99 < M < 6.49$ and $0 \text{ km} < R < 31$ km); a range of M and R must be allowed because records associated with $M = 6.24$ and $R = 21$ km are very few. The considerable scatter among the 20i spectra is obvious, but if the sample was large and the GMPM robust, the mean of ln A spectra would be close to the predicted $\mu_{\ln A}(M, R, T_n)$. One of the spectra, shown in heavier line, has $A(0.5 \text{ sec})$ approximately equal to the target value of 0.41 g. This implies that GMs with such large spectral acceleration have been recorded. Although the spectrum for such a GM has a large ordinate (relative to the median) at 0.5-sec period, it is not equally large at all periods. Clearly, the UHS, which by definition has "equally large" values at all periods, is overconservative and is not representative of individual GM spectra associated with an earthquake of magnitude equal to the mean value occurring at the mean distance. Thus, the UHS is not an appropriate target for selecting GMs to be used in dynamic analysis of structures.

13.1.3 Conditional Mean Spectrum

The conditional mean spectrum (CMS) has been developed by researchers as a target spectrum (TS) that overcomes the drawbacks of the UHS (Baker and Cornell 2005; Baker 2011). The CMS is constructed† for a selected value of the conditioning period, denoted by T^*, where the spectral acceleration is specified. Typically T^* is selected as the fundamental vibration period of the structure and $A(T^*)$ as the UHS value. Shown in Figure 13.1.4 is the CMS for the Pine Flat Dam site

† MATLAB implementation of a method to compute the CMS can be downloaded from: http://www.stanford.edu/~bakerjw/gm_selection.html.

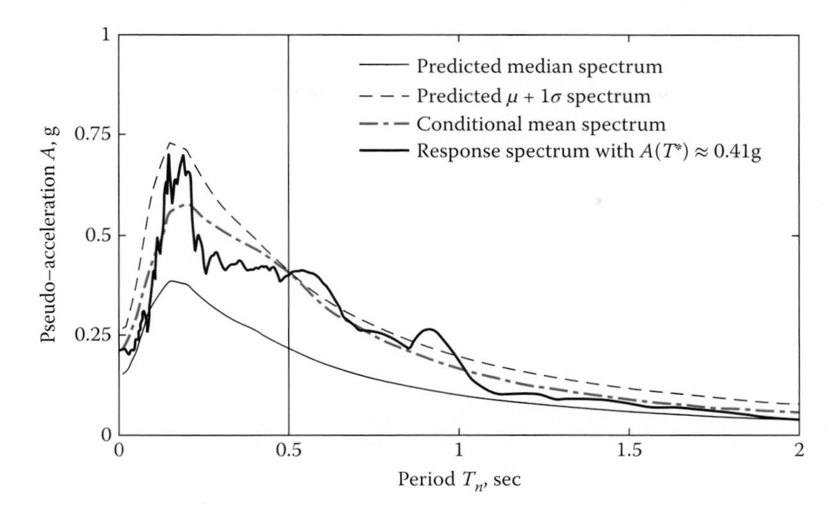

Figure 13.1.4 Conditional mean spectrum given A (0.5 sec) and response spectrum for an individual record. Also shown are the predicted median and $\mu + 1\sigma$ spectra.

and $T^* = 0.5$ sec. It has a (slight) hump near the conditioning period of 0.5 sec where it matches the UHS and then drops off on both sides. As the period decreases below, and increases above, the 0.5-sec period, the CMS usually tends to approach the median (or mean of ln A) spectrum for which $\varepsilon = 0$. The CMS is representative of recorded GMs, as evident from its similarity to one of the response spectra; hence, it is appropriate as a target spectrum for dynamic analysis of structures. This new spectrum is called the CMS because it provides the expected (mean) values of the natural logarithms of spectral accelerations at all periods given the spectral acceleration value at the conditioning period, T^*.

13.1.4 CMS-UHS Composite Spectrum

Researchers have demonstrated several concerns with using the CMS as the target spectrum to select and modify GMs. First, an intensity-based assessment that is based on the CMS conditioned on a single conditioning period, T^*, almost always underestimates significantly the demand from a risk-based assessment, which is more accurate. Intensity-based assessments underestimate the demand because the nonlinear dynamics of a complex structure depends on many more features of the GM beside spectral acceleration at a single period; the latter only characterizes the response of a linear single-degree-of-freedom (SDF) system with that natural vibration period. Second, the demands from an intensity-based assessment with a single CMS vary with the choice of T^*.

A common solution to address these concerns about the CMS is to analyze the structure with several, say three, CMSs, each conditioned on a different vibration period with the target spectral ordinate set at the UHS value (see Figure 13.1.5), and then take the maximum of the mean demands from all CMSs as the final estimate. Researchers have demonstrated that this approach provides an improved estimate of demand compared to the single CMS approach. This improvement comes at the expense of considerably larger number of nonlinear RHAs (say, by a factor of 3) of the structure.

An alternative is provided by the generalized CMS (GCMS), which conditions upon an arbitrary number of vibration periods and determines the target spectral ordinates via a reliability assessment (Loth and Baker 2015). The spectral values at the conditioning periods may be specified in several ways. One option is to specify them such that the joint probability

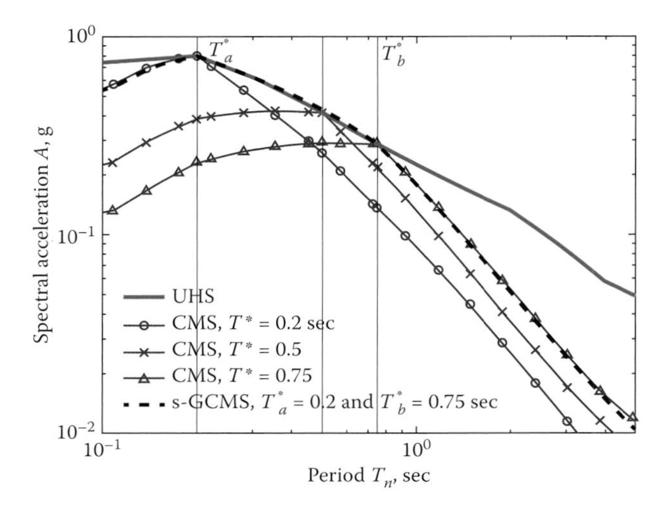

Figure 13.1.5 Comparison of s-GCMS against the UHS and three CMSs conditioned on three different periods.

of simultaneously exceeding these values agrees with the selected return period (Ni et al. 2012). Although this method leads to a ground-motion hazard that is consistent with the specified hazard level, it requires vector-valued PSHA, does not lead to a unique choice of spectral values, and all choices are either equal to or less than the UHS values. Another option is to specify spectral values as those that are most likely to cause failure of the structure, called the "design point" in structural reliability analysis (Loth and Baker 2015). This method requires a failure function that is very difficult to determine for many structures, and the "design point" is not unique; it varies with the response quantity or engineering demand parameter (EDP) of interest.

Given these challenges in practical application, the GCMS has been simplified by conditioning simultaneously upon two spectral accelerations that share a common hazard level; this spectrum will be referred to as s-GCMS (Figure 13.1.5). Specifying two spectral accelerations equal to the respective UHS values is straightforward and leads to a unique choice; however, it implies that the GM is no longer consistent with the specified return period, a topic that will be discussed later. Starting from the spectral values specified at two periods, a methodology for computing spectral values at all other vibration periods has been developed (Kwong and Chopra 2017). An example of the s-GCMS is presented in Figure 13.1.6 in both arithmetic and logarithmic scales.

The shape of the s-GCMS is controlled by the choice of the two conditioning periods: T^*_a and T^*_b. The s-GCMS will be nearly as intense as the UHS at all periods between the two conditioning periods but less intense at periods outside this range (Figures 13.1.5 and 13.1.6). This implies that as T^*_a and T^*_b approach each other, the s-GCMS reduces to the standard CMS with a single $T^* = T^*_a = T^*_b$; and as T^*_a and T^*_b become more separated, the s-GCMS becomes increasingly similar to the UHS.

For most practical applications, specifying only two conditioning periods should suffice so long as the period range defined by T^*_a and T^*_b includes the structural periods of interest. The ordinates of the s-GCMS at vibration periods between T^*_a and T^*_b are very close to the UHS ordinates (Figures 13.1.5 and 13.1.6), even though the periods in this range were not utilized in constructing the spectrum. Hence, if intense values of spectral acceleration at several vibration periods are desired, then it should be adequate to choose T^*_a as T_{min} (the shortest) and T^*_b as T_{max} (the longest) among these periods.

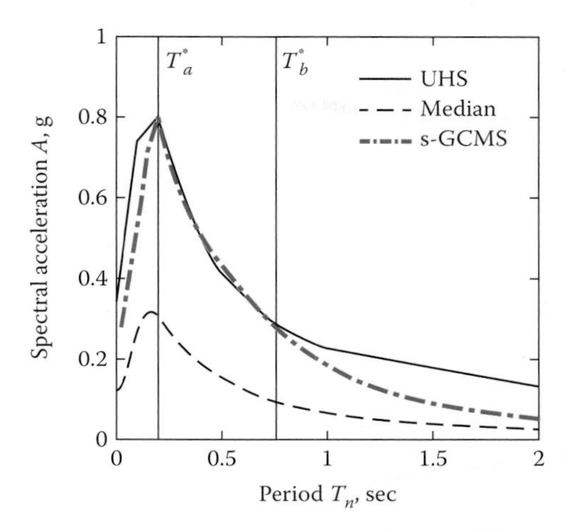

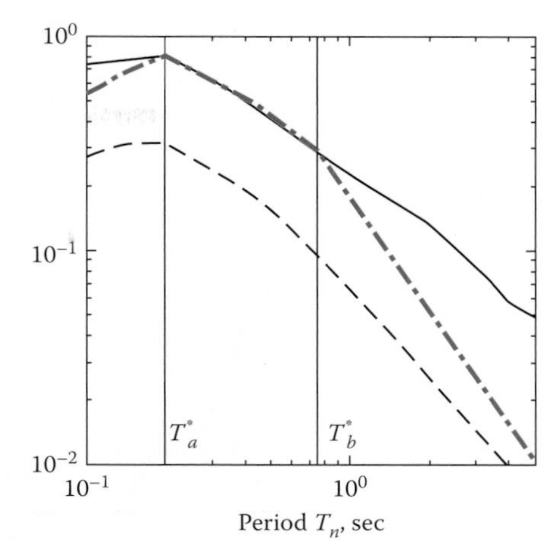

Figure 13.1.6 Example of the s-GCMS with two values of T^*, T^*, $T^*_a = 0.25$ sec, and $T^*_b = 0.75$ sec: (a) arithmetic scale; and (b) logarithmic scale. The median spectrum is constructed using a GMPM with mean values and km, determined by disaggregation of the vector-valued seismic hazard.

We now return to the implications of matching spectral ordinates at two conditioning periods to the UHS values. The return period for the ground-motion hazard has traditionally been chosen to be equal to the return period for the seismic risk. Given a return period of v_f^{-1} (e.g. 10,000 years corresponding to $v_f = 10^{-4}$ events per year), the "implicit" performance goal for a particular EDP is to determine the demand so that the annual rate of exceeding this demand is precisely v_f [see Eqs. 3 and 9 in Loth and Baker (2015)]. This demand can be obtained from the seismic demand hazard curve (SDHC) by performing a risk-based assessment of the structure, which requires a large number – say, a few hundred nonlinear RHAs of the structure (Shome et al. 1998). To avoid this onerous requirement, it is common to estimate the demand from an intensity-based assessment and use the estimate for an "explicit design check" [see Eq. 4 in Loth and Baker (2015)]. In such intensity-based assessments, the spectral acceleration at a single conditioning period T^* is often chosen equal to the UHS value so that the annual rate of exceeding this acceleration is also v_f.

By matching the spectral acceleration at two vibration periods to the UHS values for the specified return period, the s-GCMS represents a ground-motion event that has a longer return period. Such is the case because the annual rate of $A(T^*_a)$ and $A(T^*_b)$ simultaneously exceeding the respective UHS values is obviously lower than that of $A(T^*_a)$ exceeding the UHS value at T^*_a or $A(T^*_b)$ exceeding the UHS value at T^*_b; both of the latter rates are equal to v_f. Although inconsistent relative to PSHA, the more intense motions – corresponding to a longer return period – compensate for the underestimation of demand using a single CMS, as demonstrated in the context of multistory buildings (Kwong and Chopra 2017).

Although the probability of occurrence is lower, recorded GMs that naturally have intense spectral values at two significantly different periods do in fact exist. This is demonstrated for the Pine Flat Dam site for which the 10,000 years UHS was presented in Figures 13.1.5 and 13.1.6 where $A(T^*_a = 0.2 \text{ sec}) = 0.80\,\text{g}$ and $A(T^*_b = 0.75 \text{ sec}) = 0.29\,\text{g}$. Consider the case where the hazard is specified by simultaneous occurrence of these spectral values. By performing vector-valued disaggregation for this hazard (see Eqs. 9 and 4b in Kwong and Chopra (2017)), the mean controlling earthquake is defined by $\bar{M} = 6.3$ and $\bar{R} = 19$ km. Figure 13.1.7 shows response spectra for 172 GMs that have been recorded during earthquakes with $M = \bar{M} \pm 0.2$ and $\bar{R} = R \pm 10$ km. One of these spectra, highlighted by a heavy line weight, possesses intense spectral accelerations

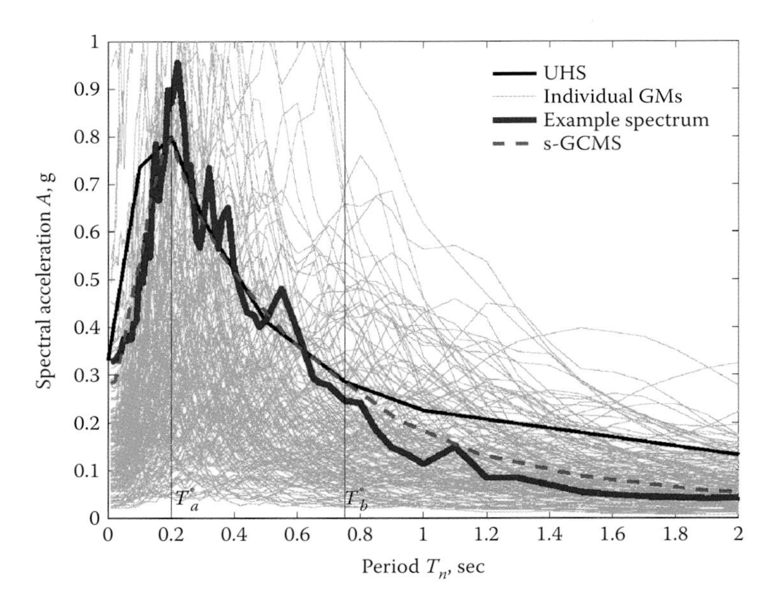

Figure 13.1.7 Comparison of the s-GCMS against the recorded GMs with $M = \bar{M} \pm 0.2$ and $R = \bar{R} \pm 10$ km, where $\bar{M}$ and $\bar{R}$ are determined by disaggregation of the vector-valued seismic hazard. A recorded GM with intense values for both $A(T_a^*)$ and $A(T_b^*)$ is highlighted.

at both conditioning periods [i.e. $A(T_a^*) = 0.87$ g and $A(T_b^*) = 0.25$ g]. Although this particular spectrum is similar to the UHS within the period range T_a^* to T_b^*, it is less intense than the UHS outside of this period range. In contrast, it is similar to the shape of the s-GCMS over the entire range of periods.

Construction of the s-GCMS requires disaggregation of the vector-valued seismic hazard to determine a single controlling earthquake scenario. This non-standard analysis, which may be argued to be impractical, can be avoided. A closer examination of the UHS, individual CMSs, and the s-GCMS enables the development of a target spectrum that can be constructed readily. Figure 13.1.5 shows that the s-GCMS is nearly as intense as the UHS at periods between T_a^* and T_b^*, but is less intense than the UHS outside of this period range. At periods shorter than T_a^*, the s-GCMS is very similar to the CMS with $T^* = T_a^*$, whereas, at periods longer than T_b^*, it is very similar to the CMS with $T^* = T_b^*$. These observations suggest that the s-GMCS might be well-approximated by a composite spectrum that combines features from both the CMS and the UHS. Formally the CMS–UHS composite spectrum is defined as

$$A_{\text{Composite}}(T_n) = \begin{cases} A_{\text{CMS}}(T^* = T_{\text{min}}) & T_n \leq T_{\text{min}} \\ A_{\text{UHS}} & T_{\text{min}} < T_n < T_{\text{max}} \\ A_{\text{CMS}}(T^* = T_{\text{max}}) & T_n \geq T_{\text{max}} \end{cases} \tag{13.1.2}$$

where, for reasons discussed earlier, T_a^* and T_b^* have been replaced by T_{min} and T_{max}, the shortest and longest structural periods of interest. This composite spectrum is essentially identical to the s-GCMS as shown in Figure 13.1.8. As T_{min} and T_{max} approach each other, the composite spectrum reduces to a single CMS with $T^* = T_{\text{min}} = T_{\text{max}}$; but when the two periods are far apart, the composite spectrum is close to the UHS. Recall from Figure 13.1.5, only two conditioning periods – T_{min} and T_{max} – enter into defining the composite spectrum, even though several conditioning periods may be necessary for the multiple-CMS approach.

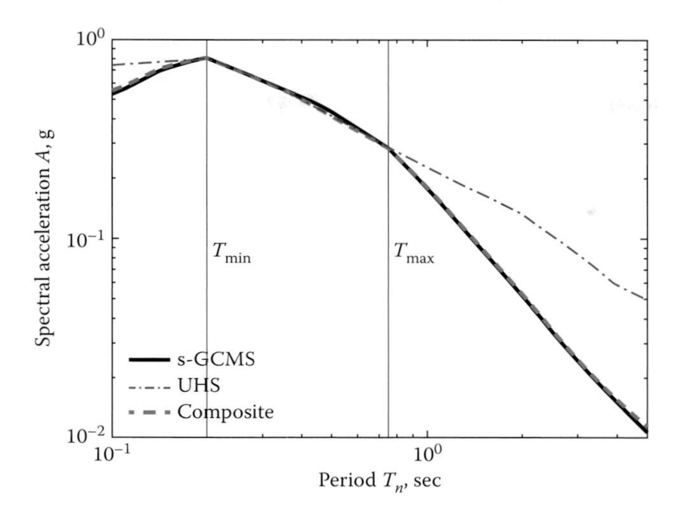

Figure 13.1.8 Comparison of composite spectrum against s-GCMS; the UHS is included for reference.

13.2 GROUND-MOTION SELECTION AND AMPLITUDE SCALING

The simplest goal of ground-motion selection and modification (GMSM) procedures is to estimate the median (or mean) values of seismic demands. For this purpose, we want to select GMs whose response spectra are similar – in some sense – to the target spectrum in amplitude and in shape. The number of recorded GMs that satisfy both requirements simultaneously is often insufficient. For example, the large majority of recorded GMs are weaker than the intensity represented by the UHS and CMS in highly seismic regions. Furthermore, the response spectra for many of the records with the desired intensity $A(T^*)$ may not be similar in shape to the CMS. Given this background, selection of GMs usually proceeds in two stages. First, every record in the database is scaled to make its spectral amplitude(s) similar to the target amplitude(s). Second, the scaled records whose response spectra are similar in shape to the target spectrum over a specified period range are selected.

This period range for selecting GMs should include the vibration periods of all modes that are significant in the response of the structure. For concrete gravity dams, the first two modes of dominantly lateral vibration and the first two modes of dominantly vertical vibration are most significant. If the fundamental vibration of the dam alone (with rigid foundation and empty reservoir) is T_1, the range 0.25 T_1 to T_1 should cover the significant modes. However, the fundamental vibration period may be lengthened by a factor of up to 2.5 because of dam–water–foundation interaction effects (Chapter 4). Thus, the relevant period range is 0.25 T_1 to 2.5 T_1. In the response of arch dams, the first four modes of "symmetric" vibration and the first four modes of "anti-symmetric" vibration are most significant. The periods of the first "symmetric" and first "anti-symmetric" vibration modes of arch dams are generally similar. If the fundamental vibration period of the dam alone is T_1, the range of 0.25 T_1 to T_1 should cover the significant modes. However, dam–water–foundation interaction may lengthen the fundamental vibration period by a factor of up to 2.0 (Chapter 9). Thus, the desirable period range is 0.25 T_1 to 2.0 T_1.

Although GMs, whose response spectra are similar in shape to the target spectrum over the period range 0.20 T_1 to 2.5 T_1, may be appropriate for estimating some EDPs for concrete dams, choosing a wider period range is prudent to ensure good estimates of a wide range of

EDPs; displacements, accelerations, stresses, cracking of concrete, sliding displacements at the base and lift lines, opening and closing of joints, etc. Therefore, we recommend a wide period range of 0.01–10 sec, which has been demonstrated to result in improved estimates of seismic demands for multistory buildings (Chopra 2017).

Three methods to scale GMs are described next. The simplest method is to scale each GM so that its $A(T^*)$ matches $A_{TS}(T^*)$, the target value from the target spectrum (TS), where we recall that T^* is the conditioning period. Thus, the *scale factor* (SF) is the ratio of the target $A_{TS}(T^*)$ and the unscaled GM's $A(T^*)$:

$$\mathrm{SF}_{T^*} = \frac{A_{\mathrm{TS}}(T^*)}{A(T^*)} \tag{13.2.1}$$

An alternative method is to scale each GM so that the average spectral amplitude over the range of periods of interest is equal to the average amplitude of the TS over the same period range; thus, the scale factor is

$$\mathrm{SF}_{\mathrm{avg}} = \frac{\sum_{i=1}^{n_p} A_{\mathrm{TS}}(T_i)}{\sum_{i=1}^{n_p} A(T_i)} \tag{13.2.2}$$

where the periods T_i, $i = 1$ to n_p, span the selected range. The third method is to scale each GM so that the selected SF minimizes the difference between the response spectrum of the scaled GM and the TS over the period range. This difference is described quantitatively as the sum of squared difference (SSD) between the two spectra:

$$\mathrm{SSD} = \sum_{i=1}^{n_p} \left\{\ln\left[A(T_i)\right] - \ln\left[A_{\mathrm{TS}}\left(T_i\right)\right]\right\}^2 \tag{13.2.3}$$

where ln(.) denotes the natural logarithm of the quantity in (.), $A(T_i)$ and $A_{TS}(T_i)$ are the spectral accelerations at period T_i of the scaled GM and of the TS, respectively. This "optimal" SF is given by (Kwong and Chopra 2017):

$$\mathrm{SF}_{\mathrm{opt}} = \left[\prod_{i=1}^{n_p} \frac{A_{\mathrm{TS}}(T_i)}{A(T_i)}\right]^{1/n_p} \tag{13.2.4}$$

where the periods T_i, $i = 1$ to n_p, span the selected range, and Π is the symbol for the product.

Now that all GMs in a large database of records have been scaled appropriately, we identify those GMs that most closely agree with (or match†, for brevity) the TS in shape. For this purpose, the SSD between the response spectrum of each scaled GM and the TS is computed from Eq. (13.2.3) and the GMs with smallest values of SSD are selected. The T_i values should cover the period range identified earlier in this section. Experience suggests 50 values of T_i per order of magnitude of periods are sufficient to ensure a good match between the two spectra. Thus for the period range of 0.05–10 sec, which spans over two orders of magnitude, more than 100 values of T_i should be chosen to cover this range.

Figure 13.2.1 shows the CMS (chosen, for example, as the TS) and the response spectra for 11 scaled GMs selected to match the CMS by the procedure described above. Parts (a) and (b) of the figure are for GMs scaled according to Eqs. (13.2.1) and (13.2.4), respectively. The response spectra of the first set of GMs all pass through $A(T^*)$ of the CMS, as enforced by the scaling criterion of Eq. (13.2.1). In contrast, the response spectra for the second set of GMs vary around $A(T^*)$ because the scaling criterion of Eq. (13.2.4) minimized the difference between the spectra

† The word "match" used here refers to a notion distinct from the concept of "spectral matching" in Section 13.4.

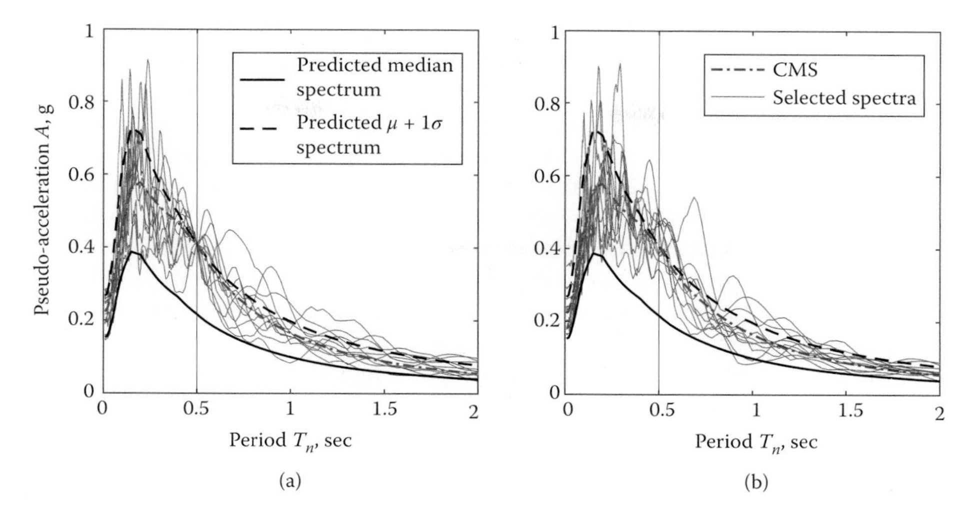

Figure 13.2.1 Response spectra for 11 scaled GMs selected for similarity with the CMS using two scale factors: (a) SF_{T^*} and (b) SF_{opt}; the period range for selection of GMs is 0.05–10 sec (not shown in figure).

over the period range of 0.05–10 sec. As a result, the latter set of GMs achieve a better overall match with the CMS.

Which of the three scale factors provides the best estimates of median seismic demands? Researchers have attempted to answer this question in the context of nonlinear RHA of multistory buildings (Kwong and Chopra 2017). However, such research specific to concrete dams has not been reported.

What is the number of GMs required? In the interest of reducing the number of nonlinear RHAs of the dam–water–foundation system, we would like to know the number of GMs required to estimate the median demands to the desired accuracy and precision. *Accuracy* of an estimate of the median demand is the degree of closeness to its true value. *Precision* of an estimate, related to *repeatability*, is the degree to which repeated estimates – using different sets of n GMs of the same type: CS-based (Section 13.3); amplitude-scaled (Section 13.2); or spectrally-matched (Section 13.4) – show the same results.

The building engineering profession has arrived at a consensus that 11 GMs are adequate to estimate the median demands to the desired accuracy and precision. This conclusion has been based on nonlinear RHAs of a range of buildings (Chopra 2017). However, similar research in the context of concrete dams remains to be accomplished. In the meantime, we recommend the use of 11 GMs.

13.3 GROUND-MOTION SELECTION TO MATCH TARGET SPECTRUM MEAN AND VARIANCE

The preceding approach of selecting GMs to match a target mean (of $\ln A$) spectrum may be sufficient if the objective is to estimate the median values of seismic demands. However, this approach ignores the inherent variance that exists in the response spectrum; recall that variance is equal to the square of standard deviation, which we have deliberately reduced in the GM selection process. As a result, the selected GMs will show smaller than "actual" variance in spectral values and hence smaller dispersion in structural response. Thus, the preceding approach is not sufficient if the goal is a complete risk assessment that leads to the probability distribution of seismic demand.

For this purpose, we introduce the *conditional spectrum* (CS) (Jayaram et al. 2011), which defines both the mean and variance of the (natural logarithm of the) target spectrum and GMs are selected and scaled, as necessary, to match both statistics; the mean spectrum is identical to the CMS. When matching a target response spectrum mean and standard deviation, GMs cannot be selected individually, as described in the preceding section. Required instead are comparisons of both statistics of GM ensembles to their target values. A GM selection algorithm has been developed based on the empirically verified observation that the set of ln A at various periods is a vector of random variables that follows a multivariate normal distribution. The selection algorithm probabilistically generates multiple realizations of the response spectra from the target distribution, and then selects those recorded GMs whose response spectra match the simulated response spectra.†

Application of the aforementioned algorithm is illustrated next for the Pine Flat Dam site. Corresponding to $M = 6.24$, $R = 21$ km, $T^* = 0.5$ sec, and $\varepsilon\ (T^*) = 1.02$, the median and dispersion of the CS – or equivalently the mean and standard deviation of the natural logarithm of the CS – are presented in Figure 13.3.1. The values of $\mu_{\ln A(T_j)}$ and $\sigma_{\ln A(T_j)}$ were obtained from a GMPM. The median and dispersion‡ of the 40 simulated response spectra are computed and compared to the target values of these statistics in Figure 13.3.1, where good agreement is apparent.

Forty GMs are selected from the Next Generation Attenuation (NGA) database to individually match the simulated response spectra. Prior to selection, each of the available 6364 GM records in the database is scaled so that its $A\ (T^*)$ matches the target $A\ (T^*)$ in Figure 13.3.1. In this selection, no constraints on magnitudes and distance of the records considered are imposed, but these can be introduced. The response spectra of the selected motions are shown in Figure 13.3.2 together with the 2.5 and 97.5 percentiles of the CS.

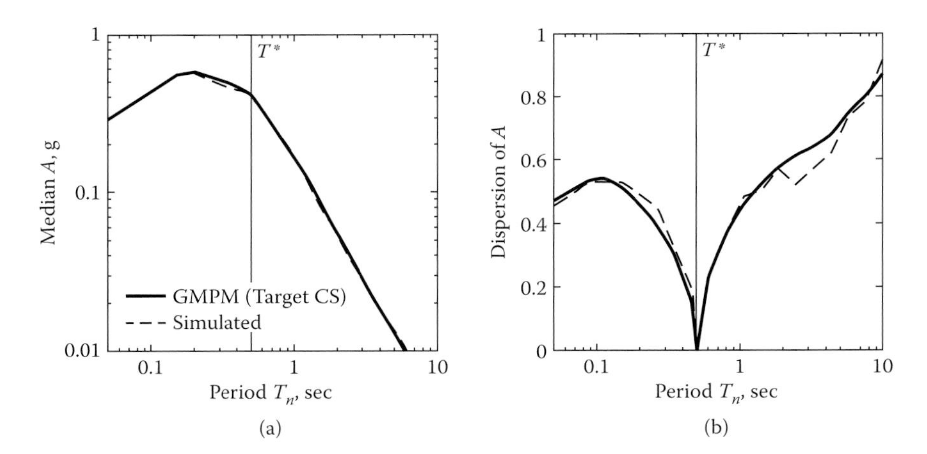

Figure 13.3.1 Comparison of median and dispersion of response spectra from (a) 40 simulated GMs and (b) GMPM.

† MATLAB implementation of the algorithm can be downloaded from http://www.stanford.edu/~bakerjw/gmselection.html.

‡ The median and dispersion of data x_i ($i = 1, 2, \ldots n$) that follows a lognormal probability distribution is defined as

$$\text{Median, } e^{\mu_{\ln x}} = \left(\prod_{i=1}^{n} x_i\right)^{1/n} \quad \text{Dispersion, } \sigma_{\ln x} = \left[\frac{1}{n-1}\sum_{i=1}^{n}(\ln x_i - \mu_{\ln x})^2\right]^{1/2}.$$

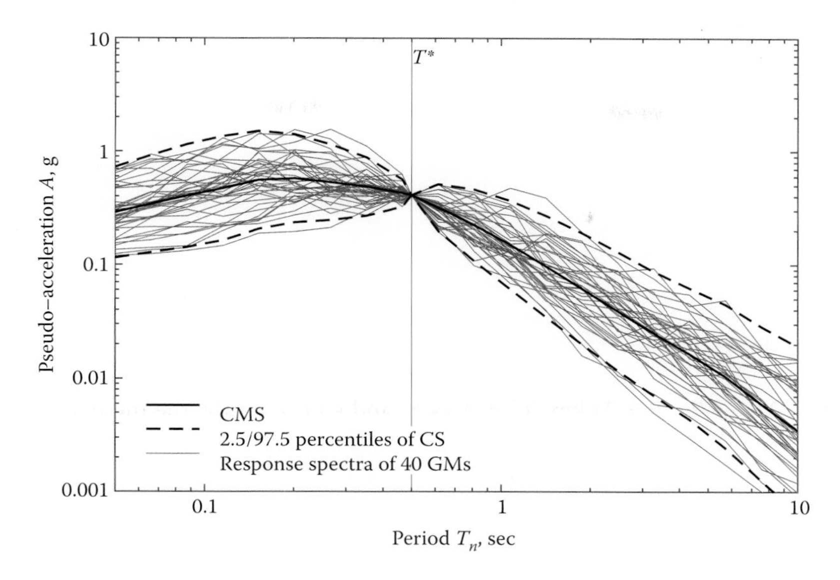

Figure 13.3.2 Fifty, 2.5, and 97.5 percentiles of the CS and response spectra of selected ground motions; CMS is the same as the 50 percentile.

13.4 GROUND-MOTION SELECTION AND SPECTRAL MATCHING

"Spectral matching" refers to the process of modifying a GM, denoted as a seed GM, so that the response spectrum for the modified GM matches the target spectrum closely. This process can be implemented in the frequency domain or the time domain. In the former approach, the Fourier amplitude spectrum of the seed GM is iteratively modified – while keeping the Fourier phase spectrum fixed – until the response spectrum of the modified GM matches the target spectrum. In the latter approach, wavelets are iteratively added to the seed GM in the time domain until its response spectrum matches the target spectrum. Implementation of the time domain approach is facilitated by RspMatch, a freely available software.† In this presentation, we do not include the frequency domain approach for two reasons; the modified GM often does not resemble GMs recorded from earthquakes, and the iterative process may not converge. The theory underlying the time domain method for spectral matching of GMs is available in references at the end of this chapter. This presentation is limited to overall concepts that are illustrated by examples.

The use of ground acceleration histories that are compatible to a design spectrum remains controversial for two reasons. First, the function that matches the entire design spectrum, such as UHS, would not be representative of a single earthquake event for reasons discussed in Section 13.1.2. This problem can be overcome by defining the CMS as the target spectrum, which, as discussed in Section 13.1.3, is representative of recorded GMs. Second, spectrum compatible acceleration-time functions have a smooth response spectrum, whereas recorded GMs display jagged response spectra. This disparity is shown in Figure 13.4.1 where the response spectra for the seed GM and the spectrally matched excitation are compared. Although this disparity or jaggedness may not seem large when spectral ordinates are plotted on logarithmic scale (part a), the true disparity becomes evident on linear scale (part b).

Because several decisions in implementation of the spectral matching method are based on experience and judgment, the method is not foolproof; thus, the spectrally matched GM should

† RspMatch can be downloaded: http://nees.org/resources/rpsmatch09.

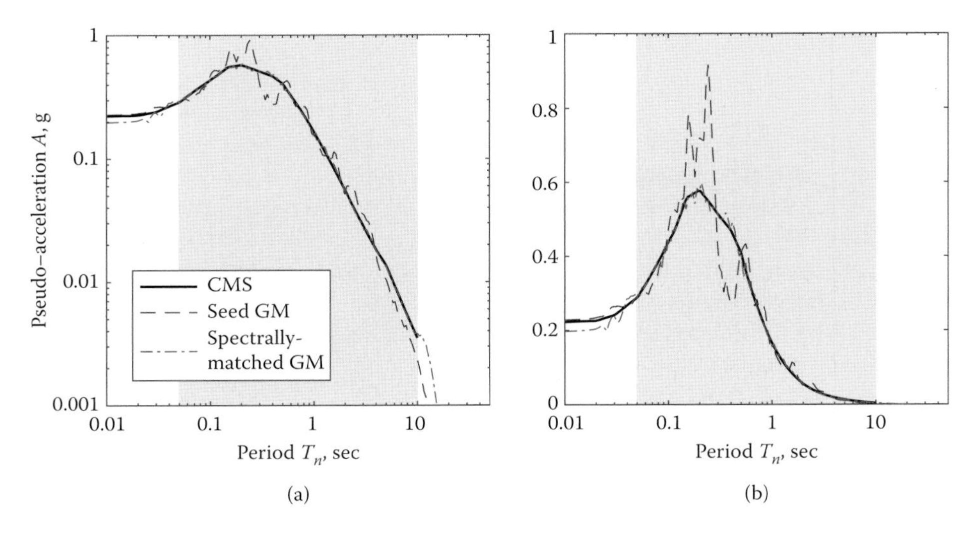

Figure 13.4.1 Conditional mean spectrum; response spectra for a seed GM and spectrally matched GM; period range used for spectral matching is shown in gray: (a) log-log scale and (b) log-linear scale.

be examined carefully. The importance of visual examination is illustrated in Figure 13.4.2, where the acceleration, velocity, and displacement histories for the seed GM and the spectrally matched GM are compared. Note that the disparity in the acceleration functions is large, indicating that the seed GM had to be adjusted greatly. Whether this is acceptable or not depends on the user's point of view. A purist would argue that drastic modification of a recorded GM is unacceptable for geological and seismological reasons. On the other hand, a pragmatist would argue that any degree of modification is acceptable so long as the acceleration, velocity, and displacement functions of the spectrally matched GM are consistent with those that could be observed in a real GM. This requirement is satisfied by the GM on the left part of the figure but not by the one on the right. In the latter case, the displacement function drifts away from the zero position with time, resulting in permanent displacement of the ground that did not exist in the seed GM. Thus, this example of spectral matching would be judged as unsuccessful. Researchers have developed experience on how to adjust the input parameters for RspMatch to achieve a successful result. If such interventions do not achieve the desired result, the particular seed GM is usually discarded.

One of the decisions to be made is the range of periods over which the seed GM is to be spectrally matched to the target spectrum. Spectral matching GMs over a wide-period range is expected to result in improved estimates of median seismic demands with minimal dispersion. However, convergence of the iterative process in spectral matching over a wide-period range is difficult to achieve.

One approach to selecting seed GMs is to ensure that they have the same earthquake source, source-to-site path, and site properties (including soil conditions) as the design earthquake. These properties include earthquake magnitude M, source-to-site distance R, fault style, directivity condition (for sites located close to large faults), and site conditions. Starting from a large database of GM records, the goal is to search for a subset that conforms to the design earthquake. Magnitude and distance are the first two parameters that should be included in defining the search window. To the extent possible, magnitude should be restricted to within 0.25 magnitude units, and the source-to-site distance to within ± 10 km, of the design earthquake. From this reduced database, the GMs recorded on unusually soft soil conditions are deleted. Finally, the desired number of seed GMs are selected randomly from the remaining dataset.

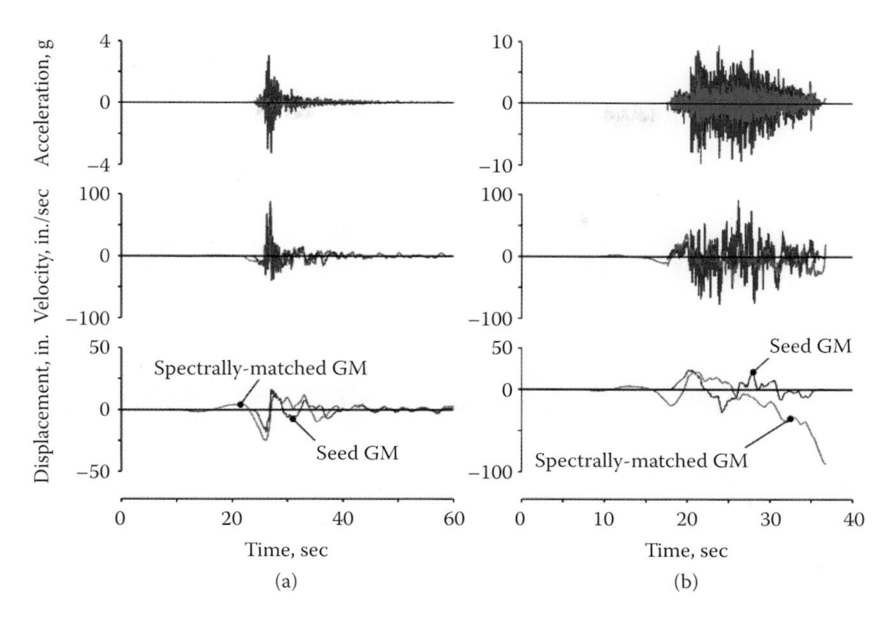

Figure 13.4.2 Ground acceleration, velocity, and displacement histories of a seed GM and spectrally matched GM. Two examples are shown: (a) successful and (b) unsuccessful.

Some researchers prefer to alternatively select seed GMs with spectral shape similar to the target spectrum, say, the CMS, as described in Section 13.1.3, to minimize the modifications necessary to achieve spectral matching.

Spectral matching can be directly implemented on each GM record, but researchers have concluded it should be scaled in amplitude to obtain a rough match with the target spectrum before applying the wavelet adjustments.

For sites close to the earthquake source, it is important that the search process identify records with directivity conditions appropriate for the site and pulse-like features arising from directivity be maintained in spectrally matched GMs.

The first selection approach described above to obtain seed GMs was implemented for the Pine Flat Dam site with the magnitude range restricted to $5.99 < M < 6.49$, and the distance range to $0 < R < 31$ km. Each selected GM was scaled so that its spectral ordinate at 0.5 sec matched that of the target spectrum. The response spectra for the resulting 40 seed GMs are shown in Figure 13.4.3a, all of which pass through $A(T_1)$ of the target spectrum, but there is considerable dispersion in the spectral ordinates at other periods, in some cases the two spectra differ by an order of magnitude. We will refer to this ensemble of GMs as the MR ensemble.

Spectral matching essentially eliminates this dispersion, as shown in Figure 13.4.3b, by the response spectra of the GMs matched over the period range of 0.05–10 sec; the response spectrum for each matched GM is essentially identical to the CMS, which was chosen as the target spectrum. Consequently, spectral matching will essentially eliminate the variability in responses of linearly elastic systems and greatly reduce the dispersion in seismic demands on nonlinear systems.

Consider another ensemble of 40 seed GMs, the CMS ensemble using the scale factor SF_{T_1}. The response spectra for these GMs and for their spectrally matched counterparts are shown in Figure 13.4.4. Because the CMS-based method selects GMs that have spectral shapes similar to the target spectrum, the dispersion in their response spectra is smaller compared to the MR ensemble (compare Figures 13.4.3a and 13.4.4a). After spectral matching, the response spectra

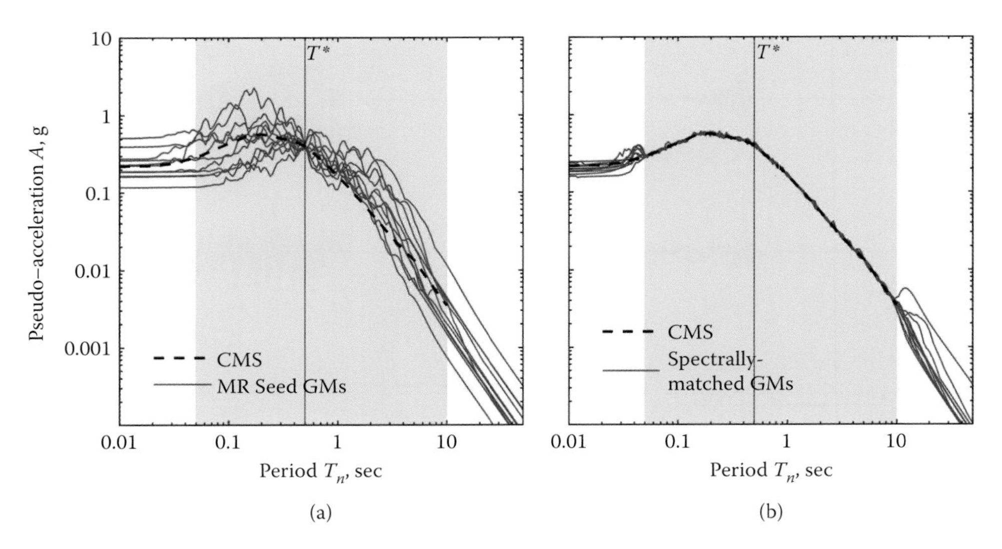

Figure 13.4.3 Response spectra for (a) MR-based ensemble of seed GMs; and (b) spectrally matched GMs; period range of 0.05–10 sec for matching is shown in gray.

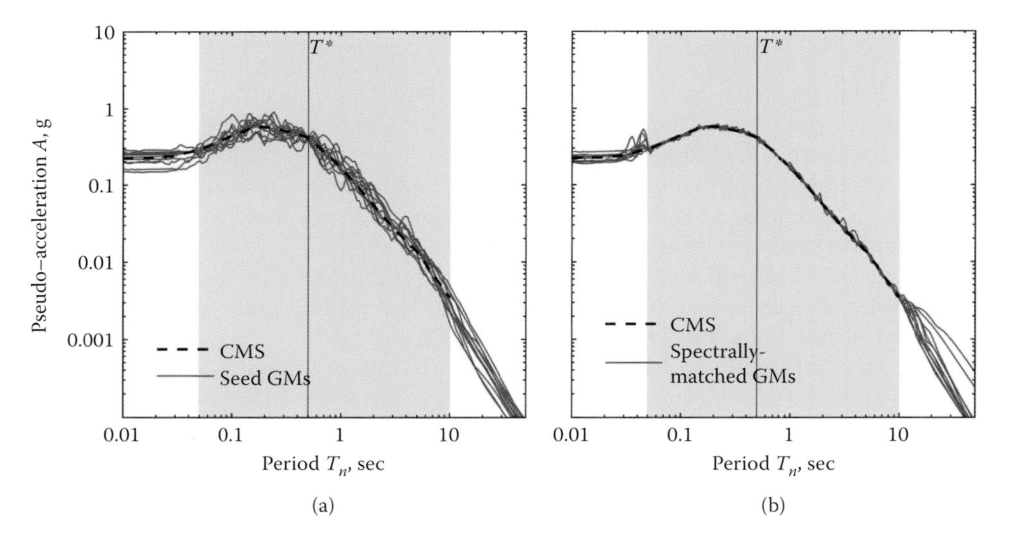

Figure 13.4.4 Response spectra for (a) CMS-based ensemble of seed GMs; and (b) spectrally matched GMs; period range of 0.05–10 sec for matching is shown in gray.

for the ensemble is virtually identical – over the period range of 0.05–10 sec – to the target spectrum. Therefore, we would expect the median seismic demands due to the two ensembles of spectrally matched GMs to be essentially identical for a linearly elastic system. However, they would differ for nonlinear systems because the response of systems responding significantly into the nonlinear range cannot be completely characterized by the response spectrum; other factors such as duration and detailed time variation of the GM are influential.

The expectations stated in the preceding two paragraphs have been confirmed by research studies in the context of nonlinear response of multistory buildings (Chopra 2017). However, similar investigations for concrete dams remain outstanding.

How many spectrally matched GMs are required to accurately determine median values of seismic demands? This issue has been systematically examined in the context of multistory buildings, leading to the conclusion that median demands estimated from nonlinear RHA for 3 GMs are unacceptable, and 11 GMs are adequate to satisfy the requirements of accuracy and precision (Chopra 2017). However, such research focused on concrete dams remains to be done.

13.5 AMPLITUDE SCALING VERSUS SPECTRAL MATCHING OF GROUND MOTIONS

Finally, we comment on the accuracy and precision of the median seismic demands estimated from amplitude-scaled GMs and spectrally matched GMs. It has been demonstrated that amplitude-scaled GMs provide good estimates of median seismic demands for multistory buildings, and accuracy is not enhanced by spectral matching; furthermore, estimates for some seismic demand parameters may be less accurate than those from amplitude-scaled GMs (Chopra 2017). We digress to observe that spectrally matched GMs do not lead to increased accuracy, although the response spectrum of every GM matches the target spectrum, a property that suggests essentially exact results for a linearly elastic system. This is yet another reminder that the response spectrum is an incomplete characterization of the response of nonlinear systems. Because amplitude-scaled GMs provide equally good, if not better, estimates of demands, amplitude scaling is the preferred method, especially because spectral matching is not foolproof; in the hands of an inexperienced user, the results can be unsatisfactory.

PART B: TWO HORIZONTAL COMPONENTS OF GROUND MOTION

13.6 TARGET SPECTRA

Developed in this section are the target spectra $A_x(T_n)$ and $A_y(T_n)$ for the x and y components of GM. Recall that the target spectrum for a single component of GM was presented in Section 13.1.4, where the seismic hazard was defined by spectral accelerations at two conditioning periods $T^* = T_a$ and T_b, henceforth denoted as T_a^* and T_b^*. For two components of GM, the spectral acceleration at one period comes from the x-component, and that for the second period from the y-component. Thus, the seismic hazard is now defined by two spectral accelerations that share a common hazard $A_x(T_a^*)$ and $A_y(T_b^*)$, i.e. both of these values are given by the UHS. Given values of $A_x(T_a^*)$ and $A_y(T_b^*)$, the expected spectral ordinates $A_x(T_n)$ and $A_y(T_n)$ – for both components of GM – at all other vibration periods, T_n, can be determined by extending the s-GCMS theory developed for one component of GM (Kwong and Chopra 2017) to two components.

Presented in Figure 13.6.1 is the UHS corresponding to the return period of 9950 years for the Pine Flat Dam site; $A_x(T_a^* = 0.2 \text{ sec}) = 0.9\,g$ and $A_y(T_b^* = 0.75 \text{ sec}) = 0.32\,g$. Consider the case where the hazard is specified by simultaneous occurrence of these spectral accelerations. By performing disaggregation of this vector-valued seismic hazard, the mean controlling earthquake is defined by $\bar{M} = 6.3$ and $\bar{R} = 15.3$ km. Computed from a GMPM, the predicted median spectrum associated with an earthquake of this magnitude at this distance is also shown in Figure 13.6.1. For this site, the s-GCMS, which now refers to a pair of spectra, was constructed by the methodology developed in Kwong and Chopra (2018). Shown in Figure 13.6.1, the s-GCMS permits several observations. First, in the period range T_a^* to T_b^*, the s-GCMS is nearly as intense as the UHS. Second, the s-GCMS is less intense than the UHS at periods shorter than T_a^* and

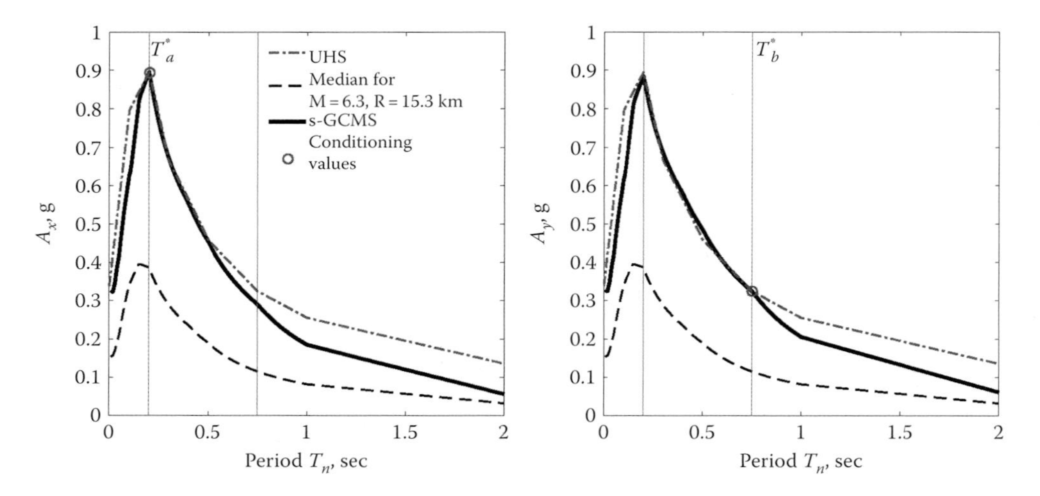

Figure 13.6.1 Comparison of s-GCMS for two components of ground motion against the UHS and corresponding median spectrum for return period of 9950 years at Pine Flat Dam site.

longer than T_b^*. Third, the ordinates of the s-GCMS at vibration periods that are far away from T_a^* and T_b^* tend to approach the median spectrum. Fourth, the s-GCMS for the two horizontal components of GM appear to be similar in shape and amplitude because these components are strongly correlated. We next examine if recorded GMs are consistent with these observations.

Although the probability of occurrence is lower, recorded GMs that naturally have intense values of $A_x(T_a^*)$ and $A_y(T_b^*)$ do, in fact, exist. Figure 13.6.2 shows the response spectra for the two components of an example GM (recorded during an earthquake with magnitude $\bar{M} \pm 0.4$ and distance $\bar{R} \pm 10$ km) that possesses intense spectral accelerations at both conditioning periods [i.e. $A_x(T_a^* = 0.2\ \text{sec}) = 0.88$ g and $A_y(T_b^* = 0.75\ \text{sec}) = 0.35$ g]. For reference, the UHS (from Figure 13.6.1) and two CMSs for conditioning periods T_a^* and T_b^*, respectively, are also shown.

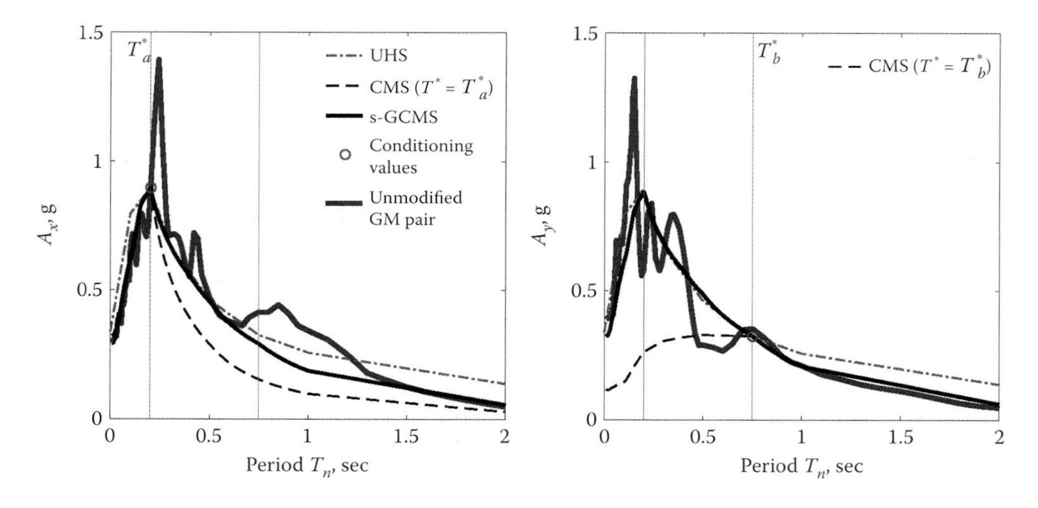

Figure 13.6.2 Comparison of s-GCMS against alternative target spectra – UHS and CMS – and spectra for an unmodified ground motion pair: (a) x-component; and (b) y-component.

The x-component of the example GM is more intense at T_b^* than the CMS with T_a^*; similarly, the y-component of GM is more intense at T_a^* than the CMS with T_b^*. Both components of the example GM are less intense than the UHS at vibration periods much shorter than T_a^* or much longer than T_b^*. Since the s-GCMS is more intense than individual CMSs but less intense than the UHS, it is a desirable target spectrum when intense values for both $A_x(T_a^*)$ and $A_y(T_b^*)$ are of interest.

For far-field sites, the response spectrum of one horizontal component is similar, on average, to the response spectrum of the orthogonal component. This is demonstrated in Figure 13.6.3a where the ratio $A_{H2}(T_n) \div A_{H1}(T_n)$ for all as-recorded ground-motion pairs in the NGA-West2 database with rupture distance greater than 30 km is plotted; note that H1 and H2 denote the axes of the recording accelerograph, which, in general, differ from the structural axes, x and y. For a randomly selected ground-motion pair, the response spectra for the two components of GM may differ significantly in both amplitude and shape; on average, however, the two spectra are very similar, especially in the period range 0–4 sec.

Figure 13.6.3b illustrates that the s-GCMS is consistent with this empirical observation; for a given choice of T_b^* and $T_a^* = T_b^* \div 3$, the s-GCMS for the two horizontal components of GM are nearly the same in the period range 0–0.5 sec, which covers vibration periods of concrete dams. Because the differences between the two spectra are less than 8%, it seems appropriate to develop a single target spectrum that is applicable to both components.

Although the s-GCMS appears to be consistent with recorded GMs, for several reasons, it is not convenient for practical applications (Kwong and Chopra 2018). First, vector-valued disaggregation for two horizontal components of GM is required to determine the controlling rupture scenario, which, in turn, is required to determine the median spectra. Second, the logarithmic standard deviation of $A_x(T_a^*)$ and $A_y(T_b^*)$ is needed to construct the s-GCMS, but most GMPMs do not directly provide this quantity. Third, many different correlation coefficients must be calculated to determine the generalized epsilon required in the equation to determine s-GCMS (Kwong and Chopra 2018). Fourth, there is no rational way of assigning a particular conditioning period, T_a^* or T_b^*, to a specific H component, because the natural vibration modes of a three-dimensional structural system are not limited to motion along any one structural axes.

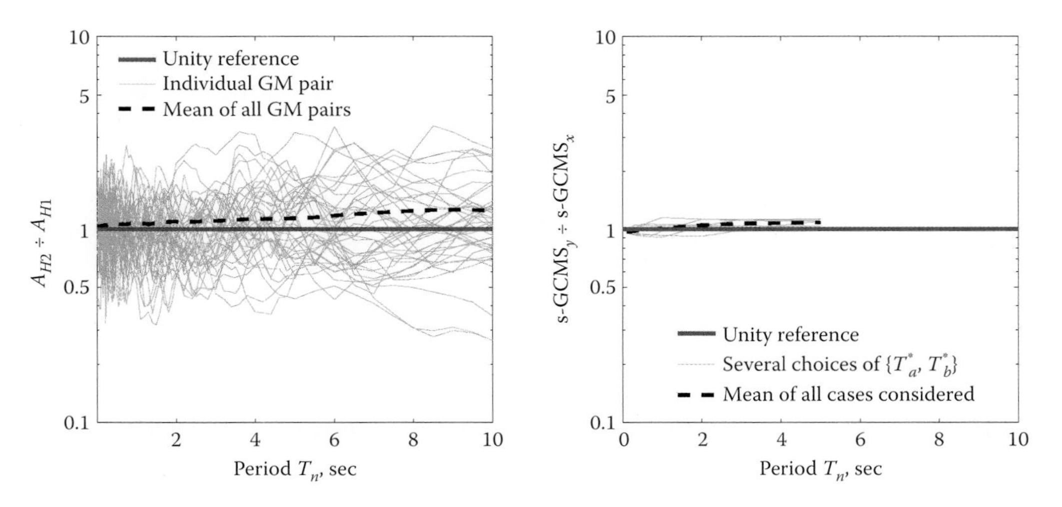

Figure 13.6.3 Ratios of response spectra for ground motion in two orthogonal directions: (a) recorded ground motion pair; and (b) several s-GCMSs, each with $T_a^* = T_b^*/3$.

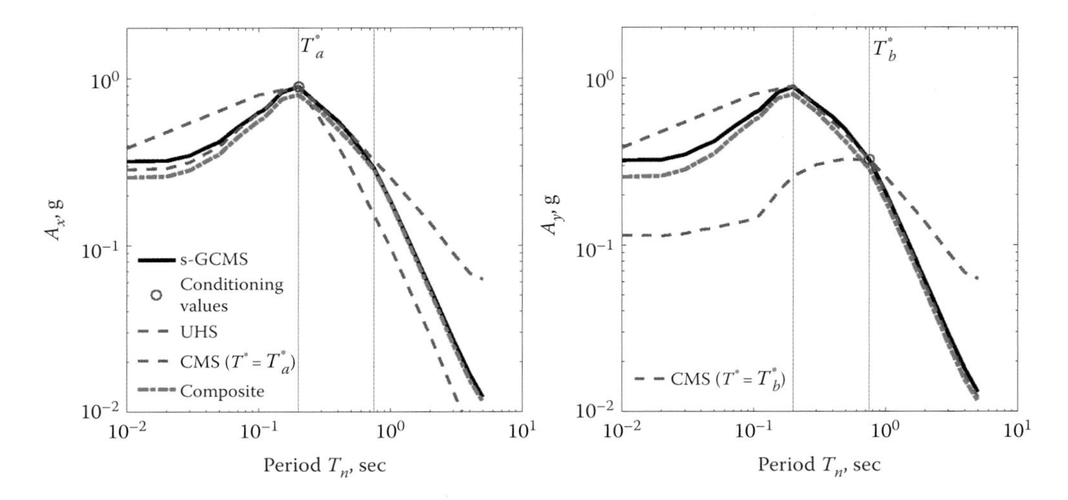

Figure 13.6.4 Example of a CMS–UHS composite spectrum: (a) x-component, and (b) y-component.

In order to overcome these drawbacks without sacrificing the benefits of s-GCMS, an alternative target spectrum is presented next.

Comparison of the UHS, s-GCMS, and individual CMSs (Figure 13.6.4) shows that at periods between T_a^* and T_b^*, the s-GCMS is nearly as intense as the UHS but outside this range it is less intense than the UHS. At periods shorter than T_a^*, it is very similar to the CMS for $T^* = T_a^*$, whereas at periods longer than T_b^*, it is very similar to the CMS for $T^* = T_b^*$. Finally, Figure 13.6.3b suggests that the target spectra along the two axes are very similar.

For most practical applications, specifying only two conditioning periods should suffice so long as the period range defined by T_a^* and T_b^* includes the structural periods of interest. As can be seen in Figure 13.6.4, the ordinates of the s-GCMS at vibration periods between the two T^* are very close to the UHS ordinates even though the periods in this range were not utilized in constructing the target spectra. Hence, if intense values of spectral acceleration at several periods are desired, then it should be adequate to choose T_a^* as $T_{\min}$ (the shortest) and T_b^* as $T_{\max}$ (the longest) among the n periods.

Based on the preceding observations, the s-GCMS is approximated by a pair of identical spectra that combine features of both the CMS and the UHS. Formally, the CMS-UHS composite spectrum is defined as

$$A_{\text{Composite}}\left(T_n\right) = \begin{cases} A_{\text{CMS}}\left(T^* = T_{\min}\right) & T_n \leq T_{\min} \\ A_{\text{UHS}} & T_{\min} < T_n < T_{\max} \\ A_{\text{CMS}}\left(T^* = T_{\max}\right) & T_n \geq T_{\max} \end{cases} \tag{13.6.1}$$

This composite spectrum is essentially identical to the rigorously derived s-GCMS, as shown in Figure 13.6.4.

13.7 SELECTION, SCALING, AND ORIENTATION OF GROUND-MOTION COMPONENTS

The third of three scaling methods for a single component of GM presented in Section 13.2 is extended to two horizontal components of GM. In particular, the SSD metric of Eq. (13.2.3) for quantifying the misfit between the GM spectrum and the target spectrum (TS) [Eq. (13.2.3)] is generalized to two components:

$$\mathrm{SSD}_{xy} = \sum_{j=1}^{2} \sum_{i=1}^{n_p} \left\{ \ln \left[A_j \left(T_i \right) \right] - \ln \left[A_{\mathrm{TS},j} \left(T_i \right) \right] \right\}^2 \tag{13.7.1}$$

where SSD denotes the sum of the squared differences; periods T_i, $i = 1$ to n_p, span the selected range, 0.01–10 sec; $A_j\,(T_i)$ denotes the spectral accelerations at period T_i of the two components ($j = 1,2$) of the scaled GM pair; and $A_{\mathrm{TS},j}\,(T_i)$ denotes the target spectral ordinate at period T_i along component j.

To preserve the relative intensity of the two horizontal components in the original recorded GM pair, both components are scaled by the same scale factor (SF). The SF that minimizes SSD_{xy} is given in Eq. (13.7.2a) (Kwong and Chopra 2018):†

$$\mathrm{SF}_{\mathrm{opt}} = \left[\prod_{i=1}^{n_p} \frac{A_{\mathrm{TS},x}\left(T_i\right) \cdot A_{\mathrm{TS},y}\left(T_i\right)}{A_x\left(T_i\right) \cdot A_y\left(T_i\right)} \right]^{1/\left(2n_p\right)} \qquad \mathrm{SF}_{\mathrm{opt}} = \left[\frac{A_{\mathrm{TS},x}\left(T_a^*\right)}{A_x\left(T_a^*\right)} \cdot \frac{A_{\mathrm{TS},y}\left(T_b^*\right)}{A_y\left(T_b^*\right)} \right]^{1/2} \tag{13.7.2a,b}$$

For the special case where it is important to minimize the misfit only at $A_x(T_a^*)$ and $A_y(T_b^*)$, the SF in Eq. (13.7.2a) reduces to Eq. (13.7.2b). After the SF for each GM pair in a given database has been determined, prospective GMs that require unreasonably large SF, say, larger than 4, can be eliminated.

Recorded along the axes H1 and H2 of the accelerograph (Figure 13.7.1a), the selected GM pair is one of many ways to describe the ground shaking at the site. Since the recording accelerograph could have been oriented in a different azimuth, all transformed (to a different set of orthogonal axes) versions of the as-recorded GM are plausible realizations of the GM at the site from the earthquake event. This implies that in selecting recorded GMs for RHA of 3D models of dams, ideally both a unique scale factor and a unique azimuth should be determined for each GM pair. The optimal azimuth is the value that leads to the best agreement between the spectra of the transformed GM and the target spectra. First, a discrete set of azimuths is specified (periodicity of 180°). For each azimuth, the scaled GM pair is transformed and the corresponding

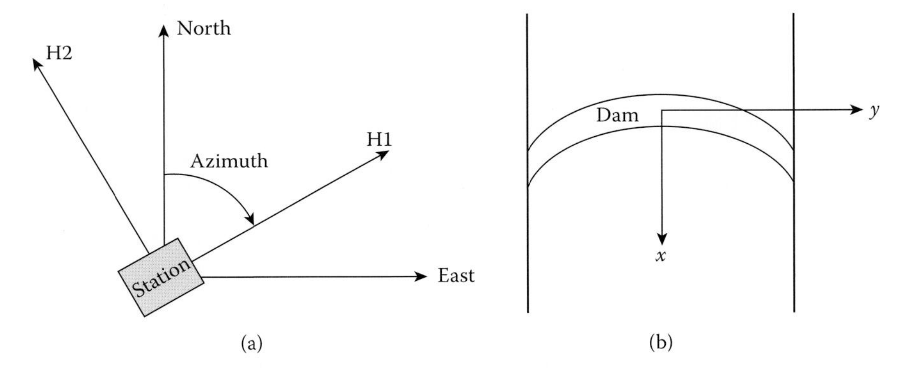

Figure 13.7.1 (a) Example of recording accelerograph with as-recorded components H1 and H2 identified; and (b) plan view (schematic) of dam with structural axes shown.

† Taking the natural logarithm of Eq. (13.7.1) to replace the product by a summation offers a computationally convenient way to compute $\mathrm{SF}_{\mathrm{opt}}$.

value of SSD_{xy} is computed. The azimuth associated with the smallest value of SSD_{xy} is defined as the optimal azimuth.

For dynamic analyses of the dam, ground acceleration histories are specified along the x- and y-axes of a 3D FE model of the dam (Figure 13.7.1b). The "best" choice would be the GM components along the direction defined by the optimal azimuth and the orthogonal direction. Alternatively, the as-recorded GM components H1 and H2 may be applied along the structural axes. The latter choice avoids calculation of the optimal azimuth, with the implication that the GM will not match as well with the target spectrum. Observe that in both cases, the selected pair of horizontal components is applied directly along the structural axes without any transformation.

Spectral matching of GMs offers an alternative that is conceptually much simpler than amplitude-scaling, especially for multi-component GMs. In this approach, first introduced in Section 13.4 for one component of GM, the response spectra of both horizontal components of a GM record are matched to the pair of target spectra (Grant 2011). Although conceptually simpler, spectral matching is not foolproof; in the hands of an inexperienced user, the results can be unsatisfactory.

PART C: THREE COMPONENTS OF GROUND MOTION

13.8 TARGET SPECTRA AND GROUND-MOTION SELECTION

A methodology to develop target spectra for two horizontal components of GM, and then to select, scale, and orient GMs to match the target spectra was presented in Sections 13.6 and 13.7. This methodology has been extended to three components of GM. Derived was a theory to develop the target spectrum for the vertical component of the GM and the earlier theory for selecting scaling and orienting horizontal components was extended to consider simultaneously the vertical component (Kwong and Chopra 2019). This methodology requires the user to specify the relative importance of the horizontal and vertical components of GM in the response of the structure.

Since horizontal GM is usually much more important in the response of concrete dams, a simpler approach is to first select these two components without considering the vertical component. The target spectrum for the horizontal components of the GM defined by Eq. (13.7.1) remains valid, and a target spectrum for the vertical component is not required. The vertical component accompanying a selected record is scaled by the same SF that was applied to the horizontal components. It has been demonstrated that this simpler approach for defining the vertical component of GM provides satisfactory results so long as GMs are matched to the target spectra by calculating SSD_{xy} [Eq. (13.7.1)] over a wide range of periods (Kwong and Chopra 2019).

14

Application of Dynamic Analysis to Evaluate Existing Dams and Design New Dams

PREVIEW

Methods for dynamic analysis of concrete dams, including effects of dam–water–foundation interaction, gradually found their way into practical applications. They have been utilized in seismic design of new dams and in seismic safety evaluation of existing dams. Conducted by government agencies and private companies, these investigations are usually not disseminated in the open literature. As a result, the scope of this chapter is limited to an overview of the dynamic analyses employed in investigations for four dams: two gravity dams and two arch dams. The first category includes the seismic evaluation of Folsom Dam (1989)[†] and seismic design of Olivenhain Dam (2003). Examples in the second category are seismic evaluation of Hoover Dam (1998) and seismic design of Dagangshan Dam (2015).

14.1 SEISMIC EVALUATION OF FOLSOM DAM

Located near Sacramento, California, Folsom Dam is a 340-ft-high gravity dam with a crest length of 1400 ft (Figure 14.1.1), consisting of 28 50-ft-wide monoliths. Designed by the Corps of Engineers using the traditional procedure outlined in Section 1.3 with a seismic coefficient of 0.05, the dam was completed in 1956. The cross section of the tallest non-overflow monolith is shown in Figure 14.1.2.

Seismic evaluation of the dam was conducted in the late 1980s (Hall et al. 1989). The earthquake for which the dam should be evaluated was selected without conducting a probabilistic seismic hazard analysis (PSHA). The maximum credible earthquake (MCE) affecting this

[†] Denotes year the investigation was completed.

Figure 14.1.1 Folsom Dam.

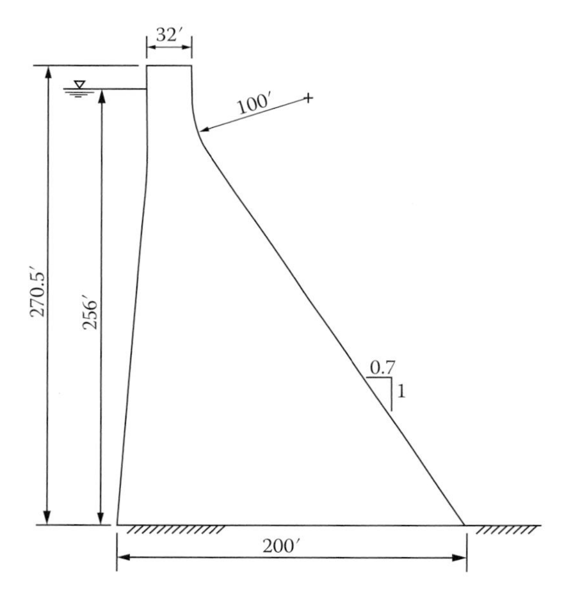

Figure 14.1.2 Cross section of the tallest non-overflow monolith.

dam was established as an earthquake of local magnitude 6.5 at a distance of about 10 miles on the East Branch of the Bear Mountain fault zone. Parameters for the horizontal component of ground motion (GM) were established as peak ground acceleration (PGA) = 0.35 g, peak ground velocity (PGV) = 20 cm/s, and bracketed duration (accelerations exceeding 0.05 g) = 16 sec. The response spectra for two selected ground motions consistent with these parameters are shown in Figure 14.1.3, where the fundamental period of the dam is noted. Corresponding vertical components of ground motions were defined by compressing the time scale of the horizontal components by a factor of 1.5 and scaling down the accelerations by a factor of 0.6.

A comprehensive program of testing concrete cores (some cores were 6 in. in diameter whereas others were 12 in. in diameter) led to the following properties for use in dynamic analysis

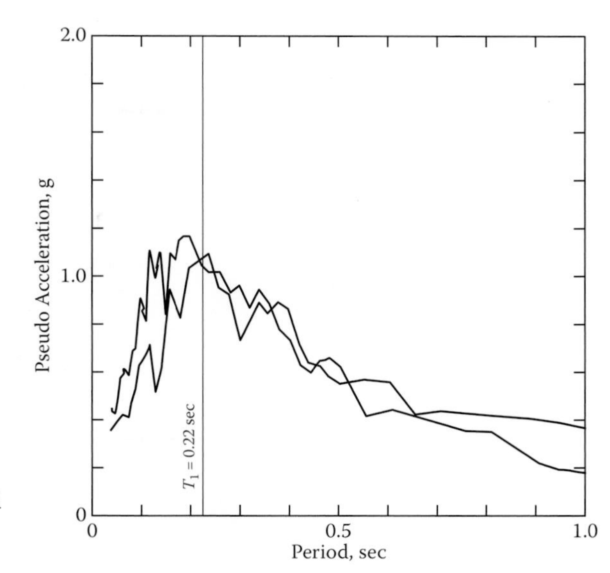

Figure 14.1.3 Pseudo-acceleration response spectra for horizontal components of selected ground motions. Source: Adapted from Hall et al. (1989).

of the dam: modulus of elasticity = 5.9×10^6 psi; Poisson's ratio = 0.19; unit weight = 158 pcf; tensile strength = 700 psi for lean concrete, and 840 psi for rich concrete provided over a thickness of 2–10 ft along the upstream and downstream faces of the dam. Based on field investigations, experience, and well-known empirical equations, properties for rock were estimated for use in dynamic analysis of the dam as modulus of elasticity = 7.9×10^6 psi, Poisson's ratio = 0.25, and unit weight = 171 pcf. Because of uncertainty in the estimates, lower and upper bound values for the modulus of elasticity of rock were selected as 5.8×10^6 psi and 11.0×10^6 psi.

Energy dissipation in the dam concrete and foundation rock was modeled as constant hysteretic damping with damping factors $\eta_s = 0.1$ and $\eta_f = 0.1$, respectively, which correspond to 5% viscous damping ratio, individually for the two substructures – dam and foundation. The wave reflection coefficient at the reservoir bottom was selected as 0.9. These damping values for the two substructures would combine to provide damping in the overall dam–water–foundation system that is much larger than damping values measured from field tests on several dams, as discussed in Section 11.3. As a result, in retrospect, the dynamic response to the selected ground motions was underestimated.

Using the EAGD-84 computer software, the earthquake response analysis of the dam–water–foundation system resulted in the envelope values of the maximum principal stresses presented in Figure 14.1.4; initial static stresses were included. The largest tensile stress of 871 psi occurs on the downstream face near the elevation where its slope changes; this value exceeds slightly the estimated tensile strength of 840 psi for the rich concrete in this region. Over a small region penetrating only a few feet from the downstream face, the stresses exceed 700 psi, the tensile strength of lean concrete, as surmised from the distribution of instantaneous stresses at the time instant when the largest stress value is attained (Figure 14.1.5). A plot of the stress history (Figure 14.1.6) indicates that the tensile strength is exceeded only once for less than 1/20th of a second. Therefore, any cracking will be very limited in extent as well as depth. Thus, it was concluded that the dam need not be retrofitted.

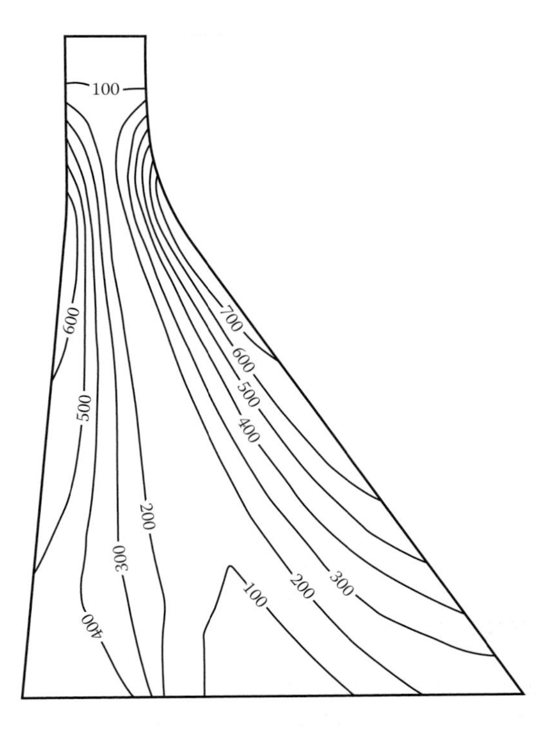

Figure 14.1.4 Envelope values of maximum principal stresses; initial static stresses are included. Source: Adapted from Hall et al. (1989). Stresses over a small zone near the downstream face exceeded 800 psi, but this contour was missing in the original figure.

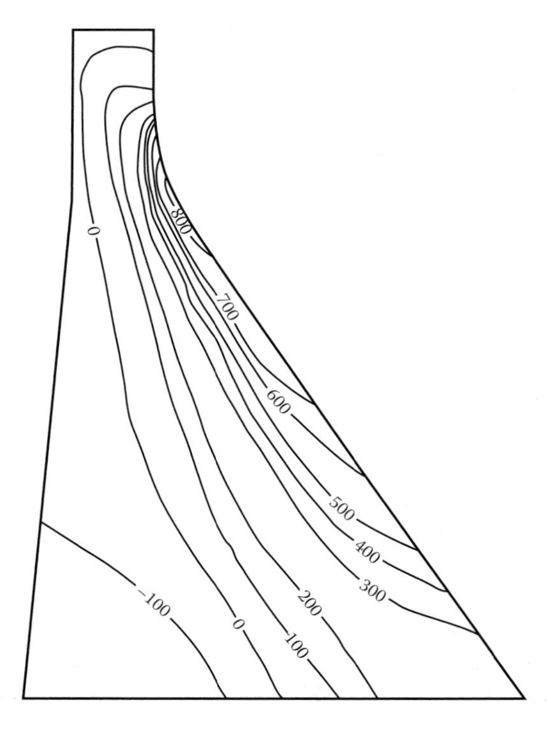

Figure 14.1.5 Maximum principal stresses at 4.69 sec after start of ground motion. Source: Adapted from Hall et al. (1989).

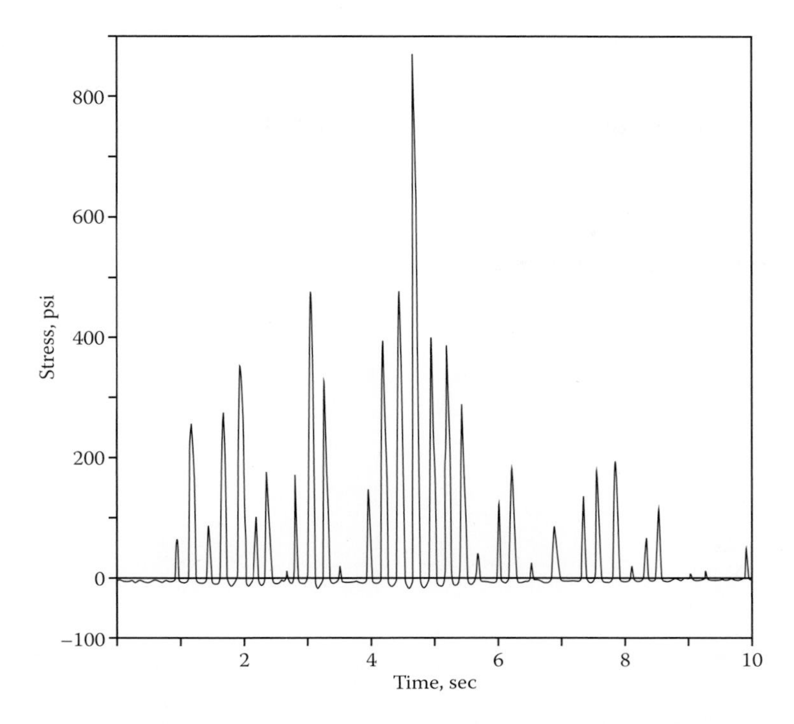

Figure 14.1.6 Variation of maximum principal stress in finite element 120 with time. Source: Adapted from Hall et al. (1989).

14.2 SEISMIC DESIGN OF OLIVENHAIN DAM

Located in San Diego County in California, Olivenhain Dam is a 318-ft-high roller-compacted concrete (RCC) dam with a crest length of 2552 ft (Figure 14.2.1). Designed by Parsons-Harza using modern dynamic analysis procedures, the dam was completed in 2003. The summary presented next is based on technical memoranda prepared by Parsons-Harza (2000a,b).

Figure 14.2.1 Olivenhain Dam, California. Source: Courtesy of Kleinfelder.

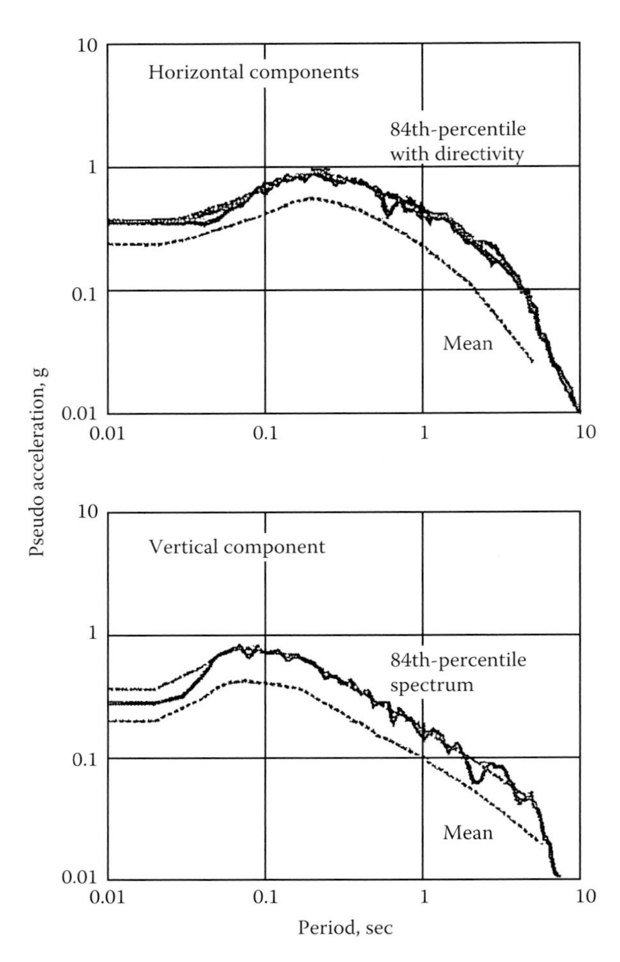

Figure 14.2.2 Design spectra, mean and 84th percentile, and response spectrum for a spectrally matched ground motion (Adapted from Parsons-Harza 2000a).

Design earthquakes were selected without conducting a PSHA. They were chosen as the MCEs for each of two faults: Magnitude $M_w = 7.5$ at a distance of 18 km for the Rose Canyon fault; and $M_w = 7.25$ at a distance of 35 km on the Elsinore fault. For each scenario earthquake, the target design spectrum was selected as the 84th-percentile spectrum determined from ground motion prediction models (GMPMs) corresponding to the estimated magnitude and distance of the design earthquake. The mean and 84th-percentile design spectra for horizontal and vertical components of ground motion are shown in Figure 14.2.2, together with the response spectrum for a recorded ground motion modified to match the target spectrum. In addition to the spectrally matched motion, a simulated motion was developed using a finite-fault composite source model. Response spectra for the simulated motions are compared against the 50th- and 84th-percentile design spectra in Figure 14.2.3.

Preliminary design of the dam using traditional procedures with a seismic coefficient of 0.2 led to the cross section shown in Figure 14.2.4.

Based on data obtained from testing of concrete cores, the following properties were selected for use in dynamic analysis of the dam: modulus of elasticity = 3.9×10^6 psi; Poisson's ratio = 0.20; and unit weight = 150 pcf. Foundation rock properties were selected as follows: modulus of elasticity = 2.6×10^6 psi; Poisson's ratio = 0.33; and unit weight = 62.4 pcf. The

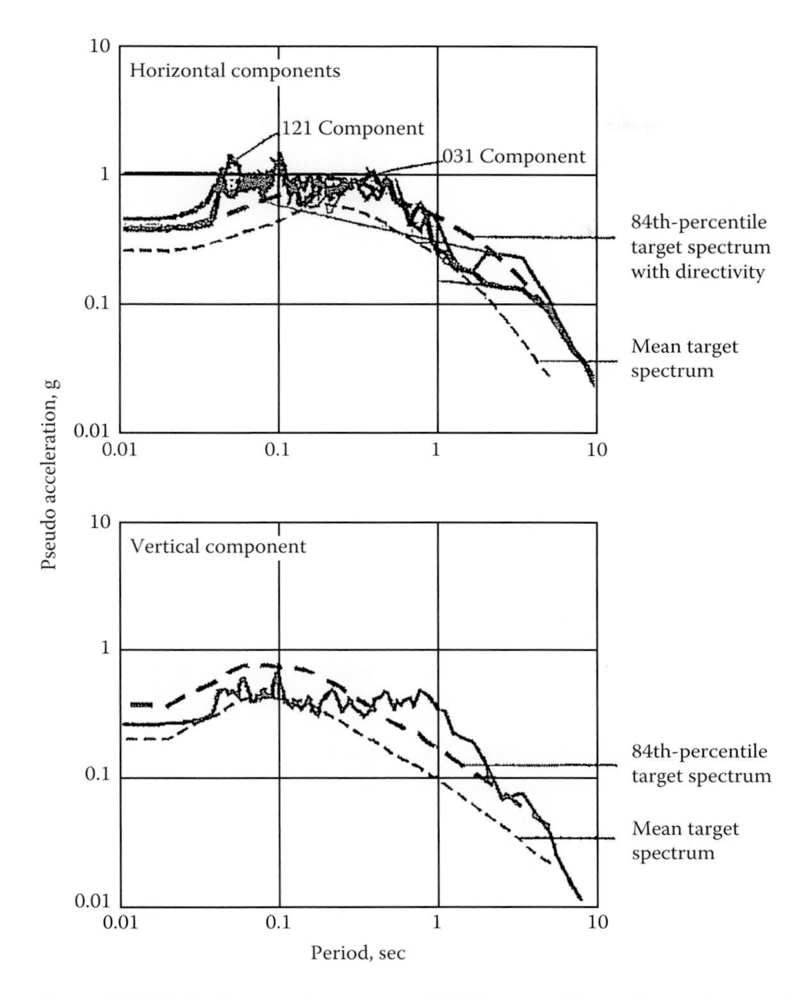

Figure 14.2.3 Design spectra, mean and 84th percentile, and response spectrum for a simulated motion developed using a finite-fault composite source model (Adapted from Parson-Harza 2000a).

dynamic compressive strength $f_c' = 4500\,\text{psi}$, which includes a 50% increase over the design strength specified for the project. Without any direct tension or splitting tension testing of concrete, the dynamic tensile strength was assumed to be 10% of f_c'.

Energy dissipation in the dam concrete and foundation rock was modeled as constant hysteretic damping with damping factors $\eta_s = 15\%$ and $\eta_f = 10\%$, which correspond to viscous damping ratio 7.5% and 5%, individually for the two substructures. There damping values for the two substructures would combine to provide excessive damping in the overall dam–water–foundation system that is much larger than the damping values measured from field tests of several dams, as discussed in Section 10.3. As a result, in retrospect, the dynamic response to the selected ground motions was underestimated.

Using the EAGD-84 computer program, earthquake response analysis of the dam–water–foundation system assuming plane strain conditions, led to the response results presented in Table 14.2.1 and Figures 14.2.5–14.2.7, initial static responses are included.

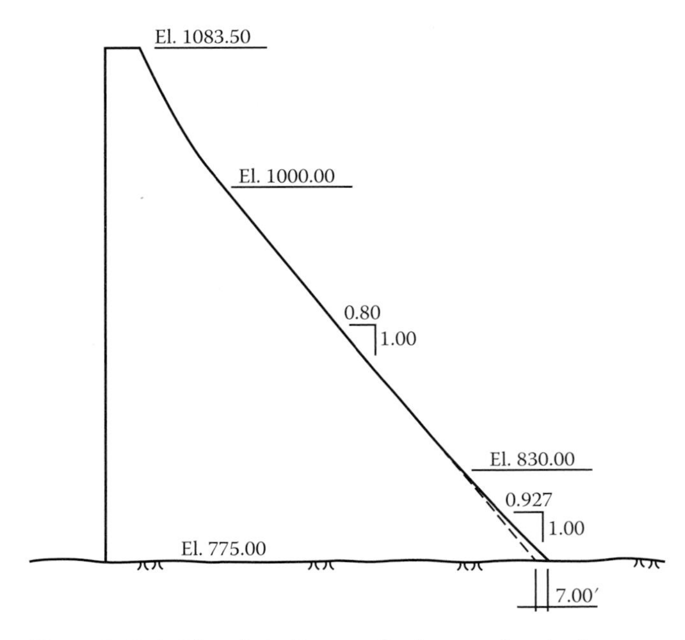

Figure 14.2.4 Olivenhain Dam: preliminary and revised cross section (Adapted from Parsons-Harza 2000b).

Table 14.2.1 Peak values of selected stresses (psi) during three ground motions (GMs).

	Rose Canyon GM	Loma Prieta GM	Landers GM
Vertical (tensile) stress at base	106	301	132
Maximum principal (tensile) stress	311	414	313
Minimum principal (compressive) stress	803	1012	824

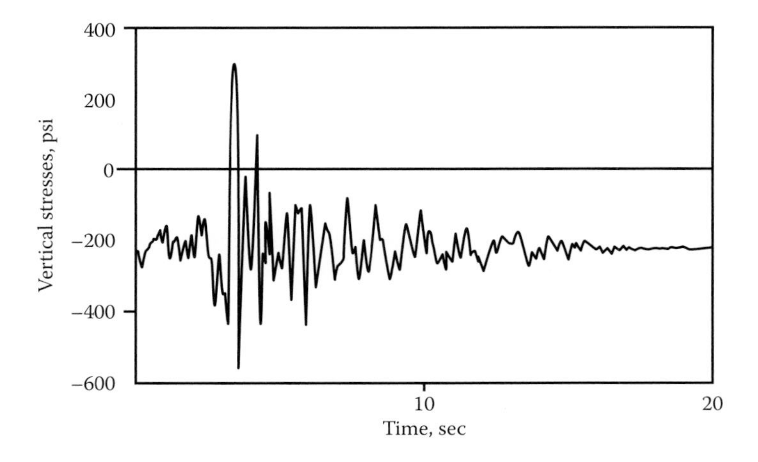

Figure 14.2.5 Vertical stress history at the location of largest stress on the dam base (Adapted from Parsons-Harza 2000b).

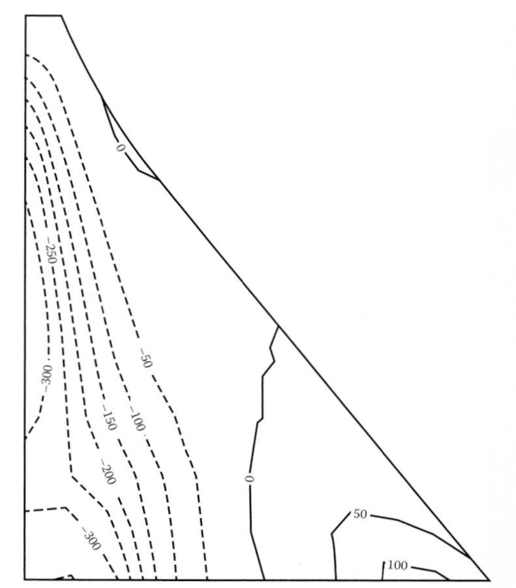

Figure 14.2.6 Maximum principal stresses in psi at the time instant of peak tensile stress at the base; contours in solid and dashed lines denote compression and tension, respectively (Adapted from Parsons-Harza 2000b).

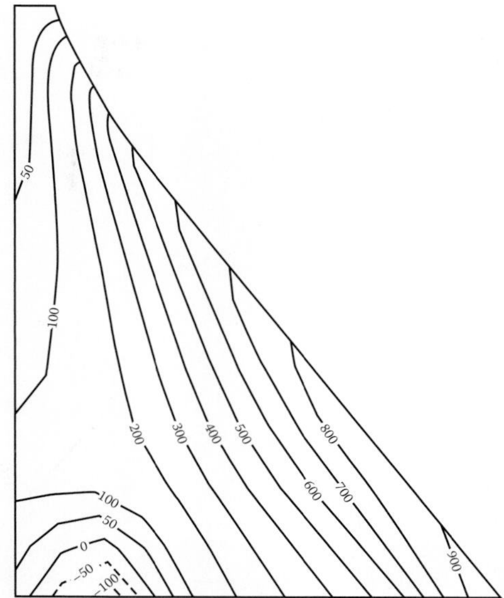

Figure 14.2.7 Minimum principal stresses in psi at the time instant of peak tensile stress at the base; contours in solid and dashed lines denote compression and tension, respectively (Adapted from Parsons-Harza 2000b).

Among the three excitations, the largest stresses are caused by the Loma Prieta ground motion. Caused by this excitation, the history of vertical stress at the location on the dam base where the largest stress occurs is presented in Figure 14.2.5. The distribution of maximum principal stresses over the dam cross section at the time instant when the peak value of vertical tensile occurs are shown in Figure 14.2.6 and corresponding results for the minimum principal stresses in Figure 14.2.7.

The peak value of maximum principal stress for all three ground motions is less than the estimated tensile strength, 450 psi, of the RCC. The peak value of minimum principal stress is well below the estimated compressive strength of 4500 psi. If the tensile strength at horizontal lift joints, which was not reported, is less than that of the parent RCC, they may open and close at the upstream and downstream edges, sequentially over a cycle of motion. The dam designers concluded that the dam will be able to withstand the MCE without permitting a sudden, uncontrolled release of impounded water.

14.3 SEISMIC EVALUATION OF HOOVER DAM

Starting in 1996, the U.S. Bureau of Reclamation embarked on a major program to evaluate the seismic safety of their dams. Among the several dams investigated was the iconic Hoover Dam, completed in 1935. Located on the border of the states of Arizona and Nevada in the U.S., it is a 725-ft (221-m)-high dam, curved in plan, with a crest thickness of 44 ft and base width of 660 ft (Figures 14.3.1 and 14.3.2).

Geologic and seismologic investigations determined the MCE as a magnitude $M_s = 6.75$ earthquake on the Mead Slope fault, centered 1.5 miles from the dam. PSHA conducted in 1993

Figure 14.3.1 Hoover Dam. Source: Arnkjell Løkke.

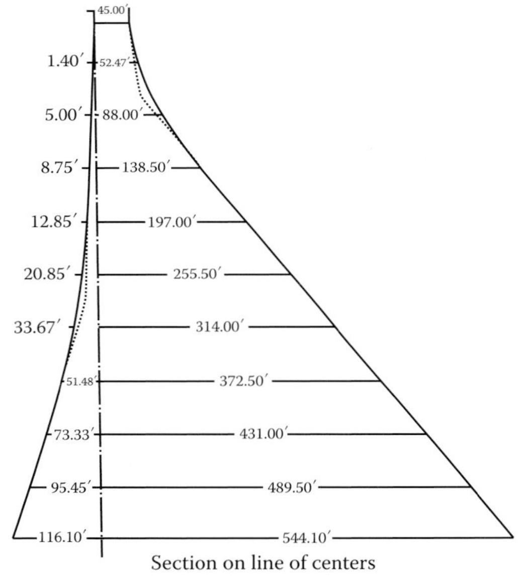

Figure 14.3.2 Cross section of Hoover Dam.

established the uniform hazard spectra for a return period of 50 000 years, associated with the Mead Slope fault (Figure 14.3.3).

For response history analysis (RHA) of the dam, the ground motion recorded at Corralitos, California, during the 1989 $M_w = 7.0$ Loma Prieta earthquake was modified to include a velocity pulse associated with near-fault effects. The response spectra for two components of this excitation exceeded the uniform hazard spectrum for a 50,000-year return period over the period range of 0.1–0.3 sec, which includes the important vibration periods of the dam (Figure 14.3.3).

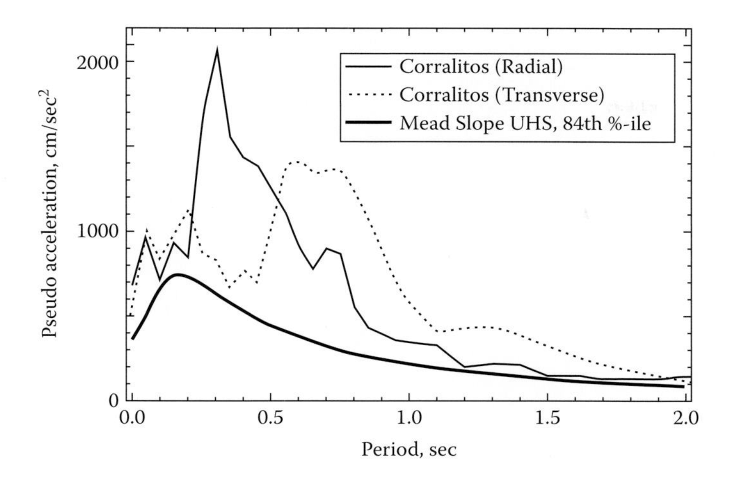

Figure 14.3.3 Uniform hazard spectrum for a 50,000-year-return period and response spectra of selected ground motions (Adapted from Bureau of Reclamation 1998d).

Based, in part, on data from testing of 6-in-diameter cores, the properties of concrete chosen for use in dynamic analysis of the dam were modulus of elasticity $E_s = 6.875 \times 10^6$ psi, Poisson's ratio = 0.20, and unit weight = 156 lb/cu ft. Laboratory tests and visual examination of the cores indicated that the intact concrete was very strong, resulting in an estimated tensile strength = 970 psi; the lift lines were bonded so well that they were difficult to locate in the cores. Properties for foundation rock were specified as follows: $E_f = 3.1 \times 10^6$ psi; Poisson's ratio = 0.26; and unit weight = 139 lb/cu ft; note that $E_f/E_s = 0.45$. For such a small value of E_f/E_s the results presented in Chapter 9 would suggest that dam–foundation interaction effects should be especially significant.

Energy dissipation in the dam concrete and foundation rock was modeled as constant hysteretic damping with damping factors $\eta_s = 0.1$ and $\eta_f = 0.1$, respectively, which correspond to 5% viscous damping ratio, individually for the two substructures – dam and foundation. The wave reflection coefficient at the reservoir bottom was selected as 0.9. These damping values for the two substructures would combine to provide overall damping in the dam–water–foundation system that is much larger than damping values measured from field tests on many dams (see Section 10.3). As a result, in retrospect, the dynamic response of the dam to the selected GMs was underestimated.

Initially, the popular finite element analysis procedure (summarized in Section 1.6) was employed wherein the mass of the foundation rock was ignored and hydrodynamic effects were modeled by an added mass of water moving with the dam, i.e. water compressibility was ignored. This analysis was implemented in SAP IV, a commercial finite-element analysis program. The aforementioned simplifications in modeling were necessitated by the limitations of SAP IV, which were common to most typical finite element analysis programs of that time.

This linear analysis led to very large tensile stresses, especially at the elevation 200 ft below the dam crest (Figure 14.3.4b); arch stresses on both upstream and downstream faces exceeded the tensile strength of concrete of 970 psi by a factor of 2, and the cantilever stresses calculated on both faces were only slightly smaller. Arch tension existed simultaneously on both faces, indicating loss of arch action. Although this mechanism was obviously not modeled in the linear analysis, it would result in redistribution of loads to the cantilevers. This scenario indicated that the dam would crack through the thickness over a large area in the upper 200 ft of the dam.

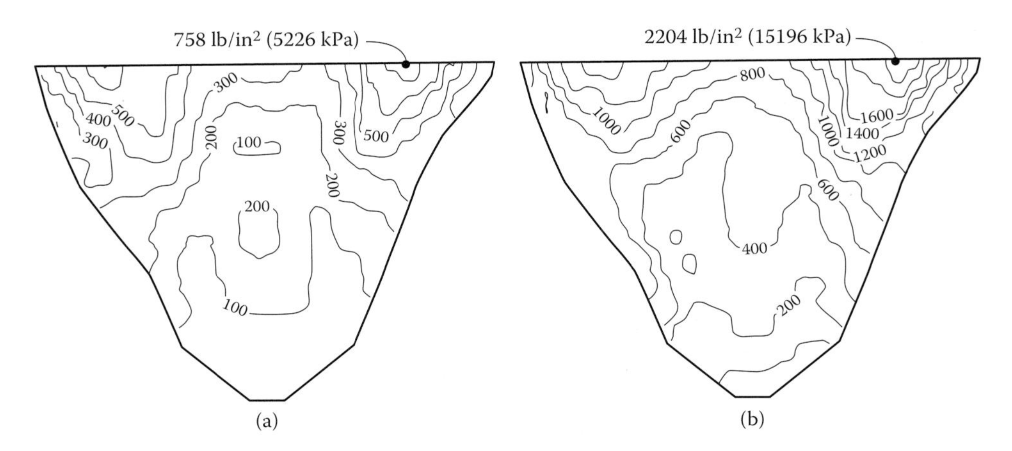

Figure 14.3.4 Peak values of tensile arch stresses in Hoover Dam for two cases: (a) including all effects of dam–foundation interaction; and (b) assuming rock to be massless. (Adapted from Bureau of Reclamation, 1998a,c.)

These results did not seem credible to Reclamation engineers, who believed that this dam should perform much better because it is curved in plan, confined in a narrow canyon (Figure 14.3.1), and its cross section is similar to gravity dams (Figure 14.3.2). To resolve this apparent inconsistency between results of computer analyses versus expectations based on engineering experience and judgment, a comprehensive investigation into the significance of dam–water–foundation interaction effects was undertaken (see Section 9.2.1).

Using the EACD-3D-96 computer program (Tan and Chopra 1996a), Hoover Dam was reanalyzed, now rigorously modeling dam–water–foundation interaction, considering the semi-unbounded geometry of the fluid and foundation domains, and including the mass of rock and compressibility of water. Reduced by a factor of 3, the stresses were now well below the estimated tensile strength of 970 psi (Figure 14.3.4a). Reclamation validated these results, which were obtained using the then newly developed computer program EACD-3D-96, by supplementary investigations. Thereafter they concluded that Hoover Dam "should be able to withstand the ground shaking from the Maximum Credible Earthquake without significant cracking …." (Bureau of Reclamation, 1998a,c).

The preceding case history of seismic evaluation of a major dam has profound implications for engineering practice. Had Reclamation engineers stopped after implementing the popular finite element analysis, which was common in engineering practice at that time, they would have been dealing with a monumental challenge of designing a remediation strategy for the highest concrete dam in the United States at a cost of perhaps several hundred million U.S. dollars. The important point, however, is that such a retrofit was unnecessary.

As mentioned earlier, the strength of Hoover Dam greatly exceeds that of (i) typical arch dams because it is confined in a narrow canyon (Figure 14.3.1), and its cross section is much thicker, in fact it is similar to gravity dams (Figure 14.3.2); and (ii) straight gravity dams because it is significantly curved in plan. Thus, if Hoover Dam were judged to be seismically deficient, the same erroneous conclusion would apply to most gravity and arch dams. The above-summarized investigation of Hoover Dam by Reclamation engineers provides a compelling case for modeling rigorously the effects of dam–water–foundation interaction including the mass of rock and compressibility of water, and by recognizing the semi-unbounded geometry of the fluid and foundation domains. The additional cost of such "realistic" analyses is negligible compared to the enormous benefits in arriving at responsible conclusions.

14.4 SEISMIC DESIGN OF DAGANGSHAN DAM

Located in Southwest China, Dagangshan Dam is a 210-m-high, double curvature arch dam with a crest length of 610 m (Figure 14.4.1). The dam consists of 29 blocks with shear keys at the contractions joints. The thickness of the crown cantilever varies from 52 m at the base to 10 m at the crest. Designed by Chengdu Engineering Corporation using modern dynamic analysis procedures, the dam was completed in 2016. The summary presented next is based on a technical report prepared by Wang et al. (2015).

A PSHA for the site corresponding to a 2% probability of exceedance in 100 years resulted in PGA = 0.557 g for the stream and cross-stream components of ground motion. With this PGA, the shape of the design spectrum for the horizontal components of GM was defined by the Chinese regulations (State Economic and Trade Commission 2000) and scaling it by a factor of ⅔ provided the design spectrum of the same shape for the vertical component; the spectrum for the horizontal GM is shown in Figure 14.4.2. In contrast, PSHA would have provided the complete

Figure 14.4.1 Dagangshan Dam, Shichuan Province, China.

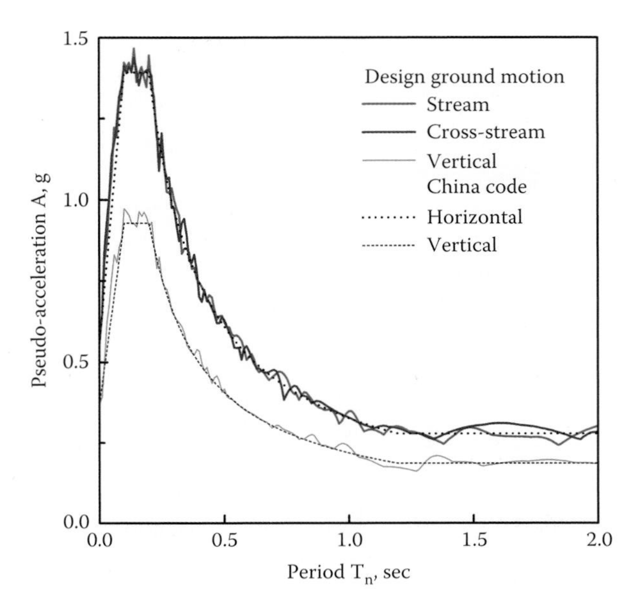

Figure 14.4.2 Design spectra and response spectra of simulated ground motion. (Adapted from Wang et al. 2015.)

design spectra for each component of GM; this standard approach would lead to uniform hazard spectra with shapes different than those selected, and the shapes of these spectra for the horizontal and vertical components of GM would not be the same. Acceleration-time functions simulated to match closely the design spectra provided the earthquake excitation for dynamic analysis of the dam (Figure 14.4.3).

Material properties, selected based on the design data, are listed next. Concrete: dynamic modulus of elasticity = 31.2 GPa; Poisson's ratio = 0.17; unit weight 2400 kg/m^3; coefficient of thermal expansion = 1×10^{-5}/°C; cohesion = 3 MPa; angle of friction = 54.46; dynamic tensile strength = 3.25 MPa; and fracture energy = 325 N/m. The foundation is composed of two types of rock (Figure 14.4.4), and their material properties are listed next. Class II rock: modulus of deformation = 20 GPa; Poisson's ratio = 0.25; and unit weight = 2650 kg/m^3. Class III rock: modulus of deformation = 7 GPa; Poisson's ratio = 0.27; and unit weight = 2650 kg/m^3.

Material damping in the concrete was modeled as Rayleigh damping with damping ratio = 5% at the first and fifth natural frequencies of the dam–foundation system; however, the rock was assumed to be undamped.

The dam–water foundation system was idealized as an assemblage of 3D solid finite elements with viscous springs to model the wave-absorbing boundaries (Figure 14.4.5). Compressibility of water was ignored but dam–foundation interaction was included, and the semi-unbounded extent of the foundation domain was recognized. Contraction joints were modeled, and plastic damage model from Lee and Fenves (1998) was selected to define the nonlinear properties of concrete.

The free-field motion at the flat surface of the foundation model (Figure 14.4.4) was defined by the ground motion described earlier. Ground motion at the bottom boundary of the foundation domain was defined as half of the surface motion, which provided the excitation for a

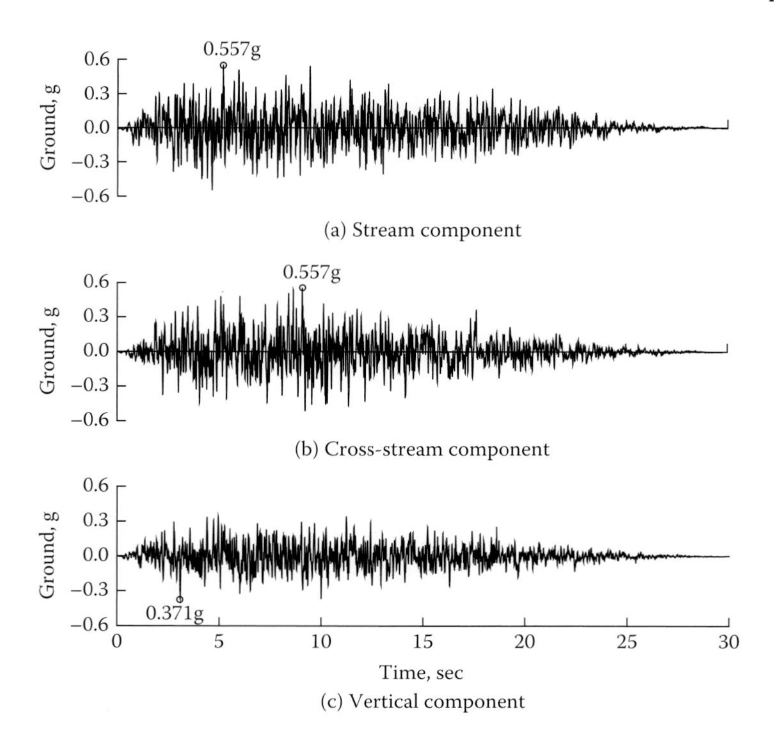

Figure 14.4.3 Simulated ground motion. (Adapted from Wang et al. 2015.)

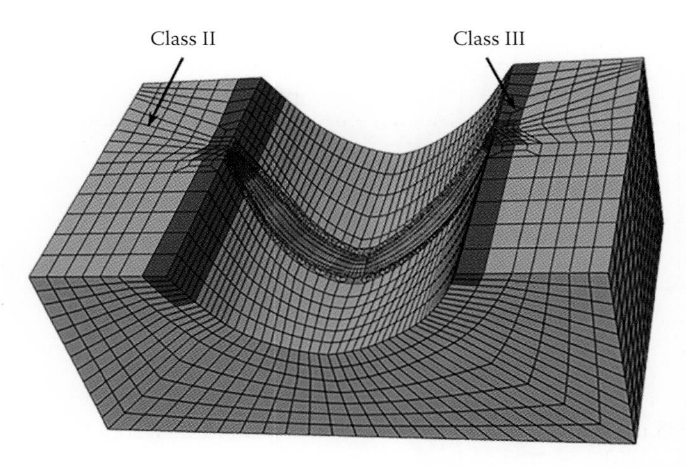

Figure 14.4.4 Foundation showing Class II and Class III rock. (Adapted from Wang et al. 2015.)

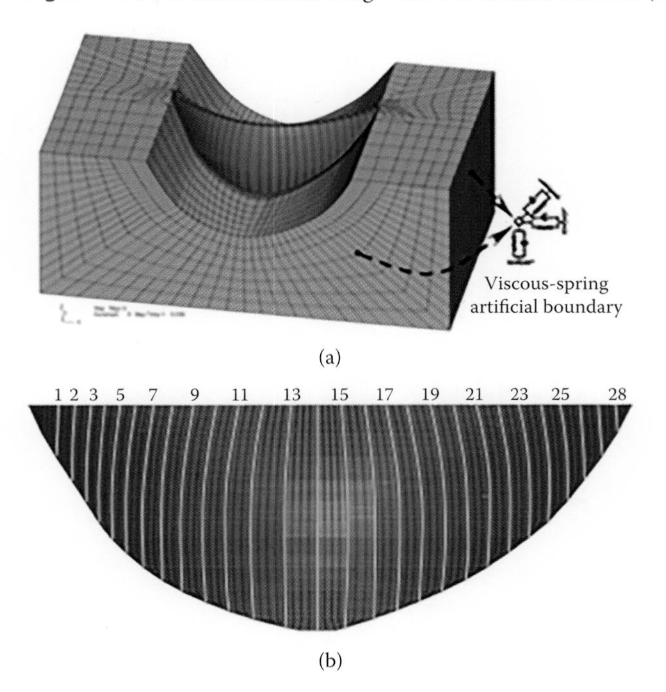

Figure 14.4.5 (a) Finite-element model of dam–foundation system; and (b) the dam and its contraction joints. (Adapted from Wang et al. 2015.)

one-dimensional analysis of the foundation to determine free-field motions at the side boundaries. Effective earthquake forces at the absorbing boundaries were determined by a procedure similar to that described in Section 11.8.

Determined by nonlinear RHA of the dam–water–foundation system, the distribution of damage for two cases – with and without steel reinforcement – is shown in Figures 14.4.6 and 14.4.7. Although the reinforcement does not influence the size of the damaged zone, it tends to distribute cracks uniformly and reduces their spacing and width. Arch and cantilever reinforcement included in construction of the dam is shown in Figure 14.4.8.

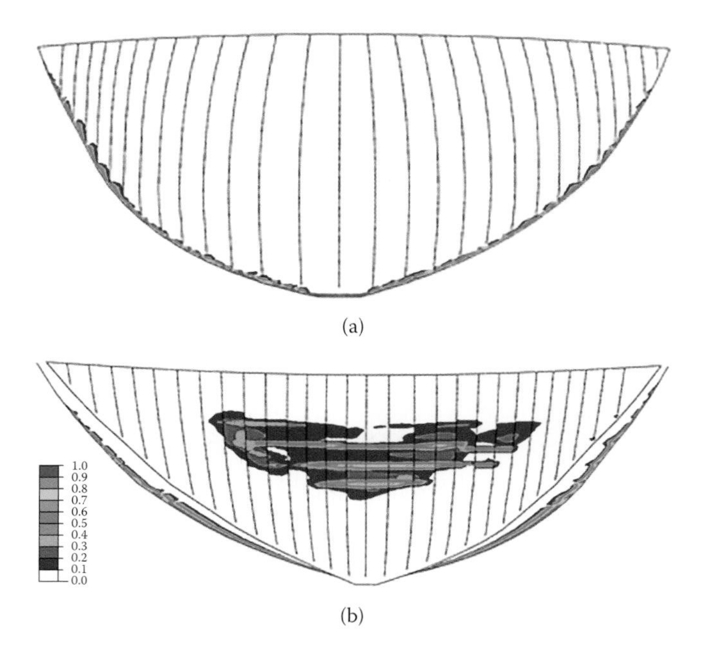

Figure 14.4.6 Damage distribution in dam without reinforcement: (a) upstream face; and (b) downstream face. (Adapted from Wang et al. 2015.)

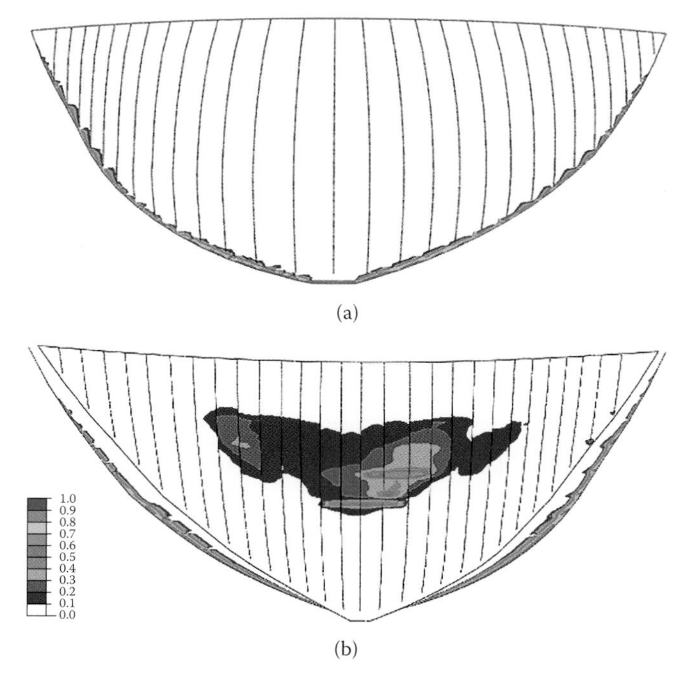

Figure 14.4.7 Damage distribution in dam with reinforcement: (a) upstream face; and (b) downstream face. (Adapted from Wang et al. 2015.)

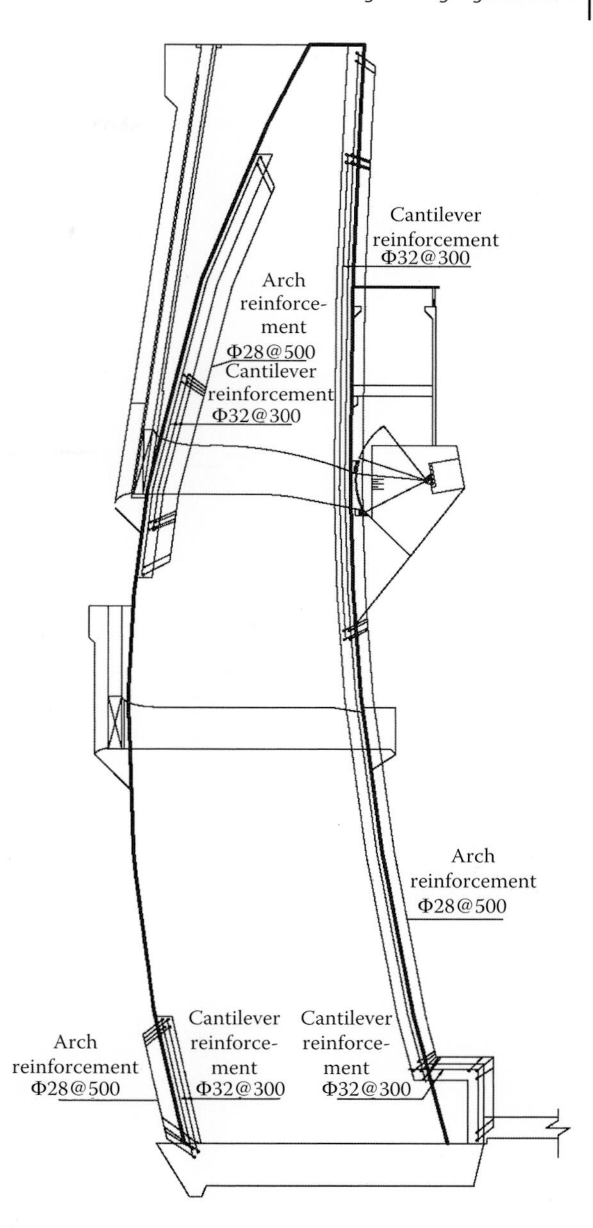

Figure 14.4.8 Arch and cantilever reinforcement in Dagangshan Dam; the symbol Φ denotes diameter in mm, and the number following @ is spacing in mm, e.g. Φ32@300 means 32-mm rebar at 300 mm spacing. (Adapted from Wang et al. 2015.)

References

Alves, S. W. (2004). Nonlinear analysis of Pacoima dam with spatially non-uniform ground motion, *Report No. EERL 2004-11*, California Institute of Technology, Pasadena, CA.

Alves, S.W. and Hall, J.F. (2006a). Generation of spatially nonuniform ground motion for nonlinear analysis of a concrete arch dam. *Earthquake Eng. Struct. Dyn.* 35 (11): 1339–1357.

Alves, S.W. and Hall, J.F. (2006b). System identification of a concrete arch dam and the calibration of its finite element model. *Earthquake Eng. Struct. Dyn.* 35 (11): 1321–1337.

ANCOLD (2017). Guidelines for Design of Dams and Appurtenant Structures for Earthquakes, Draft document not released for use (March).

Ayari, M.L. and Saouma, V.E. (1990). A fracture mechanics based analysis of concrete gravity dams using discrete cracks. *Eng. Fract. Mech.* 35 (1–3): 587–598.

Baker, J.W. (2011). Conditional mean spectrum: Tool for ground-motion selection. *ASCE J. Struct. Eng.* 137 (3): 322–331.

Baker, J.W. and Cornell, C.A. (2005). A vector-valued ground motion intensity measure consisting of spectral acceleration and epsilon. *Earthquake Eng. Struct. Dyn.* 34 (10): 1193–1217.

Bao, H., Bielak, J., Ghattas, O. et al. (1998). Large-scale simulation of elastic wave propagation in heterogeneous media on parallel computers. *Comput. Methods Appl. Mech. Eng.* 152 (1–2): 85–102. https://doi.org/10.1016/S0045-7825(97)00183-7.

Basu, U. (2004). *Perfectly Matched Layers for Acoustic And Elastic Waves: Theory, Finite-Element Implementation and Application to Earthquake Analysis of Dam–Water–Foundation Rock Systems*, PhD thesis, Department of Civil and Environmental Engineering, University of California, Berkeley, CA.

Basu, U. and Chopra, A.K. (2004). Perfectly matched layers for transient elastodynamics of unbounded domains. *Int. J. Numer. Methods Eng.* 59 (8): 1039–1074. https://doi.org/10.1002/nme.896.

Earthquake Engineering for Concrete Dams: Analysis, Design, and Evaluation, First Edition. Anil K. Chopra.

Belytschko, T. and Black, T. (1999). Elastic crack growth in finite elements with minimal remeshing. *Numer. Methods Eng.* 45 (5): 601–620.

Bhattacharjee, S.S. and Léger, P. (1992). Concrete constitutive models for nonlinear seismic analysis of gravity dams—State of the art. *Can. J. Civ. Eng.* 19: 492–509.

Bhattacharjee, S.S. and Léger, P. (1993). Seismic cracking and energy dissipation in concrete gravity dams. *Earthquake Eng. Struct. Dyn.* 22 (11): 991–1007. https://doi.org/10.1002/eqe.4290221106.

Bielak, J. (1976). Modal analysis for building-soil interaction. *J. Eng. Mech. Div.*, 771–785.

Bielak, J. and Christiano, P. (1984). On the effective seismic input for non-linear soil-structure interaction systems. *Earthquake Eng. Struct. Dyn.* 12 (3): 107–119.

Bielak, J., Loukakis, K., Hisada, Y., and Yoshimura, C. (2003). Domain reduction method for three-dimensional earthquake modeling in localized regions, Part I: Theory. *Bull. Seismol. Soc. Am.* 93 (2): 817–824.

Biot, M.A. (1956). Theory of propagation of elastic waves in fluid-saturated porous solid, Part I. Low-frequency range. *J. Acoust. Soc. Am.* 28 (2): 168–178.

Brühwiler, E. (1990). Fracture of mass concrete under simulated seismic action. *Dam Eng.* 1 (EPFL-ARTICLE-181083): 153–176.

Brühwiler, E. and Wittmann, F.H. (1990). The wedge splitting test, a new method of performing stable fracture mechanics tests. *Eng. Fract. Mech.* 35 (1–3): 117–125.

Bureau of Reclamation (1965). Concrete dams, Chapter 1: Arch dams. In: *Design Standards No. 2*. Denver, CO: U.S. Department of the Interior.

Bureau of Reclamation (1966). Concrete dams, Chapter 2: Gravity dams. In: *Design Standards No. 2*. Denver, CO: U.S. Department of the Interior.

Bureau of Reclamation (1977). *Design of Arch Dams*. Denver, CO: U.S. Department of the Interior.

Bureau of Reclamation (1983). Morrow Point Dam and power plant. In: *Technical Record of Design and Construction*. Denver, CO: U.S. Department of the Interior.

Bureau of Reclamation (1998a). Linear elastic dynamic structural analysis including mass in the foundation for Hoover Dam, *Technical Memorandum HVD-MDA-D8110-97-2*, Denver, CO: U.S. Department of the Interior.

Bureau of Reclamation (1998b). Linear elastic analysis addressing structural issues for the modification decision analysis for Deadwood Dam, *Technical Memorandum DEA-D8110-MDA-TM-98-1*, Denver, CO: U.S. Department of the Interior.

Bureau of Reclamation (1998c). Executive summary of the static and dynamic stability studies of Hoover Dam, Modification design analysis, *Technical Memorandum No. HVD-MDA-D8110-97-1*, Denver, CO: U.S. Department of the Interior.

Bureau of Reclamation (1998d). Probabilistic ground motions for Hoover Dam, Bolder Canyon Project, *Technical Memorandum No. HVD-MDA-D8-330-99-012*, Denver, CO: U.S. Department of the Interior.

Bureau of Reclamation (1999). Linear elastic static and dynamic structural analysis of Monticello Dam, *Technical Memorandum No. MONT-IE-D8110-99-2*, Denver, CO: U.S. Department of the Interior.

Bureau of Reclamation (2002). Investigation of the failure modes of concrete dams – Physical model tests, *Report No. DSO-02-02*, Denver, CO: U.S. Department of the Interior.

Bureau of Reclamation (2013). *State-of-Practice for Nonlinear Analysis of Concrete Dams*. Denver, CO: U.S. Department of the Interior.

Bybordiani, M. and Arici, Y. (2017). The use of 3D modeling for the prediction of the seismic demands on the gravity dams. *Earthquake Eng. Struct. Dyn.* 46 (11): 1769–1789. https://doi.org/10.1002/eqe.2880.

Chakrabarti, P. and Chopra, A.K. (1974). Hydrodynamic effects in earthquake response of gravity dams. *ASCE J. Struct. Div.* 100: 1211–1224.

Chopra, A.K. (1967). Hydrodynamic pressures on dams during earthquakes. *ASCE J. Eng. Mech. Div.* 93: 205–223.

Chopra, A.K. (1968). Earthquake behavior of reservoir-dam systems. *ASCE J. Eng. Mech. Div.* 94 (EM6): 1475–1500.

Chopra, A.K. (1978). Earthquake resistant design of concrete gravity dams. *ASCE J. Struct. Div.* 104 (ST6): 953–971.

Chopra, A.K. (2012). Earthquake analysis of arch dams: Factors to be considered. *ASCE J. Struct. Eng.* 138 (2): 205–214. https://doi.org/10.1061/(ASCE)ST.1943-541X.0000431.

Chopra, A.K. (2017). *Dynamics of Structures: Theory and Applications to Earthquake Engineering*, 5e, 960 pgs. Upper Saddle River, NJ: Pearson Education.

Chopra, A.K. and Chakrabarti, P. (1972). The earthquake experience at Koyna Dam and stresses in concrete gravity dams. *Earthquake Eng. Struct. Dyn.* 1 (2): 151–164.

Chopra, A.K. and Chakrabarti, P. (1973). The Koyna earthquake and the damage to Koyna Dam. *Bull. Seismol. Soc. Am.* 63 (2): 381–397.

Chopra, A.K. and Chakrabarti, P. (1981). Earthquake analysis of concrete gravity dams including dam–water–foundation rock interaction. *Earthquake Eng. Struct. Dyn.* 9 (4): 363–383.

Chopra, A. K. and Tan, H.C. (1989). Simplified earthquake analysis of gated spillway monoliths of concrete gravity dams, *Technical Report SL-89-4*, Structures Laboratory, U.S. Army Corps of Engineers Waterways Experiment Station, Vicksburg, MS, 155 pgs.

Chopra, A. K. and Wang, J.-T. (2008). "Analysis and response of concrete arch dams including dam–water–foundation rock interaction to spatially-varying ground motions," *UCB/EERC-2008-03*, Earthquake Engineering Research Center, University of California, Berkeley, CA, 127 pgs.

Chopra, A.K. and Wang, J.-T. (2010). Earthquake response of arch dams to spatially varying ground motions. *Earthquake Eng. Struct. Dyn.* 39 (8): 887–906.

Chopra, A.K. and Zhang, L. (1991). Earthquake-induced base sliding of concrete gravity dams. *ASCE J. Struct. Eng.* 117: 3698–3719.

Clough, R. W. (1980). Nonlinear mechanisms in the seismic response of arch dams, *Proceedings, International Conference on Earthquake Engineering*, Skopje, Yugoslavia.

Darbre, G.R., De Smet, C.A.M., and Kraemer, C. (2000). Natural frequencies measured from ambient vibration response of the arch dam of Mauvoisin. *Earthquake Eng. Struct. Dyn.* 29: 577–586.

Dasgupta, G. and Chopra, A.K. (1979). Dynamic stiffness matrices for viscoelastic half planes. *ASCE J. Struct. Div.* 105 (EM5): 729–745.

Dassault Systems (n.d.). Abaqus/Standard User's Manual, Version 6.13, (2013).

Domínguez, J., Gallego, R., and Japón, B.R. (1997). Effects of porous sediments on seismic response of concrete gravity dams. *ASCE J. Eng. Mech.* 123 (4): 302–311.

Dowling, M.J. and Hall, J.F. (1989). Nonlinear seismic analysis of arch dams. *ASCE J. Eng. Mech.* 115 (4): 768–789.

Duron, Z.H. and Hall, J.F. (1988). Experimental and finite element studies of the forced vibration response of Morrow Point Dam. *Earthquake Eng. Struct. Dyn.* 16 (7): 1021–1039.

El-Aidi, B. and Hall, J.F. (1989a). Non-linear earthquake response of concrete gravity dams Part 1: Modeling. *Earthquake Eng. Struct. Dyn.* 18 (6): 837–851. https://doi.org/10.1002/eqe.4290180607.

El-Aidi, B. and Hall, J.F. (1989b). Non-linear earthquake response of concrete gravity dams Part 2: Behaviour. *Earthquake Eng. Struct. Dyn.* 18 (6): 853–865. https://doi.org/10.1002/eqe.4290180608.

FEMA (2005). *Federal Guidelines for Dam Safety, FEMA 65*. Washington D.C.: Federal Emergency Management Agency.

FEMA (2014). *Selecting Analytical Tools for Concrete Dams to Address Key Events along Potential Failure Mode Paths, FEMA P-1016*. Washington D.C.: Federal Emergency Management Agency.

Fenves, G. and Chopra, A.K. (1983). Effects of reservoir bottom absorption on earthquake response of concrete gravity dams. *Earthquake Eng. Struct. Dyn.* 11 (6): 809–829.

Fenves, G. and Chopra, A. K. (1984a). Earthquake analysis and response of concrete gravity dams, *Report No. UCB/EERC-84/10,* Earthquake Engineering Research Center, University of California, Berkeley, CA, 213 pgs.

Fenves, G. and Chopra, A.K. (1984b). Earthquake analysis of concrete gravity dams including reservoir bottom absorption and dam–water–foundation rock interaction. *Earthquake Eng. Struct. Dyn.* 12 (5): 663–680.

Fenves, G. and Chopra, A. K. (1984c). EAGD-84: A computer program for earthquake response analysis of concrete gravity dams, *Report No. UCB/EERC-84/11,* Earthquake Engineering Research Center, University of California, Berkeley, CA, 78 pgs.

Fenves, G. and Chopra, A.K. (1985a). Effects of reservoir bottom absorption and dam–water–foundation rock interaction on frequency response functions for concrete gravity dams. *Earthquake Eng. Struct. Dyn.* 13 (1): 13–31.

Fenves, G. and Chopra, A.K. (1985b). Reservoir bottom absorption effects in earthquake response of concrete gravity dams. *ASCE J. Struct. Eng.* 111 (3): 545–562.

Fenves, G. and Chopra, A.K. (1985c). Simplified earthquake analysis of concrete gravity dams: Separate hydrodynamic and foundation interaction effects. *ASCE J. Eng. Mech.* 111 (6): 715–735.

Fenves, G. and Chopra, A.K. (1985d). Simplified earthquake analysis of concrete gravity dams: Combined hydrodynamic and foundation interaction effects. *ASCE J. Eng. Mech.* 111 (6): 736–756.

Fenves, G. and Chopra, A.K. (1986). Simplified analysis for earthquake-resistant design of concrete gravity dams, *Report No. UCB/EERC-85/10,* Earthquake Engineering Research Center, University of California, Berkeley, Calif., 149 pgs.

Fenves, G. and Chopra, A.K. (1987). Simplified earthquake analysis of concrete gravity dams. *ASCE J. Struct. Eng.* 113 (8): 1688–1708.

Fenves, G., Mojtahedi, S., and Reimer, R.B. (1992). Effect of contraction joints on earthquake response of an arch dam. *ASCE J. Struct. Eng.* 118 (4): 1039–1055.

Fok, K.-L. and Chopra, A.K. (1986a). Earthquake analysis and response of concrete arch dams, *Report No. UCB/EERC-85/07*, Earthquake Engineering Research Center, University of California, Berkeley, Calif., 207 pgs.

Fok, K.-L. and Chopra, A.K. (1986b). Earthquake analysis of arch dams including dam–water interaction, reservoir boundary absorption, and foundation flexibility. *Earthquake Eng. Struct. Dyn.* 14: 155–184.

Fok, K.-L. and Chopra, A.K. (1986c). Frequency response functions for arch dams: Hydrodynamic and foundation flexibility effects. *Earthquake Eng. Struct. Dyn.* 14 (5): 769–795.

Fok, K.-L. and Chopra, A.K. (1987). Water compressibility in earthquake response of arch dams. *ASCE J. Struct. Eng.* 113 (5): 958–975.

Fronteddu, L., Léger, P., and Tinawi, R. (1998). Static and dynamic behavior of concrete lift joint interfaces. *ASCE J. Struct. Eng.* 124 (12): 1418–1430. https://doi.org/10.1061/(ASCE)0733-9445(1998)124:12(1418).

García, F., Aznárez, J.J., Padrón, L.A., and Maeso, O. (2016). Relevance of the incidence angle of the seismic waves on the dynamic response of arch dams. *Soil Dyn. Earthquake Eng.* 90: 442–453. https://doi.org/10.1002/eqe.4290201104.

Ghanaat, Y. and Redpath, B.B. (1995). Measurements of reservoir-bottom reflection coefficient at seven concrete dam sites, Report No. QS95-01, U.S. Army Corps of Engineers, Waterways Experiment Station and Bureau of Reclamation on research conducted under Grant Nos. DACW39-94-C-0083 and 1425-4-CA-81-20070, Orinda, CA.

Goldgruber, M. (2015). *Nonlinear Seismic Modelling of Concrete Dams.* Graz: Technical University. https://doi.org/10.13140/RG.2.1.3001.6485.

Goodman, R.E., Taylor, R.L., and Brekke, T.L. (1968). A model for the mechanics of jointed rock. *ASCE J. Soil Mech. Found. Div.* 94 (SM 3): 637–659.

Grant, D.N. (2011). Response spectral matching of two horizontal ground-motion components. *ASCE J. Struct. Eng.* 137: 289–297.

Graves, R.W. (1996). Simulating seismic wave propagation in 3D elastic media using staggered-grid finite differences. *Bull. Seismol. Soc. Am.* 86 (4): 1091–1106.

Hall, J.F. (1988). The dynamic and earthquake behavior of concrete dams: Review of experimental and observational evidence. *Soil Dyn. Earthquake Eng.* 7: 58–121.

Hall, J.F. (ed.) (1996). Northridge earthquake reconnaissance report, *Earthq. Spectra*, Chapter 3, Geological Observations, 1: 132–138.

Hall, J.F. (1998). Efficient non-linear seismic analysis of arch dams. *Earthquake Eng. Struct. Dyn.* 27 (12): 1425–1444.

Hall, J.F. (2006). Problems encountered from the use (or misuse) of Rayleigh damping. *Earthquake Eng. Struct. Dyn.* 35 (5): 525–545. https://doi.org/10.1002/eqe.541.

Hall, J.F. and Chopra, A.K. (1980). Dynamic response of embankment, concrete-gravity and arch dams including hydrodynamic interaction, *Report No. UCB/EERC-80/39*, Earthquake Engineering Research Center, University of California, Berkeley, October 1980, 220 pgs.

Hall, J.F. and Chopra, A.K. (1982). Two-dimensional analysis of concrete gravity and embankment dams including hydrodynamic effects. *Earthquake Eng. Struct. Dyn.* 10 (2): 305–332.

Hall, J.F. and Chopra, A.K. (1983). Dynamic analysis of arch dams including hydrodynamic effects. *ASCE J. Eng. Mech. Div.* 109: 149–167.

Hall, R.L., Woodson, S.C., and Nau, J.M. (1989). Seismic stability evaluation of Folsom Dam and Reservoir project, Report No. 3: Concrete gravity dam, *Technical Report GS-87-14*, U.S. Army Engineer Waterways Experiment Station, Vicksburg, MS.

Hall, J.F., Dowling, M.J., and El-Aidi, B. (1991). *Defensive Design of Concrete Gravity Dams, EERL-91-02*. Pasadena, CA: California Institute of Technology.

Hamilton, E.L. (1971). Elastic properties of marine sediments. *J. Geophys. Res.* 76 (2): 579–604.

Hartford, D. and Baecher, G. (2004). *Risk and Uncertainty in Dam Safety*, Thomas Telford. Reston, VA: ASCE Press.

Hashash, Y.M.A., Groholski, D.R., Phillips, C.A. et al. (2011). *DEEPSOIL User Manual and Tutorial.* Version 5.0. Urbana, IL: University of Illinois.

Hatami, K. (1997). Effect of reservoir bottom on earthquake response of concrete dams. *Soil Dyn. Earthquake Eng.* 16 (7–8): 407–415.

Hillerborg, A., Modéer, M., and Petersson, P.E. (1976). Analysis of crack formation and crack growth in concrete by means of fracture mechanics and finite elements. *Cem. Concr. Res.* 6 (6): 773–781.

Hohlberg, J.M. (1992). A joint element for the nonlinear dynamic analysis of arch dams, *Report No. 186*, Institute of Structural Engineers, ETH, Zurich, Switzerland.

ICOLD (2016). Selecting seismic parameters for large dams, *Bulletin 148*, International Commission on Large Dams, Paris, France.

Jayaram, N., Lin, T., and Baker, J.W. (2011). A computationally efficient ground-motion selection algorithm for matching a target response spectrum mean and variance. *Earthquake Spectra* 27 (3): 787–815.

Kwong, N.S. and Chopra, A.K. (2017). A generalized conditional mean spectrum and its application for intensity-based assessments of seismic demands. *Earthquake Spectra* 33 (1): 123–143.

Kwong, N.S. and Chopra, A.K. (2018). Determining bidirectional ground motions for nonlinear response history analysis of buildings at far-field sites. *Earthquake Spectra* 34 (4): 1931–1954.

Kwong, N.S. and Chopra, A.K. (2019). Selecting and scaling three components of ground motions for intensity-based assessments at far-field sites, submitted for publication to *Earthquake Spectra*.

Lau, D.T., Noruziaan, B., and Razaqpur, A.G. (1998). Modelling of contraction joint and shear sliding effects on earthquake response of arch dams. *Earthquake Eng. Struct. Dyn.* 27 (10): 1013–1029.

Leclerc, M., Léger, P., and Tinawi, R. (2003). Computer aided stability analysis of gravity dams – CADAM. *Adv. Eng. Software* 34 (7): 403–420.

Lee, J. and Fenves, G.L. (1998). A plastic-damage concrete model for earthquake analysis of dams. *Earthquake Eng. Struct. Dyn.* 27 (9): 937–956.

Léger, P. (2019). Structural behavior, safety and rehabilitation of concrete dams, Compendium, Ecole Polytechnique de Montreal, Quebec, Canada (to be published).

Léger, P. and Katsouli, M. (1989). Seismic stability of concrete gravity dams. *Earthquake Eng. Struct. Dyn.* 18 (6): 889–902. https://doi.org/10.1002/eqe.4290180611.

Léger, P., Leclerc, M., and Larivière, R. (2003). Seismic safety evaluation of concrete dams in Québec. *Int. J. Hydropower Dams* 10 (2): 100–109.

Lemos, J.V. (1999). Discrete element analysis of dam foundations. In: *Distinct Element Modelling in Geomechanics* (eds. V.M. Sharma, K.R. Saxena, and R.D. Woods), 89–115. Rotterdam: Balkema.

Lemos, J.V. (2008). Block modelling of rock masses: Concepts and applications to dam foundations. *Eur. J. Environ. Civ. Eng.* 12 (7–8): 915–949. https://doi.org/10.1080/19648189.2008.9693054.

Lilhanand, K. and Teng, W.-S. (1988). Development and application of realistic earthquake time histories compatible with multiple damping response spectra, *Proceedings, 9th World Conference on Earthquake Engineering*, Tokyo, Japan.

Loh, C.-H. and Wu, T.-C. (2000). System identification of Fei-Tsui Arch Dam from forced vibration and seismic response data. *J. Earthquake Eng.* 4: 511–537.

Løkke, A. (2013). *User Manual: Pre- and Post-Processing Modules to Facilitate Analysis with EAGD-84.* Berkeley, CA: NISEE e-library, Pacific Earthquake Engineering Research Center, University of California.

Løkke, A. and Chopra, A.K. (2013). *Response Spectrum Analysis of Concrete Gravity Dams Including Dam–Water–Foundation Interaction, PEER Report 2013/17*, 72. Berkeley, CA: Pacific Earthquake Engineering Research Center, University of California.

Løkke, A. and Chopra, A.K. (2015). Response spectrum analysis of concrete gravity dams including dam–water–foundation interaction. *ASCE J. Struct. Eng.* 141 (8): 04014202. https://doi.org/10.1061/(ASCE)ST.1943-541X.0001172.

Løkke, A. and Chopra, A.K. (2017). Direct finite element method for nonlinear analysis of semi-unbounded dam–water–foundation rock systems. *Earthquake Eng. Struct. Dyn.* (8): 1267–1285. https://doi.org/10.1002/eqe.2855.

Løkke, A. and Chopra, A.K. (2018). Direct finite element method for nonlinear earthquake analysis of three-dimensional semi-unbounded dam–water–foundation rock systems. *Earthquake Eng. Struct. Dyn.* 47 (5): 1309–1328:https://doi.org/10.1002/eqe.3019.

Løkke, A. and Chopra, A. K. (2019a). A direct finite element method for nonlinear earthquake analysis of two-and three-dimensional semi-unbounded dam–water–foundation rock systems, *Report No. PEER (in preparation)*, Pacific Earthquake Engineering Research Center, University of California, Berkeley, CA.

Løkke, A. and Chopra, A.K. (2019b). Direct finite element method for nonlinear earthquake analysis of concrete dams—Simplification, modeling, and practical application. *Earthquake Eng. Struct. Dyn.* 48: 818–842.

Loth, C. and Baker, J.W. (2015). Rational design spectra for structural reliability assessment using the response spectrum method. *Earthquake Spectra* 31 (4): 2007–2026.

Lysmer, J. and Kuhlemeyer, R.L. (1969). Finite dynamic model for infinite media. *ASCE J. Eng. Mech. Div.* 95 (4): 859–878.

Maeso, O., Aznárez, J.J., and Dominguez, J. (2002). Effects of space distribution of excitation on seismic response of arch dams. *ASCE J. Eng. Mech.* 128 (7): 759–768. https://doi.org/10.1061/(ASCE)0733-9399(2002)128:7(759).

Maeso, O., Aznárez, J.J., and Dominguez, J. (2004). Three-dimensional models of reservoir sediment and effects on the seismic response of arch dams. *Earthquake Eng. Struct. Dyn.* 33 (10): 1103–1123.

McGuire, R.K. (1995). Probabilistic seismic hazard analysis and design earthquakes: Closing the loop. *Bull. Seismol. Soc. Am.* 85 (5): 1275–1284.

McGuire, R.K. (2004). *Seismic Hazard and Risk Analysis*. Oakland, CA: Earthquake Engineering Research Institute.

Medina, F., Dominguez, J., and Tassoulas, J.L. (1990). Response of dams to earthquakes including effects of sediments. *ASCE J. Struct. Eng.* 116 (11): 3108–3121.

Moczo, P., Kristek, J., Galis, M. et al. (2007). The finite-difference and finite-element modeling of seismic wave propagation and earthquake motion. *Acta Phys. Slov.* 57 (2): 177–406.

Moës, N., Polbow, J., and Belytschko, T. (1999). A finite element method for crack growth without remeshing. *Int. J. Numer. Methods Eng.* 46 (1): 131–150.

Mojtahedi, S. and Fenves, G. (2000). Effect of contraction joint opening on Pacoima Dam in the 1994 Northridge earthquake, California Strong Motion Instrumentation Program. *Data Utilization Report CSMIP/00–05 (OSMS 00–07)*, Sacramento, CA.

National Energy Administration (2015). *Code for Seismic Design of Hydraulic Structures of Hydropower Project*, NB 350-47-2015. Beijing, China: China Electric Power Press (in Chinese).

National Research Council (1990). *Earthquake Engineering for Concrete Dams: Design, Performance, and Research Needs*, 143. Washington, D.C.: National Academy Press.

Ngo, D. and Scordelis, A.C. (1967). Finite element analysis of reinforced concrete beams. *J. Am. Concr. Inst.* 64 (3): 152–163.

Ni, S.-H., Zhang, D.-Y., Xie, W.-C., and Pandey, M.D. (2012). Vector-valued uniform hazard spectra. *Earthquake Spectra* 28 (4): 1549–1568.

Nielsen, A.H. (2006). Absorbing boundary conditions for seismic analysis in ABAQUS, *Proceedings, 2006 ABAQUS Users' Conference*, Boston, MA.

Nielsen, A.H. (2014). Towards a complete framework for seismic analysis in ABAQUS, *Proc. Inst. Civ. Eng. Eng. Comput. Mech.* 167 (1): 3–12. https://doi.org/10.1680/eacm.12.00004.

Niwa, A. and Clough, R.W. (1982). Non-linear seismic response of arch dams. *Earthquake Eng. Struct. Dyn.* 10 (2): 267–281.

Nuss, L. (2001). Comparison of EACD3D96 computed response to shaker tests on Morrow Point Dam. *Technical Memorandum No. MP-D8110-IE-2001-2*, Bureau of Reclamation, Denver, CO.

Nuss, L., Chopra, A.K., and Hall, J.F. (2003). Comparison of vibration generator tests to analyses including dam–foundation–reservoir interaction for Morrow Point Dam, *Proceeding, 21st International Congress on Large Dams*, Q83-R.6, Montreal, Canada.

Nuss, L., Matsumoto, N., and Hansen, K.D. (2014). Shaken but not stirred: Earthquake performance of concrete dams, *Proceedings, U.S. Society of Dams Annual Meeting*, San Francisco, CA.

Pan, J.W., Zhang, C.H., Xu, Y., and Jin, F. (2011). A comparative study of the different procedures for seismic cracking analysis of concrete dams. *Soil Dyn. Earthquake Eng.* 31 (11): 1594–1606.

Parsons-Harza Corp. (2000a). Olivenhain Dam Design Technical Memorandum No. 2, 48 pgs.

Parsons-Harza Corp. (2000b). Olivenhain Dam Design Technical Memorandum No. 5, 199 pgs.

Pekau, O.A., Chuhan, Z., and Lingmin, F. (1991). Seismic fracture analysis of concrete gravity dams. *Earthquake Eng. Struct. Dyn.* 20 (4): 335–354.

Proulx, J. and Paultre, P. (1997). Experimental and numerical investigation of dam–reservoir–foundation interaction for a large gravity dam. *Soil Dyn. Earthquake Eng.* 7 (2): 58–121.

Proulx, J., Paultre, P., Rheault, J., and Robert, Y. (2001). An experimental investigation of water level effects on the dynamic properties of a large arch dam. *Earthquake Eng. Struct. Dyn.* 30: 1147–1166.

Proulx, J., Darbre, G.R., and Kamileris, N. (2004). Analytical and experimental investigation of damping in arch dams based on recorded earthquake. *Proceedings, 13th World Conference on Earthquake Engineering*, Paper No. 68, Vancouver, Canada.

Puntel, E. and Saouma, V.E. (2008). Experimental behavior of concrete-joint interfaces under reversed cyclic loading. *ASCE J. Struct. Eng.* 134 (9): 1558–1568.

Raphael, J.M. (1984). Tensile strength of concrete. *J. Am. Concr. Inst.* 82 (2): 151–165.

Rashid, Y.R. (1968). Analysis of prestressed concrete pressure vessels. *Nucl. Eng. Des.* 7 (4): 334–344.

Rea, D., Liaw, C.-Y., and Chopra, A.K. (1975). Mathematical models for the dynamic analysis of concrete gravity dams. *Earthquake Eng. Struct. Dyn.* 3 (3): 249–258. https://doi.org/10.1002/eqe.4290030304.

Robbe, E., Kashiwayanagi, M., and Yamane, Y. (2017). Seismic analyses of concrete dam, comparison between finite-element analyses and seismic records. *Proceedings, 16th World Conference on Earthquake Engineering*, Santiago, Chile.

Rosenblueth, E. (1968). Presión hidrodynamica en presas debida a al acceleratión vertical con refracción en el fondo, *Proceedings, 2nd Congreso National de Ingenieriá Sismica*, Veracruz, Mexico.

Saouma, V.E., Broz, J.J., Brühwiler, E., and Boggs, H.L. (1991). Effect of aggregate and specimen size on fracture properties of dam concrete. *J. Mater. Civ. Eng.* 3 (3): 204–218.

Saouma, V.E., Miura, F., Lebon, G., and Yagome, Y. (2011). A simplified 3D model for soil–structure interaction with radiation damping and free-field input. *Bull. Earthquake Eng.* 9 (5): 1387–1402.

Saouma, V., Cervenka, J., and Reich, R. (2013). *MERLIN Finite Element User's Manual*. Boulder, CO: University of Colorado.

Schnabel, P.B., Lysmer, J., and Seed, H.B. (1972). SHAKE: A computer program for earthquake response analysis of horizontally layered sites. *Report No. UCB/EERC-72/12*, Earthquake Engineering Research Center, University of California, Berkeley, CA.

Scott, R.F., Gibson, A.D., Somerville, P. et al. (1995). Geotechnical observation, Chapter 3. *Earthquake Spectra* 11 (S2): 97–142.

Shome, N., Cornell, C.A., Bazzurro, P., and Carballo, J.E. (1998). Earthquakes, records, and nonlinear responses. *Earthquake Spectra* 14: 469–500.

State Economic and Trade Commission (2000). *Specification of Seismic Design of Hydraulic Structures, DL5073-2000*. Beijing, China: China Electric Power Press (in Chinese).

Tan, H.-C. and Chopra, A.K. (1995a). Earthquake analysis and response of concrete arch dams, *Report No. UCB/EERC-95/07*, Earthquake Engineering Research Center, University of California, Berkeley, CA, 168 pgs.

Tan, H.-C. and Chopra, A.K. (1995b). Earthquake analysis of arch dams including dam–water–foundation rock interaction. *Earthquake Eng. Struct. Dyn.* 24: 1453–1474.

Tan, H.-C. and Chopra, A.K. (1995c). Dam-foundation rock interaction effects in frequency-response functions of arch dams. *Earthquake Eng. Struct. Dyn.* 24: 1475–1489.

Tan, H.-C. and Chopra, A.K. (1996a). A computer program for three-dimensional earthquake analysis of concrete dams, *Report No. UCB/SEMM-96/06*, University of California, Berkeley, Calif., 131 pgs.

Tan, H.-C. and Chopra, A.K. (1996b). Dam–foundation rock interaction effects in earthquake response of arch dams. *ASCE J. Struct. Eng.* 122: 528–538.

Tinawi, R., P. Léger, F. Ghrib, Bhattacharjee, S.S., and Leclerc, M. (1998). Structural safety of existing concrete dams: Influence of construction joints, *CEA No. 9032 G 905*, Vol. A, Vol. B., and Final Report, Canadian Electricity Association, Montreal, Quebec, Canada.

Tinawi, R., Léger, P., Leclerc, M., and Cipolla, G. (2000). Seismic safety of gravity dams: from shake table experiments to numerical analyses. *ASCE J. Struct. Eng.* 126 (4): 518–529. https://doi.org/10.1061/(ASCE)0733-9445(2000)126:4(518).

U.S. Army Corps of Engineers (1958). Gravity dam design. In: *Design Manual EM 1110-2-2200*. Washington, D.C.: U.S. Army Corps of Engineers.

U.S. Army Corps of Engineers (1994). Arch dam design. In: *Design Manual EM 1110-2-2201*. Washington, D.C.: U.S. Army Corps of Engineers.

U.S. Army Corps of Engineers (1999). Response spectra and seismic analysis for concrete hydraulic structures. In: *Design Manual EM1110-2-6050*. Washington, D.C.: U.S. Army Corps of Engineers.

U.S. Army Corps of Engineers (2016). Earthquake design and evaluation for civil works projects. In: *Design Manual EMR1110-2-1806*. Washington, D.C.: U.S. Army Corps of Engineers.

Vargas-Loli, L.M. and Fenves, G.L. (1989). Effects of concrete cracking on the earthquake response of gravity dams. *Earthquake Eng. Struct. Dyn.* 18 (4): 575–592.

Veletsos, A.S. (1977). Dynamics of structure-foundation systems. In: *Structural and Geotechnical Mechanics* (ed. W.J. Hall), 333–361. Clifton, NJ: Prentice-Hall.

Veletsos, A.S. and Meek, J.W. (1974). Dynamic behaviour of building-foundation systems. *Earthquake Eng. Struct. Dyn.* 3 (2): 121–138.

Wang, J.-T. and Chopra, A.K. (2008). A computer program for three-dimensional analysis of concrete dams subjected to spatially-varying ground motion, *Report No. UCB/EERC-2008/04*, Earthquake Engineering Research Center, University of California, Berkeley, CA, 146 pgs.

Wang, J.-T. and Chopra, A.K. (2010). Linear analysis of concrete arch dams including dam-water-foundation rock interaction considering spatially varying ground motions. *Earthquake Eng. Struct. Dyn.* 39 (7): 731–750.

Wang, J.-T., Zhang, C., and Feng, J. (2012). Nonlinear earthquake analysis of high arch dam–water–foundation rock systems. *Earthquake Eng. Struct. Dyn.* 41 (7): 1157–1176.

Wang, J.-T., Lv, D.-D., Jin, F., and Zhang, C.-H. (2013). Earthquake damage analysis of arch dams considering dam–water–foundation interaction. *Soil Dyn. Earthquake Eng.* 49: 64–74.

Wang, J.-T., Wu, M.-X., Pan, J.-W., Jim, F., Xu, Y.-J., and Zhang, C.-H. (2015). Seismic response analysis and safety evaluation of Dagangshan dam, *Technical Report of the Department of Hydraulic Engineering*, Tsinghua University, Beijing, China.

Westergaard, H.M. (1933). Water pressure on dams during earthquakes. *ASCE Trans.* 98 (2): 418–433.

Wilson, E.L. (2010). *Static and Dynamic Analysis of Structures*, Section 23.15, 4e. Berkeley, CA: Computers and Structures, Inc.

Wolf, J.P. (1988). *Soil-Structure Interaction in Time Domain*. Englewood Cliffs, NJ: Prentice-Hall.

Zangar, C.N. (1952). Hydrodynamic pressure on dams due to horizontal earthquake effects. In: *Engineering Monograph No. 11*. Denver, CO: U.S. Bureau of Reclamation.

Zhang, L. and Chopra, A.K. (1991a). Three-dimensional analysis of spatially-varying ground motions around a uniform canyon in a homogeneous half-space. *Earthquake Eng. Struct. Dyn.* 20 (10): 911–926.

Zhang, L. and Chopra, A.K. (1991b). Impedance functions for three-dimensional foundations supported on infinitely-long canyon of uniform cross-section in a homogeneous half-space. *Earthquake Eng. Struct. Dyn.* 20 (11): 1011–1028.

Zhang, C., Yan, C., and Wang, G. (2001). Numerical simulation of reservoir sediment and effects on hydro-dynamic response of arch dams. *Earthquake Eng. Struct. Dyn.* 30 (12): 1817–1837.

Zhang, C., Pan, J.W., and Wang, J.-T. (2009). Influence of seismic input mechanisms and radiation damping on arch dam response. *Soil Dyn. Earthquake Eng.* 29 (9): 1282–1293. https://doi.org/10.1016/j.soildyn.2009.03.003.

Zhang, S., Wang, G., and Yu, X. (2013). Seismic cracking analysis of concrete gravity dams with initial cracks using the extended finite element method. *Eng. Struct.* 56: 528–543.

Zienkiewicz, O.C. and Bettess, P. (1978). Fluid-structure dynamic interaction and wave forces. An introduction to numerical treatment. *Int. J. Numer. Methods Eng.* 13 (1): 1–16. https://doi.org/10.1002/nme.1620130102.

Zienkiewicz, O.C., Best, B., Dullage, C., and Stagg, K.G. (1970). Analysis of non-linear problems in rock mechanics with particular reference to jointed rock systems. In: *Proceedings, Second International Congress on Rock Mechanics*, 1–9. Belgrade: International Society for Rock Mechanics, 8–14.

Zienkiewicz, O.C., Paul, D.K., and Hinton, E. (1983). Cavitation in fluid-structure response (with particular reference to dams under earthquake loading). *Earthquake Eng. Struct. Dyn.* 11 (4): 463–481. https://doi.org/10.1002/eqe.4290110403.

Zienkiewicz, O.C., Bicanic, N., and Shen, F.Q. (1989). Earthquake input definition and the transmitting boundary conditions. In: *Advances in Computational Nonlinear Mechanics* (ed. I. St. Doltsinis). Springer.

Notation

All symbols used in this book are defined where they first appear. For the reader's convenience, this appendix, arranged in three parts to follow the organization of the text, contains the principal meanings of the commonly used notations. The reader is cautioned that some symbols denote more than one quantity, but the meaning should be clear when read in context.

PART I: CHAPTERS 2–8

Abbreviations

RHA	Response History Analysis
RSA	Response Spectrum Analysis
SDF	Single-degree-of-freedom
SRSS	Square Root of the Sum of Squares

Roman Symbols

a_g	peak ground acceleration
$a_g(t)$	horizontal component of free-field ground acceleration
$a_g^x(t)$	horizontal component of free-field ground acceleration
$a_g^y(t)$	vertical component of free-field ground acceleration
A_p	integral of $2gp(\hat{y})/wH$ over depth of impounded water for $H/H_s = 1$
$A_g^l(\omega)$	Fourier transform of $a_g^l(t)$; defined in Eq. (5.7.2)
$A(T_1, \zeta_1)$	ordinate of pseudo-acceleration response spectrum for the ground motion evaluated at period T_1 and damping ratio ζ_1
b	breadth of the dam base
$b_w(y)$	breadth of body of water moving with rigid dam; defined Eq. (2.3.28)

$B_0(\omega)$ hydrodynamic force due to horizontal ground motion; see Eq. (2.6.2a)

$B_0^l(\omega)$ added hydrodynamic force due to l-component of ground motion; see Eq. (2.4.11a)

$B_1(\omega)$ added hydrodynamic mass (real-valued component) and damping (imaginary-valued component) in the equivalent SDF system associated with the fundamental vibration mode of dam; see Eqs. (2.4.11b) and (2.6.2b)

$\mathbf{c}_c$ damping matrix for the finite-element model of the dam

C speed of hydrodynamic pressure waves in water

C_1 $= 2M_1\zeta_1\omega_1$

$\tilde{C}_1$ defined in Eq. (2.6.10b)

C_r $= \sqrt{E_r/\rho_r}$, the compression wave speed in reservoir bottom sediments

C_s $= \sqrt{E_s/\rho_s}$

$\mathcal{C}(\omega)$ compliance function for the reservoir bottom, Eq. (A2.8)

d duration of free-field ground motion

$D(T_1, \zeta_1)$ ordinate of the deformation response (or design) spectrum for the horizontal component of ground motion at period T_1 and damping ratio ζ_1

$D_1(t)$ deformation response of fundamental mode SDF system (with vibration period T_1 and damping ratio ζ_1) to $a_g^x(t)$

E_f modulus of elasticity of the foundation rock

E_r modulus of elasticity of the reservoir bottom materials: sediments or rock

E_s modulus of elasticity of the concrete in the dam; subscript "s" denotes "structure"

$f_1(y)$ equivalent static lateral forces associated with response in the first mode, as defined in Eq. (4.1.1)

$f_{sc}(y)$ equivalent static lateral forces associated with response in the higher vibration modes as defined in Eq. (4.2.2)

$f_1(x, y)$ horizontal component of equivalent static forces

$f_1^k(x, y)$ equivalent static forces associated with the peak displacement of $r^k(x, y)$ due to l-component of ground motion; $k = x, y$ denotes the x- and y-component of forces, respectively; see Eq. (2.2.13)

$\mathcal{F}(\omega)$ foundation-rock flexibility terms defined in Eq. (3.2.8)

$\bar{F}_0^l(\omega)$ frequency response function for hydrodynamic force on rigid dam due to l-component of ground motion

F_{st} $= ½\rho g H^2$, hydrostatic force on the dam

F_1 defined in Eq. (4.2.3)

g $=$ acceleration of gravity

h_1^* $= L_1^\theta/L_1^x$, effective height of a dam in its fundamental vibration mode

H depth of impounded water

H_s height of dam; subscript "s" for structure

i $= \sqrt{-1}$

$I_{0n}(\omega)$ integral defined in Eqs. (2.3.13a) and (5.5.12a)

$I_{jn}(\omega)$ integral defined in Eqs. (2.3.13b) and (5.5.12b)

I_t mass moment of inertia of the dam about the base centroid, defined in Eq. (3.2.3b)

J number of generalized coordinates included in analysis

$\mathbf{k}$, $\mathbf{k}_b$, $\mathbf{k}_{bb}$ submatrices of $\mathbf{k}_c$

$\mathbf{k}_c$ stiffness matrix for the finite-element model of the dam

K_1 $= \omega_1^2 M_1$

$K(\omega)$ dynamic stiffness terms for foundation domain, defined in Eq. (3.2.4)

l $= x$ and y denotes horizontal and vertical components of ground motion, respectively

L_1 integral defined in Eq. (4.1.3b)

L_1^l integral defined in Eq. (2.2.4)

$\tilde{L}_1$ defined in Eqs. (2.6.8c), (4.1.2b), and (4.1.7b).

$\mathbf{L}^l(\omega)$ vector with its elements defined in Eq. (5.4.13b)

L_1^θ, L_θ^x integrals defined in Eqs. (3.2.3c) and (3.2.3d)

$\tilde{\mathbf{L}}^l(\omega)$ vector with its elements defined in Eq. (5.6.2b)

$m(x, y)$ mass density of dam concrete

$m_a(y)$ "added mass" of water to model hydrodynamic pressures on a rigid dam with water compressibility ignored [Eq. (2.3.26)]; "added mass" to represent hydrodynamic effects in equivalent SDF system [Eq. (2.6.6)]

$m_a(y, \omega)$ "added mass" of water corresponding to hydrodynamic pressure $p_0^x(x, y, t)$; see Eq. (2.3.24)

$\tilde{m}_k(x, y)$ mass density of equivalent SDF system, $k = x, y$, defined in Eq. (2.6.5)

m_t total mass of the dam defined in Eq. (3.2.3a)

m_1^* $= (L_1)^2/M_1$, effective mass of the dam in its fundamental vibration mode

$\mathbf{m}$, $\mathbf{m}_b$ submatrices of $\mathbf{m}_c$

$\mathbf{m}_c$ mass matrix for the finite-element model of the dam

$M(t)$ overturning moment at base of the dam

$\bar{M}(\omega)$ frequency response function for $M(t)$

M_1 generalized mass of the dam in the fundamental vibration mode; integral defined in Eqs. (2.2.3) and (4.1.3a)

$\tilde{M}_1$ generalized mass of the equivalent SDF system, defined in Eqs. (2.6.8a), (4.1.2a), or (4.1.7a)

N number of nodal points above the base in the finite-element model of the dam

N_b number of nodal points at the base in the finite-element model of the dam

$p(x, y, t)$ hydrodynamic pressure

$p^l(x, y, t)$ hydrodynamic pressure due to l-component of ground motion

$p(y, \tilde{T}_r)$ $\text{Re}[\bar{p}_1(y, \tilde{T}_r)]$

$\bar{p}_1(y, \tilde{T}_r)$ complex-valued hydrodynamic pressure on upstream face of dam vibrating in its fundamental mode due to harmonic acceleration of dam at period $\tilde{T}_r$

$\bar{p}^l(x, y, \omega)$ frequency response function for $p(x, y, t)$ due to the l-component of ground motion

$p_o(y)$ hydrodynamic pressure on a rigid dam with water compressibility neglected

$\bar{p}_0^l(x, y, \omega)$ frequency response function for hydrodynamic pressure on a rigid dam due to the l-component of ground motion

$\bar{p}_1(x, y, \omega)$ frequency response function for hydrodynamic pressure due to dam vibrating in its fundamental mode

$\bar{p}_j^b(x, y, \omega)$ frequency response function that satisfies Eqs. (2.3.8) and (5.5.10)

$\bar{p}_j^f(x, y, \omega)$ frequency response function defined by Eq. (5.5.11c)

$q_1^l(t)$ dam response in fundamental modal coordinate to the l-component of ground motion

$\bar{q}_1^l(\omega)$ frequency response function for $q_1^l(t)$

$\bar{\tilde{q}}_1(\omega)$ frequency response function for the equivalent SDF system representing the dam, including interaction effects

$\bar{q}_h(x, \omega)$ continuous function analogous to the y-DOF elements in the vector $\mathcal{S}_{qq}^{-1}(\omega)\bar{\mathbf{Q}}_h(\omega)$

$\bar{\mathbf{q}}(\omega)$ vector of frequency response functions for displacements, relative to the free-field ground motion at the reservoir bottom

$\bar{\mathbf{Q}}_0^l(\omega)$ vector of nodal forces at the reservoir bottom statically equivalent to $-\bar{p}_0^l(x, 0, \omega)$

$\mathbf{Q}_h^l(t)$ vector of hydrodynamic forces at the reservoir bottom

$\bar{\mathbf{Q}}_h^l(\omega)$ vector of frequency response functions for $\mathbf{Q}_h^l(t)$

$\bar{\mathbf{Q}}_j^b(\omega)$ vector of nodal forces at the reservoir bottom statically equivalent to $\bar{p}_j^b(x, 0, \omega)$

$\bar{\mathbf{Q}}_j^f(\omega)$ vector of nodal forces at the reservoir bottom statically equivalent to $-\bar{p}_j^f(x, 0, \omega)$

$r(t)$ any response quantity

r_1 peak value of response due to the fundamental vibration mode

r_o peak value of $r(t)$

$r_{o,\text{ total}}$ peak value of total (static plus dynamic) response of dam

r_{sc} peak value of response due to higher vibration modes determined by static correction method

r_{st} response (e.g. state-of-stress) prior to earthquake due to self-weight of the dam, hydrodynamic pressures, ice loads, construction sequence, and thermal effects

$r^k(x, y, t)$ displacement of dam relative to its base; $k = x, y$ denotes x- and y-component of displacement, respectively

$\bar{\mathbf{r}}^l(\omega)$ vector of frequency response functions for displacements relative to the free-field ground motion of nodal points on the base due to the l-component of ground motion

$\bar{\mathbf{r}}_b^l(\omega)$ vector of frequency response functions for displacements relative to the free-field ground motion of nodal points on the base due to the l-component of ground motion

$\mathbf{r}_c(t)$ vector of nodal point displacements relative to the free-field ground displacement

$\bar{\mathbf{r}}_c^l(\omega)$ vector of frequency response functions for $\mathbf{r}_c(t)$ due to the l-component of ground motion

$\bar{\mathbf{r}}_f(\omega)$ vector of frequency response functions for displacements of nodal points at the surface of the foundation

$\mathbf{r}_p(t)$ vector of nodal displacements for finite element p

R_f period lengthening ratio due to dam–foundation interaction effects

R_r period lengthening ratio due to dam–water interaction effects

R_w $T_1^r / \tilde{T}_r$

$\bar{\mathbf{R}}_0^l(\omega)$ vector of nodal forces at the upstream face of the dam statically equivalent to $\bar{p}_0^l(0, y, \omega)$

$\mathbf{R}_b(t)$ vector of nodal forces at the base of the dam due to dam–foundation interaction

$\bar{\mathbf{R}}_b^l(\omega)$ vector of frequency response functions for $\mathbf{R}_b(t)$ due to the l-component of ground motion

$\mathbf{R}_c(t)$ vector containing hydrodynamic forces $\mathbf{R}_h(t)$ and dam–foundation interaction forces $\mathbf{R}_b(t)$

$\bar{\mathbf{R}}_c^l(\omega)$ vector of frequency response functions for $\mathbf{R}_c(t)$ due to the l-component of ground motion

$\bar{\mathbf{R}}_f(\omega)$ vector of frequency response functions for forces at the surface of the foundation

$\mathbf{R}_h(t)$ vector of hydrodynamic forces at the nodes on the upstream face of the dam

$\bar{\mathbf{R}}_h^l(\omega)$ vector of frequency response functions for $\mathbf{R}_h(t)$ due to the l-component of ground motion

$\bar{\mathbf{R}}_j^b(\omega)$ vector of nodal forces at the upstream face of the dam statically equivalent to $\bar{p}_j^b(0, y, \omega)$

$\bar{\mathbf{R}}_j^f(\omega)$ vector of nodal forces at the upstream face of the dam statically equivalent to $\bar{p}_j^f(0, y, \omega)$

$\mathcal{S}(\omega)$ matrix defined in Eq. (5.3.1)

$\mathcal{S}_f(\omega)$ dynamic stiffness matrix for the foundation substructure; defined in Eq. (5.3.4)

$\tilde{\mathcal{S}}_f(\omega)$ matrix defined in Eq. (5.4.9)

$\mathbf{S}(\omega)$ matrix whose elements are defined in Eq. (5.4.13a)

$\tilde{\mathbf{S}}(\omega)$ matrix whose elements are defined in Eq. (5.6.2a)
t time
T_1 fundamental vibration period of dam on rigid foundation with empty reservoir, given by Eq. (4.4.1)
$\tilde{T}_1$ fundamental resonant period of dam including dam–water–foundation interaction, given by Eq. (4.1.4c)
T_1^r $4H/C$, fundamental vibration period of impounded water
$\tilde{T}_f$ fundamental resonant period of dam with empty reservoir, including dam–foundation interaction, given by Eq. (4.1.4b)
$\tilde{T}_r$ fundamental resonant period of dam on rigid foundation including dam–water interaction; given by Eq. (4.1.4a)
$\mathbf{T}_p$ stress-displacement transformation matrix for finite element p
$V(t)$ shear force at the base of the dam
$\bar{V}(\omega)$ frequency response function for $V(t)$
$\bar{v}(y', \omega)$ frequency response function for vertical displacement in the layer of reservoir bottom materials
W_s total weight of dam
w unit weight of water
$w_k(x, y)$ $= gm_k(x, y)$
$w_s(y)$ weight of dam (or structure) per unit height; subscript "s" used to denote structure
x coordinate along breadth of dam
y coordinate along height of dam
$\hat{y}$ y/H
$Z_j(t)$ generalized coordinate corresponding to the jth Ritz vector
$\bar{Z}_j^l(\omega)$ frequency response function for $Z_j(t)$ due to the l-component of ground motion
$\bar{\mathbf{Z}}_j^l(\omega)$ vector whose elements are $\bar{Z}_j^l(\omega)$

Greek Symbols

α $= (1 - \xi C)/(1 + \xi C)$; wave reflection coefficient at reservoir bottom
γ dimensionless cross-sectional parameter for the dam
Γ_1 $= \Gamma_1^x$
$\tilde{\Gamma}_1$ $= \tilde{L}_1/\tilde{M}_1$
Γ_1^l $= L_1^l/M_1$
$\delta_{kl}, \delta_{xy}, \delta_{jk}$ Kronecker delta function
$\delta(x)$ Dirac delta function
ξ $= \rho/\rho_r C_r$, damping coefficient of reservoir bottom materials: sediments or rock
ζ_1 fundamental mode damping ratio of dam on rigid foundation with empty reservoir
$\tilde{\zeta}_1$ damping ratio of equivalent SDF system that models vibration of the dam in its fundamental mode, including dam–water–foundation interaction
ζ_f added damping ratio due to dam–foundation interaction; Eq. (3.4.4)
$\tilde{\zeta}_f$ damping ratio of equivalent SDF system that models vibration of the dam in its fundamental mode including dam–foundation interaction; Eq. (3.4.3)
ζ_r added damping ratio due to dam–water interaction and reservoir bottom absorption; Eq. (2.6.13)
$\bar{\zeta}_r$ damping ratio defined in Eq. (2.6.12)
η_f constant hysteretic damping factor for the foundation rock
η_s constant hysteretic damping factor for the dam concrete
$\boldsymbol{\iota}^l$ subvector of $\boldsymbol{\iota}_c^l$ corresponding to nodal points above the base
$\boldsymbol{\iota}_b^l$ subvector of $\boldsymbol{\iota}_c^l$ corresponding to nodal points at the base

$\boldsymbol{\iota}_c^l$ vector defined in Eq. (5.2.3)
λ_n nth eigenvalue for the associated dam–foundation system; see Eq. (5.4.8)
μ_n $= 2n - 1\pi/2H$; see Eq. (2.3.16)
$\mu_n(\omega)$ eigenvalue for the nth natural vibration mode of the impounded water; see Eq. (2.3.14)
ρ density of water
ρ_f density of foundation rock
ρ_r density of reservoir bottom materials
ρ_s density of dam concrete; subscript "s" denotes structure
$\sigma_p(t)$ vector of planar stress components in finite element p
$Y_n(y, \omega)$ eigenfunction for nth natural vibration mode of the impounded water; defined in Eq. (2.3.15)
$\phi(y)$ fundamental vibration mode shape of dam at its upstream face, i.e. $\phi_1^x(0, y)$
$\phi_1^k(x, y)$ fundamental natural vibration mode of dam without water; $k = x$, y denotes the x- and y-component of modal displacements, respectively
χ_n vector defined in Eq. (5.4.11)
$\chi_n(x)$ continuous function analogous to the y-DOF elements in χ_n
$\psi_n(y)$ continuous function analogous to the x-DOF elements in ψ_n^f
$\boldsymbol{\psi}_{bn}$ subvector of elements in $\boldsymbol{\psi}_n$ corresponding to nodal points on the base of dam
$\boldsymbol{\psi}_n$ nth Ritz vector of the associated dam-foundation rock system; see Eq. (5.4.8)
$\boldsymbol{\psi}_{fn}$ subvector of elements in $\boldsymbol{\psi}_n$ corresponding to nodal points at upstream face of dam
ω excitation frequency
ω_1 fundamental natural vibration frequency of dam on rigid foundation with empty reservoir
$\tilde{\omega}_1$ natural vibration frequency of the equivalent SDF system that models the fundamental mode response of the dam, including dam–water–foundation interaction
$\tilde{\omega}_f$ natural vibration frequency of equivalent SDF system that models the fundamental mode response of the dam including dam–foundation interaction
ω_n^r nth natural vibration frequency of impounded water with non-absorptive bottom, Eq. (2.3.16)
$\tilde{\omega}_r$ natural vibration frequency of the equivalent SDF system that models the fundamental mode response of the dam on rigid foundation, including dam–water interaction
Ω_r $= \omega_1^r/\omega_1$; see Eq. (2.5.1)

PART II: CHAPTERS 9–11

Abbreviations

BEM Boundary Element Method
CMS Conditional Mean Spectrum
DRM Domain Reduction Method
EDP Engineering Demand Parameter
ESI Effective Seismic Input Method
FEM Finite Element Method
UHS Uniform Hazard Spectrum
PSHA Probabilistic Seismic Hazard Analysis
PML Perfectly Matched Layer
RHA Response History Analysis

Roman Symbols

$a_g^l(t)$	l-component of free-field ground acceleration; $l = x, y, z$
$A_g^l(\omega)$	Fourier transform of $a_g^l(t)$; defined in Eq. (5.7.2)
$\mathbf{b}$	"damping" matrix for finite-element model of the fluid domain
$\mathbf{c}$	damping matrix for finite-element model of dam–foundation system in Direct FEM
$\mathbf{c}_c$	damping matrix for the finite-element model of the dam in substructure method
$\mathbf{c}_f$, $\mathbf{c}_r$	matrix of damper coefficients for nodes on absorbing boundaries Γ_f and Γ_r, respectively
C	speed of pressure waves in water
C_r	velocity of compression wave in reservoir boundary materials: sediments or foundation
d_c, d_t	damage index for concrete in compression and tension, respectively
E_f	Young's modulus of foundation rock
E_s	Young's modulus of dam concrete
f	excitation frequency, in Hz
f_t	tensile strength of concrete
$\mathbf{f}(\mathbf{r}^t)$	vector of (nonlinear) internal forces in dam–foundation system
G_F	specific fracture energy for concrete
g	the acceleration due to gravity
$\mathbf{h}$	"stiffness" matrix of finite-element model of the fluid domain
$\mathbf{H}_r$	vector of dynamic forces associated with absorbing boundary Γ_r
H	y-coordinate of the free surface of water measured from the base of the dam
i	$= \sqrt{-1}$
J	number of generalized coordinates
$\mathbf{k}$	stiffness matrix of linear part of finite-element model of dam–foundation system
$\mathbf{k}^0$	initial (linear) stiffness matrix
$\mathbf{k}^{el}(t)$	$= [1 - d(t)]\mathbf{k}^0$, the degraded elastic stiffness matrix
$\mathbf{k}$, $\mathbf{k}_b$, $\mathbf{k}_{bb}$	submatrices of $\mathbf{k}_c$
$\mathbf{k}_c$	stiffness matrix of the finite-element model of the dam in substructure method
l	direction of the free-field ground motion; $l = x, y, z$
L	length of truncated fluid domain
$\mathbf{L}^l(\omega)$	forcing vector of the dam–water–foundation system containing terms $L_n^l(\omega)$ defined in Eqs. (8.4.10b) or (8.8.5b)
$\mathbf{m}$	mass matrix of finite-element model of dam–foundation system in Direct FEM
$\mathbf{m}$, $\mathbf{m}_b$	submatrices of $\mathbf{m}_c$
$\mathbf{m}_c$	mass matrix for the finite-element model of the dam in substructure method
n	inward normal direction at the free surface, upstream dam face, or reservoir boundary as illustrated in Figure 8.5.2.
$\mathbf{n}$	outward normal vector to fluid domain
N	number of nodal points in the finite-element model of the dam in substructure method
N_b	number of nodal points at the dam–foundation interface in substructure method
$\mathbf{p}^t$	vector of (total) hydrodynamic pressures at finite-element nodes in the fluid domain
$\mathbf{p}_f^0$	free-field hydrodynamic forces at boundary Γ_r
$\mathbf{p}_r^0$	free-field hydrodynamic pressures at boundary Γ_r
$\mathbf{P}_f^0$	effective earthquake forces on boundary Γ_f

$\mathbf{P}_r^0$	effective earthquake forces on boundary Γ_r
$p^l(x, y, z, t)$	hydrodynamic pressure due to lth component of ground motion
$\overline{p}_0^l(x, y, z, \omega)$	frequency response function for hydrodynamic pressure due to the lth component of ground acceleration with a rigid dam and reservoir boundary
$\overline{p}_j^f(x, y, z, \omega)$	frequency response function for hydrodynamic pressure due to harmonic acceleration of dam corresponding to the jth Ritz vector with no motion at the reservoir boundary
q	$q = \rho / \rho_r C_r$
$\mathbf{Q}$	fluid-solid coupling matrix
$\mathbf{Q}_b$, $\mathbf{Q}_h$	fluid-solid coupling matrices at water–foundation interface Γ_b, and water–foundation interface Γ_h, respectively.
r	vector of displacements in the time domain at the dam–foundation rock interface
$\mathbf{r}^t$	vector of total displacements at finite-element nodes in dam–foundation system
$\mathbf{r}_I^0$	vector of displacements for incident seismic wave
$\mathbf{r}_c(t)$	vector of nodal point displacements relative to the free-field ground displacement; $\mathbf{r}_c^l(t)$ denotes the vector due to the l-component of ground motion (Chapter 8)
$\overline{\mathbf{r}}_c^l(\omega)$	vector of frequency response function for $r_c^l(t)$ due to the l-component of ground motion
$\mathbf{r}^s(t)$	vector of structural displacements due to the static application of the earthquake-induced free-field displacements $\mathbf{r}_b^t(t)$ at the dam–foundation interface
$\mathbf{r}_b^t(t)$	subvector of $\mathbf{r}_c^t$ corresponding to nodal points on the dam–foundation interface
$\mathbf{r}_c^t(t)$	vector of total (quasi-static plus dynamic) displacements of the dam
$\mathbf{r}_b^f(t)$	vector of the specified free-field ground motion
$\mathbf{r}_f^0$	free-field motion at boundary Γ_f
$\mathbf{R}_f^0$	free-field forces at boundary Γ_f
$\mathbf{R}_r^0$	free-field hydrodynamic pressures at boundary Γ_r
$\mathbf{R}^{st}$	vector of static forces
$\mathbf{r}_f^0$	free-field motion at Γ_f
$\mathbf{R}(t)$	vector of interaction forces at the dam–foundation interface
$\overline{\mathbf{R}}_0^l(\omega)$	vector of nodal forces at the upstream face of the dam statically equivalent to $-\overline{p}_0^l(0, y, \omega)$
$\mathbf{R}_b^l(t)$	vector of forces at the dam–foundation interface due to the l-component of ground motion
$\overline{\mathbf{R}}_b^l(\omega)$	vector of frequency response functions for $\mathbf{R}_b^l(t)$
$\mathbf{R}_c(t)$	vector of hydrodynamic forces at upstream face and interaction forces at the dam–foundation interface
$\mathbf{R}_f(t)$	vector of forces at dam–foundation interface
$\overline{\mathbf{R}}_f^l(\omega)$	vector of frequency response functions for forces at the dam–foundation interface due to the l-component of ground motion
$\mathbf{R}_h(t)$	vector of hydrodynamic forces on the upstream face of the dam; $\mathbf{R}_h^l(t)$ denotes the vector due to the l-component of ground motion
$\overline{\mathbf{R}}_h^l(\omega)$	vector of frequency response functions for $\mathbf{R}_h^l(t)$
$\overline{\mathbf{R}}_j^f(\omega)$	vector of nodal forces statically equivalent to the pressure function $-\overline{p}_j^f(0, y, \omega)$
$\mathcal{S}_f(\omega)$	complex-valued, frequency-dependent stiffness matrix for the foundation; see Eq. (8.3.1)

$\mathbf{s}$ — "mass" matrix of finite-element model of fluid domain

s,r — spatial coordinates on the upstream face of the dam; see Figure 8.5.2

s',r' — spatial coordinates on the reservoir boundary; see Figure 8.5.2

$\mathbf{S}(\omega)$ — matrix in the substructure method defined in Eqs. (8.6.1), (8.6.2), and (8.8.5a)

t — time

T_1 — fundamental vibration period of dam on rigid foundation with empty reservoir

$\tilde{T}_1$ — fundamental resonant period of the dam including dam–water–foundation interaction

$\tilde{T}_f$ — fundamental resonant period of the dam with empty reservoir, including dam–foundation interaction

$\tilde{T}_r$ — fundamental resonant period of the dam on rigid foundation including dam–water interaction; $\tilde{T}_r^s$ and $\tilde{T}_r^a$ denote periods associated with plan-wise symmetric and antisymmetric modes, respectively

$Z_j(t)$ — generalized coordinate corresponding to the jth Ritz vector

$Z_j^l(t)$ — generalized coordinate corresponding to the jth Ritz vector due to the l-component of ground motion

$\bar{Z}_j^l(\omega)$ — frequency response function for $Z_j^l(t)$

$\hat{Z}_j(\omega)$ — Fourier transform of $Z_j(t)$

$\hat{\mathbf{Z}}(\omega)$ — vector of Fourier transforms $\hat{Z}_j(\omega)$

$\bar{\mathbf{Z}}_j^l(\omega)$ — vector of frequency response functions $\bar{Z}_j^l(\omega)$

Greek Symbols

α — wave reflection coefficient at the reservoir boundary

Γ_b — water–foundation interface

Γ_f, Γ_r — absorbing boundaries at truncations of foundation and fluid domains, respectively

Γ_h — dam–water interface

δ_{nk} — Kronecker delta function

$\varepsilon^l(s,r)$ — function illustrated in Figure 8.5.2; when represented by $\varepsilon^l(s',r')$, it refers to s' and r' coordinates

$\boldsymbol{\iota}^l$ — subvector of $\boldsymbol{\iota}_c^l$ corresponding to nodal points not on the dam–foundation interface

$\boldsymbol{\iota}_b^l$ — subvector of $\boldsymbol{\iota}_c^l$ corresponding to nodal points on the dam–foundation interface

$\boldsymbol{\iota}_c^l$ — vector contains ones in positions corresponding to the l-translational degrees of freedom and zeros elsewhere; $l = (x, y, z)$

ζ_f, ζ_s — viscous damping ratio for foundation domain and dam, respectively

η_f, η_s — hysteretic (rate-independent) damping factors for foundation rock and dam concrete, respectively

λ_j — jth eigenvalue from the eigenvalue problem defined in Eq. (8.4.7)

ρ — mass density of water

ρ_f — mass density of foundation rock

ρ_r — mass density of reservoir boundary material: sediments or rock

$\boldsymbol{\sigma}_p(t)$ — stress vector for finite-element p

ω — circular or radial frequency

ξ — $\rho/\rho_r C_r$, damping coefficient for reservoir bottom materials

$\psi_j^f(s,r)$ — function representing the normal component of the jth natural vibration mode shape on the dam–water interface.

$\boldsymbol{\psi}_j$ — jth Ritz vector of the associated dam–foundation system

$\boldsymbol{\psi}_{bn}$ subvector of $\boldsymbol{\psi}_n$ that contains elements corresponding to the nodal points at the dam–foundation interface

$\boldsymbol{\psi}_{fn}$ subvector of $\boldsymbol{\psi}_n$ that contains elements corresponding to the nodal points at the dam–water interface

ν_f, ν_s Poisson's ratio of foundation rock and dam concrete, respectively

ω_1 natural circular frequency of dam with empty reservoir supported on rigid foundation

PART III: CHAPTERS 12–14

Abbreviations

CS	Conditional Spectrum
CMS	Conditional Mean Spectrum
EDP	Engineering Demand Parameter
FEM	Finite Element Method
GCMS	Generalized Conditional Mean Spectrum
GM	Ground Motion
GMSM	Ground-Motion Selection and Modification
GMPM	Ground-Motion Prediction Model
IM	Intensity Measure
MDE	Maximum Design Earthquake
OBE	Operating Basis Earthquake
PSHA	Probabilistic Seismic Hazard Analysis
RHA	Response History Analysis
s-GCMS	Simplified GCMS
SEE	Safety Evaluation Earthquake
SF	Scale Factor
SSD	Sum of the Squared Differences between two spectra defined in Eqs. (13.2.3) and (13.7.1)
TS	Target Spectrum,
UHS	Uniform Hazard Spectrum

Roman Symbols

A	Pseudo-acceleration or spectral acceleration
$A_{\text{TS}}(T_n)$	Target spectrum
$A_x(T_n)$, $A_x(T_n)$	Target spectra for x- and y-components of ground motion
$A_{\text{H1}}(T_n)$, $A_{\text{H2}}(T_n)$	Response spectra of as-recorded horizontal components H1 and H2
$A_{\text{composite}}(T_n)$	CMS-UHS composite spectrum
$A_{\text{CMS}}(T_n)$	Conditional mean spectrum
$A_{\text{UHS}}(T_n)$	Uniform Hazard Spectrum
M	Earthquake magnitude
R	Causal distance
$\text{SF}_{\text{T*}}$, SF_{avg}, SF_{opt}	Scale Factors defined by Eqs. (13.2.1), (13.2.2), and (13.2.4), respectively; also (13.7.2)
T^*	Conditioning period
T_{min}, T_{max}	Shortest and longest structural periods of interest
T_n	Natural vibration period of an SDF system

Greek Symbols

ε	Number of standard deviations by which given ln$A$ value differs from the mean predicted ln$A$
μ	Standard deviation
σ	Mean

Index

Earthquake Engineering for Concrete Dams: Analysis, Design, and Evaluation, First Edition. Anil K. Chopra.

D

E

J

K

L

M

N

O

P